Harald Riffert
Hanns Ruder
Hans-Peter Nollert
Friedrich W. Hehl

Relativistic Astrophysics

Harald Riffert
Hanns Ruder
Hans-Peter Nollert
Friedrich W. Hehl
(Eds.)

Relativistic Astrophysics

vieweg

Die Deutsche Bibliothek – CIP-Einheitsaufnahme

Relativistic astrophysics / Harald Riffert ... (ed.). –
ISBN 978-3-663-11296-9   ISBN 978-3-663-11294-5 (eBook)
DOI 10.1007/978-3-663-11294-5

Produced by Media-Print, Paderborn

ISBN 978-3-663-11296-9

Preface

This book summarizes the lectures given at the 162. WE-Heraeus Seminar which took place in the house of the German Physical Society in Bad Honnef in August 1996. Already the number 162 shows the activity and effectiveness of the WE-Heraeus Foundation. We would like to express our thanks to Jutta Adam and Dr. Volker Schäfer for the almost incredibly simple and unbureaucratical procedure of funding, organization and realization, and, of course, to the founders.

Similar to the 152. WE-Heraeus Seminar *Relativity and Scientific Computing* (Springer Verlag 1996), this seminar was a joint venture of the Astronomical Society (AG) and of the Section 'Gravitation und Relativity Theory' of the German Physical Society (DPG). Since Einstein has developed his Theory of General Relativity more than 80 years ago, the situation has changed dramatically. In the first decades main efforts were undertaken for a better understanding and for the experimental verification of the theory. Meanwhile General Relativity (GR) is one of the experimentally best confirmed theories with an accuracy better than 10^{-13}! Consequently, GR has become a powerful tool for the investigation of cosmic processes where strong gravitational fields are involved. The state of the art of our knowledge is summarized by Jürgen Ehlers in the first chapter.

An impressive example of the change of the situation is the gravitational light deflection. For about 60 years this effect has been measured at the limb of the sun more or less for its own sake. Now gravitational lensing is used almost on a "industrial" scale in the search for dark matter or planetary systems around distant stars. Almost the same is true for gravitational waves. Once an academic problem only, now gravitational wave detectors are built around the world at a cost of hundreds of millions of dollars. Probably, Einstein himself did not foresee the potential of his theory for astrophysics. Further astrophysically relevant objects which have to be described within the framework of GR are neutron stars and neutron stars binaries, accretion flows around black holes, and, of course, black holes themselves. With the exception of the physics of black holes, which has been the subject of the 179. WE-Heraeus Seminar held in Bad Honnef in August 1997, all the above topics are covered in this book by various experts in their fields. To round things up two additional contributions have been included to demonstrate the importance of GR onto the structure of space time ranging from the solar system to cosmological scales.

We are very grateful to Gabriela Meyer who did most of the editorial work for the manuscripts.

Tübingen and Köln
November 1997

H. Riffert
H. Ruder
H.-P. Nollert
F. W. Hehl

Contents

Joachim Wambsganss:
Gravitational Microlensing: Machos and Quasars

Karsten Danzmann:
Laser-Interferometric Gravitational WaveDetectors – on the Ground and in Deep Space

Ute Kraus:
Light Deflection Near Neutron Stars

Max Camenzind:
Magnetohydrodynamics of Rotating Black Holes

Harald Riffert:
Thin Accretion Disks around Black Holes

Hans-Peter Nollert:
Quasinormal Ringdown: The Late Stage of Neutron Star Mergers

Edward Seidel:
A General Relativistic Approach to Neutron Star Binary Evolution

Gustav A. Tammann and Lukas Labhardt:
A Forty-Year Search for the Hubble Constant

Michael Soffel:
Experimental Gravity

General Relativity as a Tool for Astrophysics

Jürgen Ehlers

Max-Planck-Institut für Gravitationsphysik
Albert-Einstein-Institut, Schlaatzweg 1, D-14473 Potsdam

1 Introduction: Basic Properties of Gravity

Phenomenologically the gravitational interaction may be characterized as the only (known) interaction which is both universal and long range. It affects all kinds of matter in the same way, and under quasistationary conditions its force falls off, like the Coulomb force, as r^{-2}. In contrast to electrodynamics, it is then always attractive and thus cannot be shielded. Like electromagnetism it admits waves propagating with the fundamental speed c, but in contrast to electromagnetic waves which have helicity ± 1, gravitational waves have helicity ± 2. Moreover, gravity is by far the weakest interaction (at accessible energies) as highlighted by the "fact" that between a proton and an electron Newton's force is weaker than Coulomb's by a factor of $2 \cdot 10^{39}$. Because of these well-known properties, gravity – although unmeasurably weak and totally negligible compared to the strong and weak, short range nuclear forces in the atomic and subatomic regime – dominates in the realm of celestial bodies and systems thereof. This clear-cut separation of interaction ranges also insures that one can combine in astrophysics laws of GR with laws from quantum mechanics (equations of state, transition probabilities) without difficulties, although these laws are taken from theories which are, strictly speaking, incompatible. The combination of the two properties "universality" and "long range", motivates the interpretation of gravity as a manifestation of the metric $g_{\alpha\beta}$ and its associated affine connection $\Gamma^{\alpha}_{\beta\gamma}$ and curvature $R^{\alpha}{}_{\beta\gamma\delta}$ of spacetime. According to Einstein, the field $g_{\alpha\beta}$ serves both as the *metric*, which locally defines the distinction between space and time and determines proper (clock) times and proper (ruler) distances, angles, areas and volumes, and as the *gravitational potential* whose derivatives, through the connection $\Gamma^{\alpha}_{\beta\gamma}$, govern the equations of motion of matter and non-gravitational fields. Related to this is the fact that the null cone field determined by $g_{\alpha\beta}$ determines the *causal structure* of processes, e.g., the (hyperbolic) propagation of fields. By assigning the double role to the metric, the identity ("equivalence") of geometry, gravity and inertia is built into the theory, and since so far every successful description of matter, whether classical or quantal, in terms of particles or fields, involves a metric as an indispensible ingredient, the identification of gravity with the dynamical metric structure of spacetime gives GR its fundamental significance for physics. This is very well expressed by John Wheeler's name *Geometrodynamics* for general relativity.

The purpose of this introductory survey is to review the basic concepts on which Einstein's theory of spacetime and gravity is based, to list some of its consequences and to indicate the empirical support of GR, and to hint at its increasing role in astrophysics.

2   General Relativity: Basic Concepts and Laws

As in Newtonian physics, so in GR physical processes can be represented quasi-geometrically in terms of subsets – curves, surfaces, ... – of a 4-dimensional manifold $M = \{x, y, ...\}$, "spacetime", and fields on M. The spacetime points or *events* indicate the "where and when" of processes of negligible spatial and temporal extent such as the collision of two particles. More abstractly and mathematically, M can be characterized as the common domain of definition of all fields, at least classical ones or in the "classical limit", which represent the stuff of which the physical world is thought to consist.

Piecewise, M can be covered by coordinate systems (x^α). Coordinates are labels (names) of events; they can, but need not, and in general do not have an interpretation in terms of times, distances or the like.

Fields on M include $g_{\alpha\beta}$, a Lorentz metric here taken to have the signature corresponding to $(-+++)$, and matter ("non-gravitational") fields. Among all fields, $g_{\alpha\beta}$ is distinguished by not allowing a tensorial, locally "conserved" (see below) energy-momentum distribution. (One may not like this distinction between the fields called "gravitational" and "material" and aim at a unified theory where all fields are on the same footing; but for GR this distinction *does* exist, and no really satisfactory way to overcome this dichotomy has emerged). The most important non-gravitational fields for astrophysics are the electromagnetic field $F_{\alpha\beta} = -F_{\beta\alpha} = 2A_{[\beta;\alpha]}$ derivable from a 4-vector potential A_α, and the fields ε, p, U^α – proper energy density, pressure and 4-velocity describing isotropic bulk matter such as a fluid. A more detailed model of matter is provided by kinetic theory. Its basic state variable is a distribution function $f(x^\alpha, p_\alpha)$ defined on the cotangent space (phase space) T^*M of M whose points are pairs (x^α, p_α) consisting of an event x^α and a covector p_α at x^α. Such an f may describe, e.g., a distribution of photons, atoms, or stars.

Having chosen the fields needed to describe some objects or processes, the next step is to put down the *laws* which these fields are to obey. But before doing that, let me remind you of some mathematical machinery associated with the metric $g_{\alpha\beta}$, which is used in formulating those laws.

First, the metric determines (via its Christoffel symbols $\Gamma^\alpha_{\beta\gamma}$) a derivative operator ("covariant derivative") ∇_α acting on tensor fields and, by means of that, a parallel transport of tensors along curves. Curves with a parallel-transported tangent vector are called geodesics.

Nearby geodesics, although individually as straight as possible, in general exhibit relative curvature. More precisely, let $x^\alpha(s, \epsilon)$ describe a family of geodesics, one for each value of ϵ. Then $\xi^\alpha = (\partial_\epsilon x^\alpha)_{\epsilon=0}$ may be interpreted, except for a scale factor $\epsilon > 0$, as a vector connecting $x^\alpha(s, 0)$ to $x^\alpha(s, \epsilon)$ for "infinitesimal" ϵ. It satisfies the equation of geodesic deviation ("Jacobi eq."),

$$\frac{\nabla^2 \xi^\alpha}{ds^2} = R^\alpha{}_{\beta\gamma\delta} \dot{x}^\beta \dot{x}^\gamma \xi^\delta, \tag{1}$$

where $\dot{x}^\alpha = (\partial_s x^\alpha)_{\epsilon=0}$ is the tangent vector of the geodesic $x^\alpha(s,0)$. The curvature tensor $R^\alpha{}_{\beta\gamma\delta}$ in (1) is uniquely determined by the connection of the metric $g_{\alpha\beta}$. From it, one forms the Ricci tensor $R_{\alpha\beta} = R^\gamma{}_{\alpha\gamma\beta}$, the Ricci scalar $R = R^\alpha{}_\alpha$ and the Einstein tensor $G^{\alpha\beta} = R^{\alpha\beta} - \frac{1}{2}g^{\alpha\beta}R$ which measure certain average curvatures. $G^{\alpha\beta}$ obeys the (contracted Bianchi) identity

$$G^{\alpha\beta}{}_{;\beta} = 0. \tag{2}$$

(As usual, ";" indicates covariant differentiation.)

With these tools we can state the basic laws of GR. They consist of *non-gravitational laws* and Einstein's *gravitational field equation*. The non-gravitational laws are natural generalizations of special-relativistic (SR-) laws from flat to curved spacetime, obtainable by substituting the flat-space metric $\eta_{\alpha\beta}$ and derivative ∂_α by their curved-space generalization $g_{\alpha\beta}$, ∇_α. The physical motivation behind this rule is Einstein's principle of equivalence (to be considered briefly below). Examples are "Maxwell's" equations in vacuum

$$F_{[\alpha\beta;\gamma]} = 0, \quad F^{\alpha\beta}{}_{;\beta} = 0; \tag{3}$$

"Euler's" equations for a perfect fluid[1],

$$\varepsilon_{;\alpha}U^\alpha + (\varepsilon + p)U^\alpha{}_{;\alpha} = 0, \quad U_\alpha U^\alpha = -1, \quad (\varepsilon + p)U^\alpha{}_{;\beta}U^\beta + (g^{\alpha\beta} + U^\alpha U^\beta)p_{;\beta} = 0; \tag{4}$$

and "Vlasov's" equation for a collisionless gas (of massive particles or photons),

$$p^\alpha \frac{\nabla f}{\partial x^\alpha} = p^\alpha \left(\frac{\partial f}{\partial x^\alpha} - \Gamma^\gamma_{\alpha\beta}p_\gamma \frac{\partial f}{\partial p_\beta} \right) = 0. \tag{5}$$

In these and other cases, the non-gravitational fields determine, in special relativity, a symmetric energy-momentum tensor $T^{\alpha\beta}$ in terms of which local conservation of energy and momentum is expressed as $T^{\alpha\beta}{}_{,\beta} = 0$. In the examples considered above they have the forms

$$\text{Maxwell}: \quad T^{\alpha\beta} = F^{\alpha\gamma}F^\beta{}_\gamma - \frac{1}{4}g^{\alpha\beta}F_{\gamma\delta}F^{\gamma\delta}, \tag{6}$$

$$\text{Euler}: \quad T^{\alpha\beta} = (\varepsilon + p)U^\alpha U^\beta + pg^{\alpha\beta}, \tag{7}$$

$$\text{Vlasov}: \quad T^{\alpha\beta} = \int f(x^\gamma, p_\gamma)p^\alpha p^\beta \frac{d^3p}{p^0}. \tag{8}$$

(In (8), the integral is over the mass shell.) The formal generalization of the SR-conservation law reads

$$T^{\alpha\beta}{}_{;\beta} = 0. \tag{9}$$

---

[1] Throughout, I put $c = G = 1$

This covariantly generalized energy-momentum law indeed follows from the covariantly generalized laws (3), (4), (5), respectively, in agreement with the equivalence principle. In all these laws, the effect of gravity on matter is contained in the covariant derivative operation ";". It has the consequence that (9) is *not* a local conservation law (although it is called so, by abuse of language). This must be so: Since gravity acts on matter, the latter's energy-momentum should not be conserved. Moreover, one cannot add a gravitational energy-momentum tensor to that of matter to obtain an exact, local, overall energy-momentum conservation law. According to the principle of equivalence and its mathematical expression, a tensorial gravitational field strength from which $T^{\alpha\beta}_{\text{grav}}$ could be formed, does not exist. This follows from Einstein's equivalence principle as stringently as the non-existence of exact world lines of elementary particles from Heisenberg's uncertainty relation. In both cases a price has to be paid for an empirically well-founded principle. Also, in both cases the old concepts may be used as approximations under suitable conditions, e.g., if one uses an "effective energy-momentum tensor" for gravitational waves.

The collection of basic laws of GR is completed by Einstein's gravitational field equation

$$G^{\alpha\beta} = 8\pi T^{\alpha\beta} - \Lambda g^{\alpha\beta}. \tag{10}$$

It is constructed such that it implies the energy-momentum law (9) via the identity (2). In fact, (10) can be characterized (in 4 dimensions) as the only tensorial equation of the form $L^{\alpha\beta} = \kappa T^{\alpha\beta}$ where $L^{\alpha\beta}$ depends only on $g_{\alpha\beta}$ and its first and second derivatives and which implies (9) identically [Lovelock's theorem (1972)]. (The symmetry of $L^{\alpha\beta}$, and thus of $T^{\alpha\beta}$, and the mathematically significant fact that $L^{\alpha\beta}$ is linear in $g_{\alpha\beta,\gamma\delta}$, both follow from the theorem.)

The field equation (10) is more than an identity only if $T^{\alpha\beta}$ is specified in terms of matter fields (and $g_{\alpha\beta}$)which have to obey laws such as (3), (4) or (5). Eq. (10) does not determine the metric in terms of matter nor vice versa; it interrelates metric and matter. Since its left-hand-side is non-linear in $g_{\alpha\beta}$, (10) by itself implies interactions of the gravitational field with itself and interactions of parts of matter mediated by $g_{\alpha\beta}$.

To summarize: In GR, a model for some physical situation consists of a space-time $(M, g_{\alpha\beta})$ and a collection of non-gravitational fields such as $F_{\alpha\beta}, \varepsilon, p, U^{\alpha}, f$, which obey (10) with the appropriate $T^{\alpha\beta}$ exemplified by eqs. (6) – (8). Such a model is dynamically complete if (10), coupled to the pertinent non-gravitational laws such as (3), (4), or (5) and possibly equations of state like $p = p(\varepsilon)$, admits a well-posed initial-value problem. Physically plausible and mathematically useful properties of $T^{\alpha\beta}$ include several energy-positivity conditions which state in different variations that energy density is non-negative and dominates pressures and stresses. Such conditions enter, e.g., singularity theorems. It appears to be unknown whether such conditions follow from microscopic models of bulk matter.

Instead of assuming field equations directly, one may (for most non-dissipative systems) base the theory on an *action functional* and obtain the field equations as Euler-Lagrange equations of that. Since this is useful for setting up evolution equations and in contexts where one needs to relate the macroscopic description to an underlying microscopic model of matter, e.g., to the standard model of particle physics, this will be briefly mentioned here. The action functional has the form

$$A_D[g, \varphi] = \frac{1}{16\pi} \int_D (R - 2\Lambda)\eta + \int_D L_m \eta \tag{11}$$

where D is an arbitrary compact domain of M, g denotes the metric, R its Ricci scalar, and φ indicates matter fields on which L_m depends. η denotes the volume element determined by the metric. The gravitational part of the action is the only one whose density, $R - 2\Lambda$, is an invariant function of the metric and its derivatives which leads to second-order field equations (Vermeil 1917, Weyl 1991); this fact provides another characterization of the field equation (10). Taking L_m from the standard model allows to relate it, at least formally and semi-classically, to gravity. The domain of definition of the fields φ in this case is no longer spacetime, but a suitable principal bundle over M.

3 Remarks on Interpretation, Observables, and Correspondences with other Theories

To connect the mathematical formalism of GR with observations and experiments, one needs to relate some of its mathematical constructs to things which are known (or taken for granted) and which can be pointed out in the lab. For the case of GR, the *interpretational dictionary* includes:

Particle	$\longleftrightarrow$	timelike curve,
Freely falling particle	$\longleftrightarrow$	timelike geodesic
Extended body	$\longleftrightarrow$	timelike world tube
Light ray	$\longleftrightarrow$	null geodesic.

These correspondences are consistent with the laws formulated above. The geodesic law for test particles can be "derived" (more or less) from a particle model and the field equations, and the identification of light rays with null geodesics can be justified by applying the WKB method to Maxwell's equations.

The rules above are qualitative. To account for measurements, one needs the following *basic observables*:

a) Proper times,

$$\int ds = \int \sqrt{-g_{\alpha\beta}dx^\alpha x^\beta}, \tag{12}$$

taken along timelike world lines;

b) angles between light rays which meet in an observer's eye or at a telescope;

c) energy fluxes of electromagnetic radiation or particle beams in various spectral regions, given in theory by

$$-\int_\Sigma T^{\alpha\beta}N_\alpha U_\beta d\Sigma, \tag{13}$$

where Σ denotes a timelike 2-surface representing, e.g., the opening of a telescope during some measurement, U^α being the 4-velocity of the telescope and N^α the unit normal to Σ.

Note that these three basic observables refer to the nearly flat vicinity of the observer's world line. Distances of stars, galaxies etc. are not directly measured, but are "theoreticals" computed from the observables by means of the theory. (This is particularly relevant to cosmology where several "distances" are employed, all inferred from values of observables and assumptions on the sources and the cosmic metric.)

For interpretations and in order to relate general relativity approximately to theories which are not part of GR, e.g., the quantum mechanics of atoms, one may use generalized Fermi coordinates (τ, ξ^a) based on a timelike world-line l which represents some point of a laboratory. τ denotes proper time on l, and ξ^a are geodesic normal coordinates on the hypersurfaces Σ_τ generated by geodesics intersecting l orthogonally at a point on l corresponding to time τ. In the neighborhood of l, these coordinates approximate rectangular coordinates of a rigid, possibly accelerated and rotating frame like that of an observatory on earth. If l is geodesic and the vectors $\frac{\partial}{\partial \xi^a}|_l$ are parallel along l, these coordinates represent a frame of reference which is "locally inertial" on l (Einstein's lift). In either case one may apply, locally near the origin, "ordinary" physics as if one had a rigid, Minkowskian or Newtonian frame, and so fit into GR whatever is needed. This device is used, e.g., to treat atoms at neutron star surfaces or to use ordinary thermodynamics in a GR-description of gravity.

Applying the field equation (10) in a local inertial frame shows how Newtonian gravity arises as a local approximation to GR. In fact, one may even use this correspondence between local inertial frames in a curved spacetime with Newtonian inertial frames to "derive" (10) from Poisson's law of Newtonian gravity.

Note that the whole theory of GR including its physical interpretation is generally covariant: Two collections $(M, g_{\alpha\beta}, ...)$ and $(M', g'_{\alpha\beta}, ...)$ represent the same physics if and only if they are related by a diffeomorphism mapping M onto M', $g_{\alpha\beta}$ onto $g'_{\alpha\beta}$ etc.

4   Some Developments, Implications and Problems of GR

From the role of gravity indicated in the introduction there arise astrophysical questions such as

(i) What are the equilibrium states of compact bodies containing some specified amount of matter?
(ii) If perturbed, how do such bodies react, how much gravitational radiation do they emit?
(iii) How do the components of a nearby isolated system move, does such a system radiate?
(iv) How does gravitational radiation act on bodies (detectors)?
(v) How is electromagnetic radiation affected by a gravitational field?
(vi) What does the theory say about the structure and evolution of the universe?

In trying to answer these and similar questions to be considered later during this seminar, the relativity theorist has to study the implications of Einstein's equations, coupled to matter models, under various circumstances – if possible by proving theorems, more often by approximation methods, and in most realistic cases, by numerical methods.

Here I shall but briefly consider a number of approaches, which have been used and which are under further development in order to find out whether and how GR can answer such questions.

4.1 Initial-Value Problems

One way, and perhaps the most useful way, to study generic properties of solutions of the field equation (10), is to analyze the initial-value problem: Which data, if any, can be chosen freely on a spacelike hypersurface such that there exists one and – up to diffeomorphisms – only one solution of (10) compatible with those data? What are the domains of dependence in which the solution is determined by the data on parts of the initial hypersurface (causality)? Do solutions develop singularities, and of which kind? Which part of the initial data contains information about (incoming/outgoing) gravitational radiation?

The first important answer which emerged from the analysis of the first of these questions is: Of the 10 components of (10), 4 impose conditions on the initial data, while 6 of them are evolution equations. If coordinates are chosen such that the initial hypersurface Σ is given by $t = t_0$, the four "constraints" are $G^0_\mu = 8\pi T^0_\mu$, while the remaining six equations $G_{ab} = 8\pi T_{ab}$ are evolution equations. The gravitational data involved in the constraints are the intrinsic metric g_{ab} of Σ and a symmetric tensor K_{ab} which, in the evolved spacetime, represents the extrinsic curvature of Σ in M. To obtain a well-posed initial-value problem, one has to (i) select appropriate matter data, (ii) choose four coordinate conditions to break the diffeomorphism invariance, (iii) supplement the six equations $G_{ab} = 8\pi T_{ab}$ by the four equations (9), (iv) add additional matter constraints and evolution equations depending on the matter present. It then turns out that, for the vacuum equations $R_{\lambda\mu} = 0$ as well as for the cases with matter mentioned in Sect. 2, the local initial-value problem is indeed well-posed. In particular – and this is often not recognized, although it has been shown several years ago by H. Friedrich (1985) – the evolution equation can be made (symmetric, first order) hyperbolic for (essentially) arbitrary coordinate (and/or frame) conditions.

Some *global* results are also available for asymptotically flat data (and solutions) and for compact Σ; i.e., for isolated systems and cosmological spactimes. The most important open question is perhaps: Which asymptotically flat initial data (on $\Re^3$, e.g.) imply that the resulting spacetime is asymptotically Minkowskian at spatial and null infinity (see Sect. 4.2). Another open problem is: How can one treat in an initial-value problem, spacetimes containing (one or several) material bodies separated by empty space? Here, the problem is to control the evolution of the boundaries of the bodies. (So far, only very special exact results have been established.) Both these problems would have to be solved as parts of the n-body problem in GR (even for $n = 1$!).

Since the initial data g_{ab}, K_{ab} of the gravitational field are subject to four constraint equations, the question arises as to which variables can be chosen freely and which, modulu boundary conditions, determine g_{ab}, K_{ab} obeying the constraints. The answer, at least under some rather general conditions, is that four functions can be chosen freely on Σ; one choice consists of the conformal metric of Σ and a tracefree, transverse part of a (suitably densitized) part of K^{ab}. These variables may be considered as the "true degrees of freedom" of the gravitational field.

To end this subsection I mention in passing that, starting with the action functional (11), one can express Einstein's equation in Hamiltonian form. The true degrees of freedom of gravity then correspond to two pairs of canonical variables per space point. Several choices of canonical variables have been used in attempts to quantize gravity. Classically, various forms of the (constraint + evolution) equations (ADM equations) form the basis of constructing solutions of (10) numerically. A hot topic of research is which formulations are most suitable for specific problems in numerical GR.

4.2 Asymptotics, Global Energy-Momentum

Several of the problems listed at the beginning of Sect. 4 refer to systems one wishes to idealize as isolated. The basic idea, due to R. Penrose, to formalize "isolation" is to require spacetime $(M, g_{\alpha\beta})$ to admit a conformal extension $(\hat{M}, \hat{g}_{\alpha\beta})$, $M \subset \hat{M}$, with $\hat{g}_{\alpha\beta} = \Omega^2 g_{\alpha\beta}$ such that on the boundary of M in $\hat{M}$, $\hat{g}_{\alpha\beta}$ remains Lorentzian while the conformal factor Ω vanishes, with $\Omega_{,\alpha} \neq 0$ there. One has to distinguish between null infinity, to be approached from M via future – or past-directed null geodesics, and spatial infinity, to be approached via spacelike geodesics. The important and difficult task to relate this geometrical concept of asymptotic flatness to the field equations has been tackled and largely solved, by H. Friedrich (1997) by means of his regular conformal field equations. His formulation permits to represent physically infinite parts of spacetime, e.g., a part of the source's history and a slice between two retarded times, reaching out arbitrarily far into the radiation zone near future null infinity, as a compact part of the unphysical, extended spacetime $\hat{M}$. The use of this approach for numerical GR is only in its beginning stage.

The most difficult remaining problem is to control the behavior of Ω and $\hat{g}_{\alpha\beta}$ (or equivalent variables) near spatial infinity where the physical conformal curvature, the Weyl tensor, unavoidably blows up if the spacetime has a non-vanishing total (ADM-) mass. Promising work on this is progressing.

Within this asymptotic framework clear, covariant definitions of a (conserved) total energy-momentum 4-vector at spatial infinity (ADM-energy-momentum) and of a null-energy-momentum (Bondi), a functional on spherical sections of null infinity, have been possible; and it has been shown that the corresponding energies are positive for all non-flat spacetimes satisfying some regularity conditions. What remains is to relate these two kinds of global 4-momenta to each other, a problem clearly related to the "spatial infinity problem" mentioned above.

The results and problems touched in the previous and in this subsection are closely related to each other. In particular, they are part of a rigorous formulation of the n-body problem of GR (even for n = 1). Even as long as the latter problem – unquestionably a key problem of astrophysics – can be treated by means of approximation methods only, the conceptual and formal background provided by initial value and asymptotic work should serve as guidance in improving these methods.

4.3 Black Holes

The "most typically relativistic" objects are undoubtedly black holes. As is well known, within pure GR – disregarding Yang-Mills and similar fields – their equilibrium states are known explicitly and characterized by their mass, angular momentum and electric charge. I shall not recall the black hole thermodynamics except for the remarks that it still appears to be a challenge to find a formulation of (phenomenological or statistical) thermodynamics which covers both ordinary and black hole thermodynamics (is there a useful gravitational field entropy?), and that, as a possible problem of semi-classical gravity, the "dissolution" of a black hole by emission of Hawking radiation is not understood quantitatively.

The interest of black holes for astrophysics is obvious; they should exist as remnants of (some) supernovae, and the presence of supermassive holes in centers of galaxies (also "our own") is observationally indicated. The challenge here is to find observable features that are truly relativistic, related, e.g., to horizons, ergo-regions or accretion disc behavior which requires relativistic thermo-hydrodynamics. Indications exist, but – as far as I am aware – no firm evidence.

4.4 Gravitational Radiation and Equations of Motion

Any relativistic theory of gravitation will predict gravitational waves, as first pointed out by H. Poincaré (1905, 1908), independently discovered for GR by Einstein in 1916 and 1918, and observationally verified, though indirectly, by J.H. Taylor, R.A. Hulse and co-workers since 1979 (Taylor 1994, Hulse 1994).

Using the field equation (10) with $\Lambda = 0$, linearized at flat spacetime as background, one easily finds that such waves propagate with the fundamental speed c and have polarization states according to helicity $s = \pm 2$. The difficult problems to determine the radiation emitted by self-gravitating bodies or systems and the radiation reaction effects on the motions of the sources have been and still are active areas of research, both by means of analytical post-Minkowskian and post-Newtonian methods and combinations thereof, and by numerical methods. The most difficult problems, in my opinion, are to interrelate the bodies and their nearzone fields to the far (radiation) field and to incorporate in such a treatment that there is no, or not much, incident gravitational radiation. Though the problem can be formulated rigorously in terms of the concepts and tools sketched in Sect. 4.1 and 4.2, for the reasons indicated it cannot yet be solved in that way, and even the most sophisticated and far-reaching approximation methods employ ad-hoc assumptions to bridge some gaps in their derivations. (As a technical point I should like to mention that in these approximation methods, the interplay between evolution equations, constraints and coordinate conditions is not clear and deserves attention.)

Pushing aside such worries, it appears that for nearly Newtonian systems (solar system) and moderately relativistic systems (PSR 1913+16, PSR 1534+12), the dynamics and the related radiation are well understood, at least up to and including terms of the order of $\left(\frac{v}{c}\right)^5$ beyond Newtonian terms in the equations of motion.

The most challenging problem in this field is perhaps to understand quantitatively inspiralling binary systems of neutron stars and/or black holes and the process of merging of their components. Here, interaction of analytic and numerical work will be especially needed. These problems are considered urgent because of the great opportunity to observe gravitational radiation around the turn of the century by means of LIGO, VIRGO and GEO.

5 Tests of the Foundations of GR

It is useful to classify experiments supporting general relativity in two classes, those which test the Einstein equivalence principle and those which test the gravitational field equation. The tests in the first category support not only GR, but all metric theories of gravity, i.e., those theories of gravity which (i) assume the existence of a Lorentz metric, which has the physical meaning embodied heuristically in the equivalence principle and formalized by the rules which state how $g_{\alpha\beta}$ and its connection, or derivative, enter nongravitational laws of physics, as exemplified by eqs. (1) – (9) and the interpretation rules of Sect. 3, and (ii) which reduce approximately to Newton's theory for quasistationary, low speed, weak gravity, nearly isolated systems. (Note the last condition which is usually not stated, but necessary; for Newton's theory was formulated and successful for such systems only. The so-called "Newtonian" cosmology, while useful in some contexts, does not have deterministic laws in the general case of unbounded, inhomogeneous matter distributions, claims in the literature notwithstanding.) Only tests of the second category can discriminate between metric theories, and thus support (or disprove) GR itself.

5.1   Tests of the Equivalence Principle

Here, all tests of SR could be listed. I shall briefly consider a few experiments only which serve to support the assumption that at each event there are local inertial frames. This implies that, given any practically rigid frame of reference, e.g., a laboratory on earth, it must be possible to measure the acceleration $\vec{g}$ and the angular velocity $\vec{\omega}$ of that frame with respect to a local inertial frame; the results must be unique whatever the means to determine them.

For a laboratory on earth, $\vec{g}$ and $\vec{\omega}$ have indeed been determined (i) mechanically, using directly freely falling particles or, indirectly, a Foucault pendulum or a similar device or torsion balance experiments; ii) electrodynamically, via the (local), so-called gravitational redshift and the Michelson/Sagnac effects; (iii) quantum mechanically, using the effects of nearly homogeneous acceleration or (dynamical) rotation fields, relative to the lab, on the phase of wave functions of neutrons (or other particles) propagating through crystals. Within the experimental uncertainties, which vary between about 10^{-11} and 10^{-2}, $\vec{g}$ and $\vec{\omega}$ do agree, thus supporting the equivalence principle and, as part of it, the universality of free fall and of "dynamical irrotationality".

Such experiments, if carried out at different places, also show – within the Einsteinian interpretational framework which I here adopt – that SR is not "globally" valid, since there are, in fact, true, i.e., inhomogeneous, gravitational ($\equiv$ curvature) fields. Indeed, nearby local inertial frames exhibit relative accelerations determined by the tidal field, as formalized in (1).

Note that, in i) one is testing the "test-body limit" of (9), in ii) one is testing the " ; " in (3), and in iii) one is testing the admissibility of "inertial potentials" in Schrödinger's equation.

The equivalence principle also claims that there is but one, "locally" Euclidian, spatial metric. The so-called space-isotropy experiments can be interpreted as verifying, with the staggering accuracy of 10^{-22}, that the (spatial) metric which enters the kinetic energy, $\frac{1}{2}\delta_{ab}p^a p^b$, indeed is identical to the one which enters Coulomb's potential energy, via $(\delta_{ab}x^a x^b)^{-\frac{1}{2}}$.

Here, the Vessot-Levine experiment (gravity probe A) should also be mentioned; it supports the $g_{00} = -(1 + 2\frac{U}{c^2})$-term in the Newtonian approximation to any metric theory of gravity. In contrast to the "local" time dilation experiment ((ii) above), this experiment tests the "global" gravitational time dilation, in the sense that it refers to a domain in which the gravitational field (in Newton's sense) is inhomogeneous, or (in Einstein's sense), curvature cannot be neglected.

5.2 Tests of the Gravitational Field Equation

These tests will be treated in detail in another lecture; here I shall only indicate their "theoretical role". At the first post-Minkowskian (1 PM) level, one linearizes the field equation (10) at the Minkowskian background metric (with vanishing $T^{\alpha\beta}$), integrates them for a "first order" matter tensor $T^{\alpha\beta}$, and uses the resulting metric to study effects on test bodies. The resulting pertinent effects are

a) Light deflection by the sun;

b) Signal retardation by the solar field;

c) Geodetic (de Sitter) precession of the angular momentum of the earth-moon system, considered as a test gyro in the field of the sun;

d) Schiff precession produced by, e.g., the earth's angular momentum, acting on gyroscopes orbiting the earth;

e) Lense-Thüring precession of the line of nodes of satellites orbiting the earth, due to the earth's angular momentum.

As is well known, the effects a) – c) have been measured; they depend on the g_{00} and g_{ab} terms of the linearized metric, i.e., on the Newtonian potential. The possibility of measuring d), which involves the gravitomagnetic components g_{0a} of the metric, has been under discussion for a long time; it will hopefully be realized in the foreseeable future (gravity probe B). As to e), which also concerns gravitomagnetism, there is a recent analysis of the orbits of the satellites LAGEOS I and III by Ciufolini according to whom the effect can be extracted from the data, though only with a 30 % uncertainty.

At the next higher level (IPN), one takes into account the nonlinear term $2U^2/c^4$ in g_{00}, in addition to the previous ones, and one keeps terms of order $(v/c)^2$ beyond the Newtonian ones in the equations of motion (9). This gives the "anomalous" precessions of the perihelia of the planets. At this level, the Nordtvedt effect should also be mentioned which tests that even self-gravitational binding energy contributes in the same way to the inertial and passive gravitational masses, strengthening the law of universality of free fall.

The next, and so far last, level which has been tested concerns a mixture of the 1 PN approximation and those terms of the $2\frac{1}{2}$ PN approximation which are related to gravitational radiation reaction and which, consequently, imply secular effects in the motions of the bodies involved. These terms, of the order of $(\frac{v}{c})^5 \sim (\frac{GM}{Rc^2})^{2.5}$, have been used, with and also without use of an energy balance between sources and radiation field, to provide indirect evidence for gravity to be transmitted with the fundamental speed c and the existence of gravitational radiation. The systems in question are the binary pulsars PSR 1913+16 and PSR 1534+12, mentioned already in Sect. 4.4. In these cases the pulse arrival times have been used, in connection with the 1 PN orbits and GR-effects, to determine the Keplerian and post-Keplerian orbit elements and the masses of the components; these data then permit, with the radiation reaction terms, to predict the energy loss rate and the decay rate of the orbital period. For "the" binary pulsar the result for $\dot{P}_b$ agrees within the uncertainty of about $4 \cdot 10^{-3}$ with the measured value, obtained by Hulse (1994) and Taylor (1994).

In a semiquantitative way one may also consider the observational successes of the standard hot big bang cosmology as support for Einstein's equation; for the computation of the abundances of light elements in the early universe uses the gravitating effect of radiation which determines the cosmic scale function, and this is certainly not a Newtonian effect.

The next steps in testing GR are expected to come from the direct detection of gravitational waves and from direct tests of gravitomagnetic effects. A weak point in testing the field equation may be that, so far, the tests of GR essentially have been concerned with the vacuum equations and the relation of that to very nearly spherical sources; quantifiable tests of (10) for the interior equation, in the support of the energy-momentum tensor, appear to be missing.

6   Additional Remarks on Astrophysical Applications of General Relativity

In Sect. 4 and 5 some applications of GR to astrophysics have been touched already. "Applications" is perhaps not the appropriate term, for so far, gravitational theory as a part of physics, i.e., as a means to quantitatively represent and to "explain" real phenomena, *is* essentially a part of astrophysics. In this last section, I shall list what may be considered the main subfields of gravitational physics, and give some comments on those aspects of GR which play a major role in these fields.

6.1   Structure of Single Compact, Non-Rotating or Rotating Material Objects; their Instabilities, Oscillations, Quasi-Normal Modes; their Coupling to Modes of the Gravitational Field

Here, GR enters via the gravitating role of pressure and the space-curving role of energy and pressure, as exhibited in the Tolman-Oppenheimer-Volkov equation of gravito-hydrostatics which leads to the Buchdahl inequality for the compactness $\frac{2GM}{c^2R}$ of a static, spherical object and the Chandrasekhar and Landau limits for masses of white dwarfs and neutron stars. In the case of rotating objects, gravitomagnetic effects enter, and the interrelation of gravity and equations of state of dense matter is of particular interest. Finally, in considering oscillations, the coupling of matter to the degrees of freedom of the gravitational field and the resulting damping effects enter the scene.

6.2   Motion and Radiation of Binary Systems Containing Compact Objects

Here the changes of celestial mechanics due to GR, the modification of the central force law combined with special-relativistic effects, the propagation of changes of the metric field with c and the resulting damping as well as the emission of gravitational waves are relevant. The quasistationary motion is mainly governed by the elliptic aspects (constraints) of the field equation (10) combined with (9), while the hyperbolic aspects (true degrees of freedom) enter perturbatively only. (For results and open problems, see Sect. 4 and 5)

6.3   Black Holes as Results of Gravitational Collapses (Supernovae) and as Energy Producing Engines at Galactic Centers and in Quasars

This is the area in which strong field effects and typical GR-structures (horizons, ergo regions) are expected to show up; here relativistic effects of hydrodynamics and radiation transport in jets and accretion disks may prove important and observable.

Interesting questions concerning supernovae, which require numerical work, are under which conditions the resulting object is a neutron star or a black hole, and what kind of gravitational waves are emitted in the process.

6.4   Gravitational Waves, their Generation, Propagation and Detector

Here, the hyperbolic aspects of the field equation dominate in the far radiation zone; see remarks in previous sections.

6.5   Light Deflection, Gravitational Lensing

This, now very active, field has not been mentioned so far in this survey, except for the "classical" light deflection by the sun. Going back to the foundations, one recognizes that the mere existence of that deflection shows that spacetime is not conformally flat, the Weyl part of the curvature tensor in general does not vanish. As a consequence, the light "cones" (in spacetime, not in its tangent spaces), though they are cone-like near the vertex, in general will develop caustics and self intersections. Quantitatively, the local structure of caustics can be studied by applying (1) to connection vectors between null geodesics – light rays – which form small beams, decomposing in that equation the curvature tensor into its Weyl and Ricci parts and substituting for the latter the energy-momentum tensor according to the field equation (10). The resulting distortion of the light cones by matter leads to the phenomena of gravitational lensing, first theoretically conceived by Einstein in 1912, first observed in 1979 by Walsh et al. (1979). One speaks of macrolensing if the differential light deflection leads to several images of a source as in the case of the "double quasar" and the quadruple "Einstein cross". The differential deflection can also be so strong that a distant galaxy, lensed by a cluster of galaxies, appears, in one or several images, as an elongated, sausage-like object or even as an (Einstein) ring. The flux can thereby be enlarged by factors of 30 or more. In cases where the deflection is not strong enough to lead to resolvable images, it can nevertheless cause well observable (and observed) changes of light curves of distant sources whenever the light beam is intercepted by, e.g., stars in our galaxy; this is called microlensing. If the disturbing object is a star with a planetary companion, this can be seen in the shape of the light curve; several such cases have been observed. Thirdly, distant galaxies, lensed by a cluster, can appear as "arclets" on a CCD, and the statistical evaluation of the positions and ellipticities of those has been used to find the mass distribution including dark matter within the lensing cluster (weak lensing). Lensing observations cannot only be used to determine masses and mass distributions (and, therefore, mass to light ratios), but also – in connection with a cosmological model – to determine H_0, Λ, q_0. So far, however, the results are not yet conclusive.

6.6   Cosmology, Microwave Background, Structure Formation

The distribution of matter in the universe, at least that of the visible matter, is very inhomogeneous at least up to scales of 100 Mpc; only on still larger scales does it appear to be uniform (or not?). At present, the role of GR in cosmology is mainly to provide analytical and numerical tools to study linear and non-linear perturbations superimposed on a homogeneous and isotropic Friedmann-Lemaitre model. Whereas such background models can also be obtained by extending Newton's theory to unbounded matter distributions, only GR provides reliable laws for light propagation also for inhomogeneous models (and thus, for lensing).

Moreover, GR perturbation theory for matter gives unique evolution laws respecting relativistic causality, while "Newtonian" perturbation theory gives deterministic laws only for spatially periodic perturbations, and, of course, with instantaneous forces. As soon as propagation of disturbances or horizons or gravitational waves are significant, only GR is trustworthy. This holds also whenever radiation transport, in particular through inhomogeneities, occurs; clearly, photons do not fit well into Newtonian theory. This is not the place to enter into details of the very active field of cosmology; as a last remark I wish to point out only that, mathematically, the formation of optical caustics (in lensing) and of dynamical caustics (in structure formation) have an elegant common basis, the theory of Lagrangian singularities.

There is one conceptual point I wish to add. In astrophysics, one uses GR frequently (mostly?) as a refinement of Newtonian gravitation theory and/or special relativity, a "formalism" which provides "corrections" to the laws of these parent theories. It is, then, taken for granted that bodies have masses, angular momenta, mass centers and the like, and that there are equations of motion for these mass centers, etc. This may be adequate for many, or even most, practical purposes. It hides, however, a basic fact: The primary variables which appear in the laws of GR are local fields like $g_{\alpha\beta}$, energy density ε, 4-velocity U^α etc., and it is far from obvious whether and how integral concepts as those just mentioned can be defined exactly in GR; in extremely relativistic situations, such concepts will not apply at all. What, then, is their status? Should we be content with considering these concepts as approximate ones, definable only within particular approximation schemes, or should we make efforts to look for concepts which are truly relativistic, exactly definable within GR? Take concepts like active gravitational mass, passive gravitational mass, inertial mass, gravitational constant, which are used in describing observations. Do we understand them (sufficiently)? These questions may be considered as concerned with splitting hairs. But, sometimes hair splitting may be enlightening, perhaps even useful, as history shows.

7   Afterthought

I fear I have missed my assigned task. I did not treat "tools" for astrophysics – computational methods useful for specific, concrete purposes. Instead, I considered the role of GR in astrophysics, and that only in the form of a more or less annotated table of contents. I hope that, as a survey of thoughts, it may nevertheless be useful to some readers.

Acknowledgments: I should like to thank Professor Ruder and Professor Hehl for inviting me to give the opening talk of this seminar. Also, I am grateful to Professor Ruder for having insisted that I write up my notes, and him and his helpers to get it "latexed".

References

Einstein, A. (1916): Näherungsweise Integration der Feldgleichungen der Gravitation. *Sitzungsber. Preuß. Akad. Wiss.*, **1**, 688

Einstein, A. (1918): Über Gravitationswellen. *Sitzungsber. Preuß. Akad. Wiss.*, **1**, 154

Friedrich, H. (1985): On the Hyperbolicity of Einstein's and other Gauge Field Equations. *Commun. Math. Phys.*, **100**, 525

Friedrich, H. (1997): Einstein's Equation and Conformal Structures (eds. S. Huggett et al.). Preprint AE I-019 (Oct. 1996), to appear in *Geometric Issues in the Foundations of Science*. Oxford University Press, Oxford

Hulse, R.A. (1994): The discovery of the binary pulsar. *Rev. Mod. Phys.*, **66**, 699

Lovelock, D. (1972): The Form-Dimensionality of Space and the Einstein Tensor. *J. Math. Phys.*, **13**, 874

Poincaré, H. (1905): Sur la dynamique de l'électron. *C. R. Acad. Sci. Paris*, **140**, 1504

Poincaré, H. (1908): La dynamique de l'électron. *Revue générale des Sciences pures et appliqués*, **19**, 386

Taylor, J.H. (1994): Binary pulsars and relativistic gravity. *Rev. Mod. Phys.*, **66**, 711

Vermeil, H. (1917): Notiz über das mittlere Krümmungsmaß einer n-fach ausgedehnten Riemannschen Mannigfaltigkeit. *Nachrichten v. d. Königl. Ges. d. Wiss. zu Göttingen, Math.-Phys. Klasse*, 334-344

Walsh, D., Carswell, R.F., and Weymann, R.J. (1979): 0957+561 A, B: twin quasistellar objects or gravitational lens? *Nature*, **279**, 381

Weyl, H. (1991): *Raum, Zeit, Materie*. 8. edition. Springer, Berlin, Anhang I

Cosmological Dark Matter as Seen with Weak Gravitational Lensing

Peter Schneider

Max-Planck-Institut für Astrophysik
Karl-Schwarzschild-Str. 1, D-85740 Garching
peter@mpa-garching.mpg.de

Abstract

The distortion of images of faint high-redshift galaxies by the tidal gravitational field of mass concentrations allows to investigate the mass distribution of individual galaxy clusters, the investigation of the statistical properties of the mass distribution in galaxy halos, and the detection of dark halos without any reference to their luminosity. In addition, the statistical properties of the image distortion field on large scales can be used to infer directly the power spectrum of cosmological density fluctuations and to find (dark) matter concentrations purely from their mass properties. I will outline the basic methods of this new research field in extragalactic astrophysics, and present several recent results; in particular, a high-resolution mass map of a high-redshift cluster of galaxies is presented and compared to the light distribution.

1 Introduction

Gravitational light deflection has been one of the key tests of Einstein's Theory of General Relativity. Several authors in the 1920's have pointed out that this effect may give rise to spectacular effects, such as multiple images or ring-like images of distant sources, but no one expressed his vision so clearly as Zwicky in 1937, when he claimed that the observation of the gravitational lens effect will be 'a certainty'; he also estimated the probability of a distant source to be multiply imaged to be a few tenth of a percent, very close to modern estimates, and he predicted that the lens effect will allow the determination of the mass of distant cosmic objects and, due to the magnification effect, allow deeper looks into the universe [for an account of the history of this field and for references, see Chap. 1 of Schneider et al. (1992), see also Narayan and Bartelmann (1997) and Wu (1996) for recent reviews]. These predictions were eventually verified when Walsh et al. (1979) discovered the first lensed QSO, where two QSO images with redshift $z_s = 1.41$, separated by $6''$, have nearly identical spectra from radio to X-ray frequencies, with a giant elliptical galaxy at redshift $z_d = 0.36$, situated in a cluster of galaxies, between the images. Today, the number of multiply-imaged QSOs is about 15; in addition, 6 ring-shaped radio images have been found, in some cases with a (lower-redshift) galaxy at the ring center [for a recent review of the observational situation, see Refsdal and Surdej (1994), and the recent proceedings edited by Kochanek and Hewitt (1996)]. The discovery of giant luminous arcs in 1986 by Lynds and Petrosian (1986) and Soucail et al. (1987) has shown that clusters of galaxies can act as efficient lenses; cluster lensing today is one of the most active fields of gravitational lensing [for a recent review, see Fort and Mellier (1994)]. Finally, the impressive demonstration (Alcock et al. 1993, Aubourg et al. 1993, Udalski 1993) of the feasibility of the suggestion by Paczyński (1986) to search for compact dark objects in the halo of our Galaxy, has led to an active and successful search of Galactic microlensing events, both towards the LMC and the Galactic bulge [for a recent review, see Paczyński (1996)].

These discoveries have opened up a new road towards investigating massive structures in the Universe. Since gravitational light deflection is insensitive to the nature and physical state of the deflecting mass, it is ideally suited to study dark matter in the Universe. In this review, I will concentrate on the detection and measurement of dark halos of galaxy size or larger, using the techniques of weak lensing.

2   What is Weak Lensing?

Light rays from distant sources are deflected if they pass near an intervening matter inhomogeneity. This gravitational lens effect is responsible for the well-established lens systems like multiply-imaged QSOs, (radio) 'Einstein' rings, the giant luminous arcs in clusters of galaxies, and the flux variations of stars in the LMC and the Galactic bulge seen in the searches for compact objects in our Galaxy, as mentioned above. These types of lensing events are nowadays called 'strong lensing', to distinguish it from the effects discussed here: light bundles are not only deflected as a whole, but distorted by the tidal gravitational field of the deflector. This image distortion can be quite weak and can then not be detected in individual images. However, since we are lucky to live in a Universe where the sky is full of faint distant galaxies, this distortion effect can be discovered statistically. This immediately implies that weak lensing requires excellent and deep images so that image shapes (and sizes) can be accurately measured and the number density be as high as possible to reduce statistical uncertainties. Weak gravitational lensing can be defined as using the faint galaxy population to measure the mass and/or mass distribution of individual intervening cosmic structures, or the statistical properties of their mass distribution, or to detect them in the first place, independent of the physical state or nature of the matter, or the luminosity of these mass concentrations. In addition, weak lensing can be used to infer the redshift distribution of the faintest galaxies.

In order to describe these concepts in somewhat more detail, the basic theory of gravitational lensing should briefly be recalled. The formal description of gravitational lensing is basically simple geometry. Consider a mass distribution (the deflector) at some distance $D_{\rm d}$ from us, and some source at distance $D_{\rm s}$ (see Fig. 1). Then, draw a reference line ('optical axis') through lens and observer, define planes ('lens plane' and 'source plane') perpendicular to this optical axis through lens and source, and measure the transverse separations of a light ray in the source and lens plane by $\boldsymbol{\eta}$ and $\boldsymbol{\xi}$, respectively. Then from simple geometry, the relation between these two vectors is

$$\boldsymbol{\eta} = \frac{D_{\rm s}}{D_{\rm d}}\boldsymbol{\xi} - D_{\rm ds}\hat{\boldsymbol{\alpha}}(\boldsymbol{\xi}) \quad , \tag{1}$$

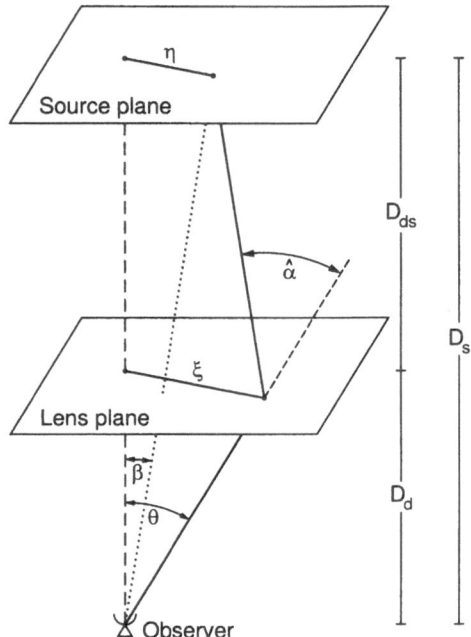

Figure 1 The geometry of a gravitational lens.

where $\hat{\alpha}(\xi)$ is the *deflection angle*. Since all deflection angles one is interested in are very small (even in clusters deflection angles one is interested in are very small (even in clusters of galaxies, the deflection angles are well below 1 arcmin), and thus the gravitational fields are weak, the *linearized field equation of General Relativity* can be employed, which implies that the deflection angle is a linear functional of the mass distribution. Since the deflection angle of a light ray passing a point mass M at separation r is $4GM/(rc^2)$, the deflection angle at position ξ caused by a mass distribution described by the *surface mass density* $\Sigma(\xi)$ becomes

$$\hat{\alpha}(\xi) = \int_{\mathbf{R}^2} \mathrm{d}^2\xi' \, \frac{4G\Sigma(\xi')}{c^2} \, \frac{\xi - \xi'}{|\xi - \xi'|^2} \quad , \tag{2}$$

where the integral extends over the lens plane.

The simple description of a gravitational lens situation can be justified much more thoroughly from Relativity; the reader is referred to Schneider et al. (1992, Chap. 4), and Seitz et al. (1994) for a rigorous treatment. Here it suffices to note that for all situations encountered in this review, the gravitational lens equations provide excellent approximations; in particular, the simple geometrical derivation of (1) remains valid in a universe if the distances are interpreted as *angular-diameter distances*.

It is convenient to replace the physical lengths in (1) by angular variables, by defining $\beta = \eta/D_\mathrm{s}$, $\theta = \xi/D_\mathrm{d}$,

$$\alpha(\theta) = \frac{D_\mathrm{ds}}{D_\mathrm{s}}\hat{\alpha}(\hat{D_d}\theta)) = \frac{1}{\pi} \int_{\mathbf{R}^2} \mathrm{d}^2\theta' \, \kappa(\theta') \frac{\theta - \theta'}{\|\theta - \theta'\|^2} \quad , \tag{3}$$

with the *dimensionless surface mass density*

$$\kappa(\boldsymbol{\theta}) = \frac{\Sigma(D_{\mathrm{d}}\boldsymbol{\theta})}{\Sigma_{\mathrm{cr}}} \quad \text{with} \quad \Sigma_{\mathrm{cr}} = \frac{c^2}{4\pi G}\frac{D_{\mathrm{s}}}{D_{\mathrm{ds}}D_{\mathrm{d}}} \quad ; \tag{4}$$

then the *lens equation* simply reads

$$\boldsymbol{\beta} = \boldsymbol{\theta} - \boldsymbol{\alpha}(\boldsymbol{\theta}) \quad . \tag{5}$$

The *critical surface mass density* Σ_{cr} is a characteristic value which separates strong from weak lenses; if $\kappa \ll 1$ everywhere (i.e., $\Sigma \ll \Sigma_{\mathrm{cr}}$), then the deflector is weak, whereas if $\kappa \sim 1$ for some $\boldsymbol{\theta}$, the lens may produce multiple images and is called strong. *Multiple images* occur if the lens equation (5) has multiple solutions $\boldsymbol{\theta}$ for the same source position $\boldsymbol{\beta}$.

Light bundles are not only deflected as a whole, but differential deflection occurs. Hence, in a first approximation, a circular light bundle acquires an elliptical cross section after passing a deflector. The differential deflection changes the solid angle subtended by a source. Since the *surface brightness* (or the specific intensity) is unchanged by light deflection – this follows from Liouville's theorem, or the fact that light deflection neither creates nor destroys photons – the change in solid angle leads to a change of observed flux from a source: the flux of an infinitesimally small source with surface brightness I and solid angle $\Delta\omega$ is $S = I\,\Delta\omega$. If $\Delta\omega_0$ is the solid angle subtended by an infinitesimally small source in the absence of a deflector, then the observed flux of an image of this source at $\boldsymbol{\theta}$ is $S = \mu(\boldsymbol{\theta})\,S_0$, where the *magnification* μ of an image of an infinitesimally small source is

$$A(\boldsymbol{\theta}) = \frac{\partial\boldsymbol{\beta}}{\partial\boldsymbol{\theta}} \tag{6}$$

is the Jacobian matrix of the lens equation; in components, $A_{ij} = \partial\beta_i/\partial\theta_j \equiv \beta_{i,j}$. The matrix A describes the *locally linearized lens mapping*. We can write the components of A as

$$A = \begin{pmatrix} 1 - \kappa - \gamma_1 & -\gamma_2 \\ -\gamma_2 & 1 - \kappa + \gamma_2 \end{pmatrix} = (1 - \kappa)\mathcal{I} - \|\gamma\| \begin{pmatrix} \cos(2\varphi) & \sin(2\varphi) \\ \sin(2\varphi) & -\cos(2\varphi) \end{pmatrix} \quad , \tag{7}$$

where γ is called *shear* and describes the tidal gravitational forces ($\mathcal{I}$ is the two-dimensional identity matrix). The components of the shear can be calculated from (3) and (6); if we write them in complex notation, $\gamma = \gamma_1 + i\gamma_2$, one finds

$$\gamma(\boldsymbol{\theta}) = \frac{1}{\pi} \int_{\mathbf{R}^2} \mathrm{d}^2\theta' \; \mathcal{D}(\boldsymbol{\theta} - \boldsymbol{\theta}')\,\kappa(\boldsymbol{\theta}') \quad , \tag{8a}$$

with the complex function

$$\mathcal{D}(\boldsymbol{\theta}) = \frac{\theta_2^2 - \theta_1^2 - 2i\theta_1\theta_2}{\|\boldsymbol{\theta}\|^4} \quad . \tag{8b}$$

The eigenvalues of A are $1 - \kappa \pm \|\gamma\|$, where $\|\gamma\| = \sqrt{\gamma_1^2 + \gamma_2^2}$. The fact that the two eigenvalues of A will be different in general implies that a circular source will be imaged, to first approximation, into an ellipse, whose axis ratio is given by the ratio of these two eigenvalues, and the orientation of the major axis is described by the angle φ. We shall later discuss the image distortion for a general source.

Note that det A can vanish, which formally implies a diverging magnification. Of course, real magnifications remain finite. A real source is extended, and the magnification averaged over an extended source is always finite. Even if we had a point source, the magnification would remain finite: in this case, the geometrical optics approximation breaks down and light propagation had to be described by wave optics, yielding finite magnifications [see Chap. 7 of Schneider et al. (1992)]. Astrophysically relevant situations involve sufficiently large sources for the geometrical optics approximation to be valid. The closed curves on which det $A = 0$ are called *critical curves*; the corresponding curves in the source plane, obtained by inserting the critical points into the lens equation, are called *caustics*. An image close to a critical curve can have a large magnification; also, the number of images of a source changes by ± 2 if and only if the source position changes across a caustic. In this case, two images merge at the corresponding point of the critical curve, thereby brightening, and disappear once the source has crossed the caustic. The caustic is not necessarily a smooth curve, but it can develop *cusps*. A source close to, and inside a cusp has three bright images close to the corresponding point of the critical curve, whereas it has one bright image if situated just ouside the cusp.

3   Cluster Mass Reconstruction from Weak Lensing

The fact that the sky is densely covered by faint galaxy images allows the statistical study of distortions of light bundles from these high-redshift sources. The basic idea here is that the shape of a galaxy image is affected by the tidal gravitational field along its corresponding light bundle. This tidal field causes a circular galaxy to form an elliptical image. Since galaxies are not round intrinsically, this effect can not be detected in individual galaxy images (except when the distortion is so strong as to lead to the formation of arcs), but since the intrinsic orientation of galaxies can be assumed to be random, a coherent alignment of images can be detected from an ensemble of galaxies. In this and the next three sections, we shall discuss several aspects of this general idea.

If one considers the line-of-sight towards a cluster of galaxies, one can assume that the main contribution to the tidal gravitational field along light bundles corresponding to galaxies behind the cluster comes from the cluster itself, unless there are other clusters near this line-of-sight. The tidal field, or the shear, is given by (8). Since the relation (8a) between shear and surface mass density is a convolution-type integral, it can be inverted, e.g., by Fourier methods, to yield (Kaiser and Squires 1993)

$$\kappa(\boldsymbol{\theta}) = \frac{1}{\pi} \int_{\mathbf{R}^2} \mathrm{d}^2\theta' \, \Re\left[\mathcal{D}^*(\boldsymbol{\theta} - \boldsymbol{\theta}') \gamma(\boldsymbol{\theta}')\right] + \kappa_0 \quad , \tag{9}$$

where the asterisk denotes complex conjugation, and $\Re(x)$ is the real part of the complex variable x. Hence, if the tidal field γ can be measured, the surface mass density of the cluster can be obtained from (9) up to an overall constant. The reason for this constant to occur is that a homogeneous mass sheet does not cause any shear.

One can think of several methods to characterize the shape of a galaxy image. A convenient method is provided by using the matrix of second brightness moments,

$$Q_{ij} = \frac{\int \mathrm{d}^2\theta \, I(\boldsymbol{\theta}) \, (\theta_i - \bar{\theta}_i) \, (\theta_j - \bar{\theta}_j)}{\int \mathrm{d}^2\theta \, I(\boldsymbol{\theta})} \quad , \tag{10}$$

where $I(\boldsymbol{\theta})$ is the surface brightness distribution, and $\bar{\boldsymbol{\theta}}$ is the center of light of the galaxy image, defined such that $\int d^2\theta\, I(\boldsymbol{\theta})\,(\boldsymbol{\theta} - \bar{\boldsymbol{\theta}}) = 0$. Defining in analogy the tensor of second brightness moments $Q_{ij}^{(s)}$ of the intrinsic brightness distribution of the galaxies, one finds from the lens equation (5) and the conservation of surface brightness, $I(\boldsymbol{\theta}) = I^{(s)}(\boldsymbol{\beta}(\boldsymbol{\theta}))$ that $Q^{(s)} = A\,Q\,A$, where A is given by (7).

In the following, we shall for simplicity restrict our attention to non-critical clusters only, i.e., we shall assume that $\det A > 0$ everywhere. The reader is referred to Schneider and Seitz (1995) and Seitz and Schneider (1995) for the treatment of critical clusters. One then defines the complex ellipticity of an image as

$$\epsilon = \frac{Q_{11} - Q_{22} + 2\mathrm{i}Q_{12}}{Q_{11} + Q_{22} + 2\sqrt{Q_{11}Q_{22} - Q_{12}^2}} \quad , \tag{11}$$

and correspondingly the ellipticity $\epsilon^{(s)}$ of the intrinsic brightness profile of the galaxy in terms of $Q_{ij}^{(s)}$. For example, if an image has elliptical contours of axis ratio $r \leq 1$, then $\|\epsilon\| = (1-r)/(1+r)$. From the relation $Q^{(s)} = A\,Q\,A$ one then derives the transformation between intrinsic and observed ellipticity (Schneider 1995)

$$\epsilon^{(s)} = \frac{\epsilon - g}{1 - g^*\epsilon} \quad , \tag{12}$$

where

$$g = \frac{\gamma}{1 - \kappa} \tag{13}$$

is the (complex) *reduced shear*. Finally, averaging over a set of galaxy images, together with the assumption that the intrinsic ellipticity distribution is isotropic, so that $\langle \epsilon^{(s)} \rangle = 0$, one finds that

$$g = \langle \epsilon \rangle \quad . \tag{14}$$

Several comments have to be made at this point:
(a) The definition (10) of the quadrupole moments cannot be applied to real images, as the integration extends to infinity. In order not to be completely dominated by noise, a weighting function has to be included in the integrals. However, with an angle-dependent weight function, the relation between Q and $Q^{(s)}$ no longer has a simple form and is only approximately given by $Q^{(s)} = A\,Q\,A$; the deviations from this law depend on the intrinsic brightness profile of the source and the weighting function. Even worse is the effect of seeing and an anisotropic point-spread-function (PSF), in particular if the latter is not known very precisely. Several methods to deal with these complications have been discussed in the literature (e.g., Bonnet and Mellier 1995, Kaiser et al. 1995). In particular, a calibration of the relation between ϵ and $\epsilon^{(s)}$ is obtained from numerical simulations and from applying these methods to degraded HST images. It is clear that HST images with their unprecedented angular resolution are best suited for this kind of work, and that ground-based images are more difficult to analyze. Future ground-based observations will make use of the calibration that can be obtained from HST images, in particular if an HST field is centered on the ground-based image.

(b) The fact that the observable g has to be obtained from averaging over an ensemble of galaxy images implies that this method has a finite resolution, i.e., the averaging process is performed over the galaxy images within a certain smoothing length from the point of interest. Several methods of smoothing have been discussed (Kaiser and Squires 1993, Seitz and Schneider 1995); we prefer smoothing with Gaussian weights. Since the number of images over which the average is performed is finite, the relation $\langle \epsilon^{(s)} \rangle = 0$ is not strictly valid due to the finite width of the intrinsic ellipticity distribution; only the expectation value of $\epsilon^{(s)}$ vanishes. The smoothing length needs not to be kept constant, but can be adapted to the local 'strength of the signal'.

(c) It is clear from (14) that only the reduced shear is an observable, but not the shear itself as needed in the inversion equation (9). If the lens is weak in the sense $\kappa \ll 1$, then $g \approx \gamma$, and (9) can be applied directly. In general, one can replace γ in (9) by $(1 - \kappa)g$, which then yields an integral equation for $\kappa(\boldsymbol{\theta})$. As shown in Seitz and Schneider (1995), this integral equation can be easily solved in a few iteration steps. If this nonlinear correction is taken into account, then $\kappa(\boldsymbol{\theta})$ is no longer determined up to an overall additive constant as implied by (9), but there exists a global invariance transformation (Schneider and Seitz 1995)

$$\kappa(\boldsymbol{\theta}) \to \lambda\kappa(\boldsymbol{\theta}) + (1 - \lambda) \quad , \tag{15}$$

which leaves all image shapes invariant. This invariance transformation is the mass sheet degeneracy discussed in a different context by Gorenstein et al. (1988). Of course, the allowed values of λ are restricted by the requirement that the resulting mass distribution is non-negative. Hence, this constraint always allows to obtain a lower limit on the mass. An alternative way to obtain a lower limit to the mass inside circular apertures has been discussed by Kaiser (1995a; see also Sect. 5 below) – the so-called aperture densitometry – which also allows a rigorous estimate of the uncertainty of this lower limit. Also, if the data field is sufficiently large, one might expect that κ decreases to near zero at the boundary of the field, which then yields a plausible range for λ; this in fact is one of the arguments to demand wide-angle fields.

(d) The integral in (9) extends over the whole sky; on the other hand, data are given only on a finite data field (CCD field) $\mathcal{U}$. If the field $\mathcal{U}$ is not sufficiently large, and the contributions of the integral (9) from outside the data field are neglected, the estimate of the surface mass density is no longer unbiased, but boundary artefacts occur. Kaiser (1995a) noticed that there exists a *local* relation between the gradient of κ and certain combinations of first derivatives of the shear components,

$$\nabla\kappa = \begin{pmatrix} \gamma_{1,1} + \gamma_{2,2} \\ \gamma_{2,1} - \gamma_{1,2} \end{pmatrix} =: \boldsymbol{u}(\boldsymbol{\theta}) \quad ;$$

performing averages over line integrations of this local relation allows the construction of unbiased finite-field inversion formulae (Schneider 1995, Bartelmann 1995, Seitz and Schneider 1996, Squires and Kaiser 1996). In Seitz and Schneider (1996), an inversion formula has been derived which filters out a particular noise component in the data which is readily identified as such: As is clear from the above equation, the vector field $\boldsymbol{u}$ should be a gradient field. However, since it is determined observationally, the resulting $\boldsymbol{u}$ will not be a gradient field, but attain a rotational component. Using the fact that the rotational component is solely due to 'noise', one then requires that its average over the data field vanishes, which then yields a unique decomposition of $\boldsymbol{u}$ into a gradient and a rotational component. On quantitative comparison with other inversion formulae, the resulting mass reconstruction formula reveals to be the most accurate direct inversion method known today [see also the lowest three panels in Fig. 6 of Squires and Kaiser (1996)].

(e) The transformation (15) leaves all image shapes invariant, but affects the magnification, $\mu \to \mu/\lambda^2$. Hence, this invariance transformation can be broken if the magnification can be measured. Two possibilities have been mentioned in the literature: Broadhurst et al. (1995) noticed that the magnification effect changes the local number density of galaxy images, $n(S) = n_0(S/\mu)/\mu$, where $n(S)$ are the cumulative number counts, and $n_0(S)$ are the counts in the absence of lensing. Assuming a local power law, $n_0(S) \propto S^{-\alpha}$, then $n(S)/n_0(S) = \mu^{\alpha-1}$. The blue galaxy counts have $\alpha \approx 1$, and so no magnification bias effect is expected. However, counts in the red have a flatter slope, $\alpha \approx 0.75$, and a number density decrease should be seen in regions of high magnifications. The number counts of galaxies with a red color has an even flatter slope, and the magnification effects become stronger. Indeed, this effect has been seen in the clusters A1689 (Broadhurst 1995) and Cl 0024+16 (Schneider et al. 1997). The magnification effect also changes the redshift distribution at fixed apparent magnitude. Bartelmann and Narayan (1995) noticed that individual galaxy images become apparently brighter, at fixed surface brightness. Assuming a sufficiently tight intrinsic magnitude - surface brightness relation, the magnification can be obtained locally. The additional information coming from the magnification effects cannot be incorporated easily in a direct inversion formula such as (9), and there are two possibilities to make use of it: one could obtain the surface mass distribution from a direct inversion, such as (9), and use the magnification information afterwards to fix the transformation parameter λ in (15). Or, one could use a reconstruction method which takes into account the *local* magnification information. One possibility for the latter is a maximum-likelihood approach (Bartelmann et al. 1996, Seitz et al. 1997, in preparation) for the reconstruction of the deflection potential ψ. For an alternative approach see Squires and Kaiser (1996).

(f) We have implicitly assumed that all sources have the same redshift, i.e., that the critical surface mass density Σ_{cr} is the same for all sources. This assumption is not too bad if the cluster is at a sufficiently low redshift, since then the ratio $D_{\mathrm{ds}}/D_{\mathrm{s}}$ can be assumed constant for faint galaxies. In general, however, the redshift distribution of galaxies has to be taken into account. In the weak lensing regime ($\kappa \ll 1$, $\|\gamma\| \ll 1$), only the mean value of $D_{\mathrm{ds}}/D_{\mathrm{s}}$ enters the reconstruction. The non-linear case is more complicated (Seitz and Schneider 1997) and requires the functional form of the redshift distribution. On the other hand, this dependence may also allow to obtain constraints on the redshift distribution of the faintest galaxies. Alternatively, Bartelmann and Narayan (1995) pointed out that the expected strong dependence of surface brightness on the redshift of galaxies, together with the dependence of the lensing strength on source redshift, may allow to determine the redshift distribution of galaxies by studying the variation of lensing strength (i.e., mean ellipticity) as a function of surface brightness. Also, the comparison of lens reconstruction of clusters at different redshifts allows conclusions about the redshift distribution as a function of magnitude – see Smail et al. (1994) and in particular Luppino and Kaiser (1997) who have discovered a strong shear signal in a cluster with redshift $z = 0.83$, implying that a large fraction of the faint galaxies which show the shear effect must have a redshift well in excess of one.

The cluster reconstruction method described above has been applied to several clusters. Fahlman et al. (1994) analyzed the shear field of the cluster MS1224 and obtained a mass-to-light ratio of $\sim 800h$, where h is the Hubble constant in units of $100 \, \mathrm{km/s/Mpc}$;

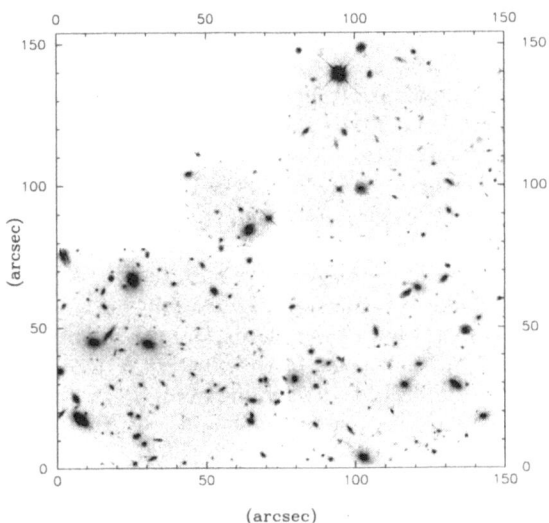

Figure 2 The WFPC2 image of the cluster Cl0939+4713 (A851); North is at the bottom, East to the right. The coordinates are in arcseconds. The cluster center is located at about the upper left corner of the left CCD, a secondary maximum of the bright (cluster) galaxies is seen close to the interface of the two lower CCDs, and a minimum in the cluster light is at the interface between the two right CCDs. In the lensing analysis, the data from the small CCD (the Planetary Camera) were not used.

in particular, the mass derived is much larger than that obtained from a virial analysis. For the cluster A1689, an M/L-ratio of about $450h$ was found by two independent groups (Kaiser 1995b, Tyson and Fischer 1995). A similar value for the M/L-ratio was found for two clusters by Smail et al. (1995).

We (Seitz et al. 1996) have recently analyzed the 'weak' lensing effects in the cluster Cl0939+4713 (A851), using WFPC2 data (Dressler et al. 1994). Since the WFPC2 field is fairly small, we have data only in the center of the cluster, where the lensing is not weak. Also, the small field requires the use of an unbiased finite-field inversion technique, and we used the one derived in Seitz and Schneider (1996). Figure 2 shows the WFPC2 image of the cluster, and the reconstructed mass distribution, together with results from a bootstrapping analysis, is shown in Fig. 3. From the latter figure, one infers that the reconstruction yields basically four significant features in the mass map: a maximum close to the position where the cluster center is predicted from optical observations, a secondary maximum roughly in the lower right CCD, an overall gradient in the lower two CCDs increasing 'to the left', and a pronounced minimum at the interface between the two right CCDs. Comparing these features with the optical image (Fig. 2) one sees that the maximum is clearly visible in the bright (cluster) galaxies, and also the secondary maximum and the minimum in the light distribution is clearly seen to correspond to similar features in the mass map. In addition, the two maxima may be traced by the X-ray emission, as indicated by the ROSAT PSPC-map (Schindler and Wambsganss 1996). Hence, in this cluster we have strong evidence of significant substructure in the mass, and that the light distribution on average follows this substructure; this has also

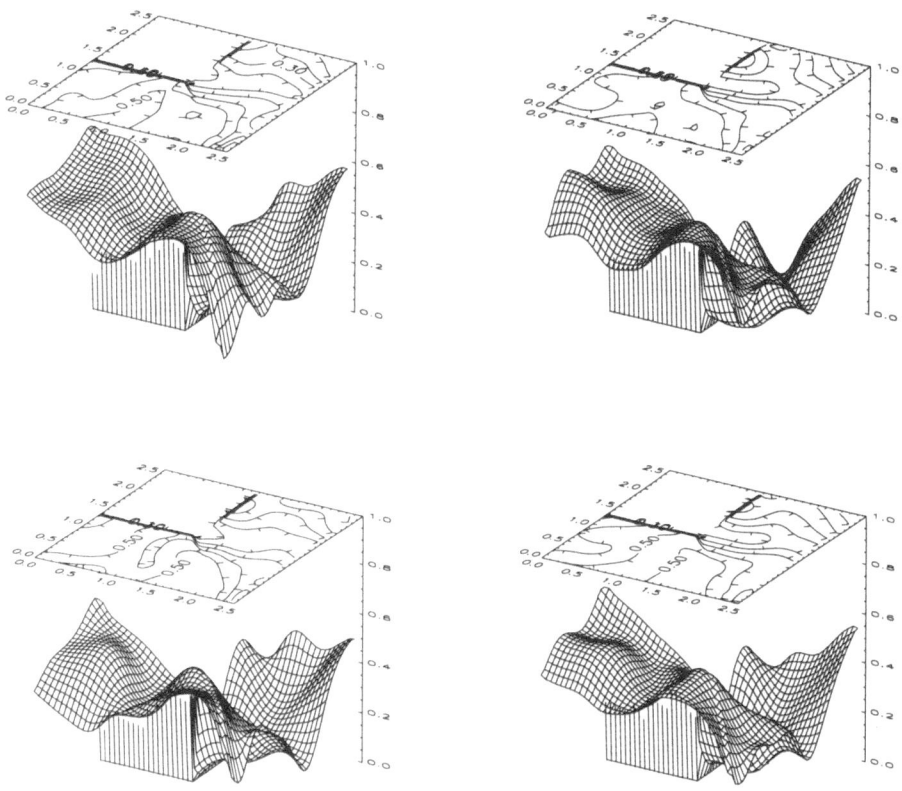

Figure 3 The lower right panel shows the reconstructed mass distribution of A851, assuming a mean redshift of the $N = 295$ galaxies with $24 \leq R \leq 25.5$ of $\langle z \rangle = 1$. The other three panels show reconstructions obtained from the same data set via bootstrapping, i.e., selecting randomly (with replacement) $N = 295$ galaxies from the galaxy sample. The similarity of these mass distributions shows the robust features of the reconstruction, i.e., a maximum, a secondary maximum, an overall gradient, and a pronounced minimum; these features can be compared with the light distribution as shown in Fig. 2.

been demonstrated quantitatively. It will be interesting to compare the mass map with a detailed HRI map which will be obtained soon (S. Schindler, private communication). The M/L-ratio of the cluster within the WFC field depends on the assumed redshift distribution of the background galaxies. Assuming that the mean redshift of galaxies with $24 \leq R \leq 25.5$ is about unity, we find that $M/L \sim 200h$, a value significantly lower than for, e.g., MS1224. However, this is not too surprising, since A851 is the highest-redshift cluster in the Abell catalog which clearly biases towards high optical luminosity. In this cluster, we also have detected the magnification effect discussed above, which has allowed us to obtain not only a strict lower limit on the mass inside the data field, but also to obtain an estimate of the mass, which led to the above value for the M/L-ratio. Note, however, that this mass calibration is uncertain due to the fact that an (unknown) fraction of the faint galaxies are cluster members which renders the estimate of the magnification effect uncertain.

Squires et al. (1996) compared the mass profiles obtained from a weak lensing analysis of the cluster A2218 with that derived from analyzing the X-ray radiation from this cluster. With the assumption that the hot X-ray-emitting intracluster gas is in thermal pressure-supported hydrostatic equilibrium, the mass profile of the cluster can be determined from equating the pressure force with the gravitational force. The mass map obtained from the mass reconstruction agrees qualitatively with the optical and X-ray light profile; using the aperture mass estimate, a mass-to-light ratio of $M/L = (440 \pm 80)h$ in solar units is obtained. The radial profile of the mass appears to be flatter than isothermal. Within the error bars, it agrees with the mass profile obtained from the X-ray analysis, with a slight indication that at large radii the lensing mass is larger than the mass obtained from X-rays. This cluster also contains a large number of arcs and multiply-imaged galaxies which have been used by Kneib et al. (1996) to obtain a detailed mass model of the central region of this cluster. In addition to the main mass concentration, there exists a secondary clump of cluster galaxies whose effects on the arcs is clearly visible. The separation of these two mass centers is $67''$. Whereas the resolution of the weak lensing mass map as obtained by Squires et al. is not sufficient to obtain a distinct secondary peak, the elongation of the central density contours extend towards the secondary galaxy clump.

4   Galaxy-Galaxy Lensing

The shear field around clusters is sufficiently strong to measure their mass distributions (see Sect. 3). One can easily show that, assuming an isothermal mass profile, the 'detection efficiency' of a lens scales like σ^4, where σ is the velocity dispersion. This scaling then implies that individual galaxies are too weak for their presence to be detected in their shear field[1], but one should be able to detect this effect from a large ensemble of galaxies, if the signals from the individual galaxies are added statistically. The signal one would expect is a slight tangential alignment of background galaxies relative to the direction connecting this background galaxy with a near foreground galaxy.

Tyson et al. (1984) have investigated this effect using ~ 60000 galaxies; they obtained a null result. More recently, Brainerd et al. (1996) have analyzed a deep field; they have divided their galaxy sample into 'foreground' and 'background' galaxies, according to the optical magnitudes, and then studied the angle between the major axis of the background galaxy and the line connecting the background galaxy with the nearest foreground galaxies. The distribution of this angle shows a deficit at small angles, and an excess at large angles, indicating the expected tangential alignment. Since an accurate measurement of image ellipticities from the ground is very difficult, only galaxies brighter than $r = 24$ were used; the effect disappears for fainter galaxies, which most likely shows the effect of the PSF on small images. Brainerd et al. have then simulated data, treating galaxies as truncated isothermal spheres, and distributing them in redshift, and they showed that the effect they observe is in accordance with expectations from their modelling. Recalling that this effect was detected (at a $3\text{-}\sigma$ level) with 'only' 506 'background' galaxies, it appears that one can use galaxy-galaxy lensing as a tool to investigate statistically the mass distribution in galaxies, since larger samples will become available soon (also, ground-based images with a smaller and/or more stable PSF will allow the use of fainter

---

[1] Assuming a number density of 50 galaxies/arcmin^2, the minimum velocity dispersion for which a $3\text{-}\sigma$ detection would be possible is about $350\,\mathrm{km/s}$ (Miralda-Escudé 1991, Schneider and Seitz 1995).

galaxies). Schneider and Rix (1997) have proposed a maximum likelihood method for the analysis of galaxy-galaxy lensing, which is very sensitive to the characteristic velocity dispersion of the galaxies, and which can also yield significant lower bounds on the halo size of galaxies. An application of this method to the many thousands of galaxies in the HST Medium Deep Survey will allow the determination of the characteristic velocity dispersion of galaxies to very high accuracy; in fact, Griffiths et al. (1996) have discovered a galaxy-galaxy lensing signal in the MDS. Dell'Antonio and Tyson (1996) have detected a galaxy-galaxy lensing signal in the Hubble Deep Field on scales below $5''$, and further detections of galaxy-galaxy lensing in the HDF on larger angular scales, using additional redshift information to separate foreground from background galaxies, have been reported by R. Blandford (private communication) and M. Hudson (private communication).

To determine the characteristic size of dark halos, one should use shallower, but wide-angle field surveys which will become available soon, e.g., the ESO Imaging Survey (EIS) and the Sloan Digital Sky Survey. Using the above mentioned Maximum Likelihood method, Erben (1997) has shown that the statistical uncertainties in the mass parameters of galactic halos are incredibly small if a survey with the depth and the angular coverage as the EIS is considered, which implies that further information can be obtained, such as the evolution of the mean redshift as a function of apparent magnitude of the galaxies. Also shown by Erben was the fact that the assumption about spherical halos is not an essential one; replacing spheres by elliptical mass distributions only slightly increases the width of the likelihood contours.

5   Detection of (Dark) Matter Concentrations

On wide-field images, one can search for (dark) mass concentrations by looking for statistically significant alignments of faint galaxy images. Let $w(x)$ be a weight function; one can then define an aperture mass

$$m(\boldsymbol{x}_0) := \int \mathrm{d}^2 x \, \kappa(\boldsymbol{x}) \, w \left(\|\boldsymbol{x} - \boldsymbol{x}_0\| \right) = \int \mathrm{d}^2 x \, \kappa(\boldsymbol{x} + \boldsymbol{x}_0) \, w \left(\|\boldsymbol{x}\| \right) \quad , \qquad (16)$$

which is the integral of κ in a circular aperture around $\boldsymbol{x}_0$, weighted by the function w. One can show (Kaiser et al. 1994, Schneider 1996) that m can be expressed directly in terms of the *tangential shear* $\gamma_t(\boldsymbol{y}; \boldsymbol{x}_0)$ *at position* $\boldsymbol{y}$ *relative to the point* $\boldsymbol{x}_0$, $\gamma_t(\boldsymbol{y}; \boldsymbol{x}_0) = -\Re \left[\gamma(\boldsymbol{y} + \boldsymbol{x}_0) \mathrm{e}^{-2\mathrm{i}\varphi} \right]$, where φ is the polar angle of $\boldsymbol{y}$:

$$m(\boldsymbol{x}_0) = \int \mathrm{d}^2 y \, \gamma_t(\boldsymbol{y}; \boldsymbol{x}_0) \, Q(\|\boldsymbol{y}\|) \quad , \qquad (17)$$

where we have defined

$$Q(x) := \frac{2}{x^2} \int_0^x \mathrm{d}x' \, x' \, w(x') - w(x) \quad , \qquad (18)$$

provided $\int \mathrm{d}x \, x \, w(x) = 0$; this last condition on $w(x)$ guarantees that the additive constant in (9) does not appear in (16). In the case of weak lensing, $\kappa \ll 1$, the image ellipticity ϵ at each point is an unbiased estimate of the local shear. Hence, the integral in (17) can be transformed into a sum over image ellipticities. The advantage of this approach is that the resulting quantity $m(\boldsymbol{x}_0)$ has well-defined statistical properties, so that the signal-to-noise ratio can be easily calculated for any chosen weight function w. The weight function w can be optimized for maximizing the signal-to-noise ratio for a given shape of the mass profiles expected. As has been shown in Schneider (1996), this method yields the possibility to reliably detect isothermal-like mass concentrations with velocity dispersion in excess of $600\,\mathrm{km/s}$, i.e., of very weak clusters of galaxies, without any reference to their luminosity. A systematic search for such (dark) mass concentrations is feasible; one only needs wide-field images of sufficient image quality. In fact, with a quite similar approach, Fort et al. (1996), Smail and Dickinson (1995) and Bower and Smail (1997) have detected a significant shear field around several high-redshift QSOs and radio galaxies, indicating a (dark) mass in their line-of-sight.

By choosing the weight function $w(x)$ in (16) to be a positive constant for $0 \leq x \leq x_1$, and a negative constant for $x_1 \leq x \leq x_2$, such that the integral condition is satisfied, then the integral in (17) only extends over the ring with $x_1 \leq x \leq x_2$; the resulting m is then proportional to the mean surface mass density in the inner circle minus the mean surface mass density in the ring; since the latter is non-negative, a lower limit on the mass inside x_1 can be derived from a measurement of the galaxy ellipticities in the ring (Kaiser 1995a).

In a similar spirit as above, one can also define mass multipoles, by supplementing the integrand in (16) by a $e^{in\varphi}$ factor. Also in this case, the resulting expression can be transformed into an integral over the shear, and in the weak lensing case can be replaced by a sum over galaxy ellipticities (Schneider and Bartelmann 1997).

6   Lensing by the Large-Scale Structure

The cosmological density fluctuations out of which the structure in the universe has formed (at least in the conventional model of gravitational instability – which has received impressive support from the detection of microwave background fluctuations by COBE) can also distort the images of high-redshift galaxies. The corresponding distortions have been calculated by Blandford et al. (1991), Kaiser (1992), and Villumsen (1996, and references therein), and are expected to be small; nevertheless, depending on the cosmological model, these distortions are measureable in principle, either by averaging the ellipticity of galaxy images over large fields, or by considering the two-point correlation function of galaxy ellipticities on large scales. If such an effect can be measured, it will allow a direct measurement of the power spectrum of the density fluctuations on the appropriate scales, very much like COBE has done. However, since the finite thickness of the recombination zone washes out fluctuations in the microwave background on scales below $\sim 5'$, comoving angular scales below $\sim 20\,\mathrm{Mpc}$ cannot be probed by them. On such small scales, the cosmic shear provides a very convenient method to measure the density fluctuations, independent on any assumptions about biasing.

The previously mentioned papers calculated the effect of cosmic shear by assuming a linear evolution of the density fluctuations. Jain and Seljak (1997) have calculated the two-point correlation function of galaxy ellipticities, using the full non-linear evolution of the power spectrum. They have shown that the resulting rms shear is significantly increased relative to the linear calculations, on scales below $\sim 10'$, so that it should be easier to measure than previously suspected. In fact, Schneider et al. (1997) probably have discovered a cosmic shear in one direction. Bernardeau et al. (1997) calculated the skewness of the projected density field as measurable from the cosmic shear, using quasilinear theory of structure formation, and Kaiser (1996) considered strategies for measuring the cosmic shear; he also pointed out that the growth of the strength of cosmic shear with source redshift is a useful discriminator between cosmological models. The same data from which galaxy-galaxy lensing was detected by Brainerd et al. (1996) have been used to search for the 'cosmic shear'; keeping in mind the difficulties to measure accurate ellipticities of very faint images from the ground, it is not surprising that Mould et al. (1994) did not find a statistically significant shear signal on a field of $4'.8$ radius. Using the same data, but a different method for analyzing the image ellipticities (basically, giving less weight to 'small' images, which are most contaminated by the PSF), Villumsen (1996) obtained a shear signal with a formal 5-σ significance. Further observations are needed to confirm this result; as mentioned before, the observations are very difficult to carry out, and the expected effects are so small that even tiny systematical effects which escape detection can mimic a significant detection.

7 Conclusions

Weak gravitational lensing has been demonstrated in the last two years to be an extremely powerful tool for extragalactic astronomy and cosmology. The fact that the theoretical concepts are so simple and well understood, and its insensitivity to the state and nature of the matter probed makes it a unique probe of (dark) mass in the universe on all scales – from MACHOs to the large-scale structure itself. The progress that has been made is intimately related to developments on the observational side. Realizing that we live in an era where wide-angle field cameras, space telescopes, and 10-meter class telescopes make their first appearance, it is clear that in weak lensing we have only scratched the surface: these new instrumental possibilities will dramatically increase the rate and quality of data, allowing surveys for dark matter concentrations. The refurbishment of the HST has enabled images of faint galaxies with unprecedented image quality and resolution. These images, together with new theoretical developments, will allow us to understand better the relation between observed image shapes and the true image shapes, before degradation with a PSF. The combination of dark matter maps from weak lensing and X-ray and dynamical studies of clusters will yield fresh insight into the structure, dynamics, and history of these systems. If the systematic effects of ground-based imaging can be understood sufficiently well, we might be able to obtain the cosmic density and the power spectrum of density fluctuations directly from lensing.

References

Alcock, C., Akerlof, C.W., Allsman, R.A., Axelrod, T.S., Bennett, D.P., Chan, S., Cook, K.H., Freeman, K.C., Griest, K., Marshall, S.L., Park, H.-S., Perlmutter, S., Peterson, B.A., Pratt, M.R., Quinn, P.J., Rodgers, A.W., Stubbs, C.W., and Sutherland, W. (The MACHO Collaboration) (1993): Possible gravitational microlensing of a star in the Large Magellanic Cloud. *Nature*, **365**, 621

Aubourg, E., Bareyre, P., Brehin, S., Maurice, E., Prevot, L., and Gry, C. (1993): Evidence for gravitational microlensing by dark objects in the Galactic halo. *Nature*, **365**, 623

Bartelmann, M. (1995): Cluster mass estimates from weak lensing. *Astron. Astrophys.*, **303**, 643

Bartelmann, M. and Narayan, R. (1995): The Lens Parallax Method: Determining Redshifts of Faint Blue Galaxies through Gravitational Lensing. *Astrophys. J.*, **451**, 60

Bartelmann, M., Narayan, R., Seitz, S. and Schneider, P. (1996): Maximum Likelihood Cluster Reconstruction. *Astrophys. J.*, **464**, L115

Bernardeau, F., van Waerbeke, L., and Mellier, Y. (1997): Weak Lensing Statistics as a Probe of Omega and Power Spectrum. *Astron. Astrophys.*, in press

Blandford, R.D., Saust, A.B., Brainerd, T.G., and Villumsen, J.V. (1991): The Distortion of Distant Galaxy Images by Large-Scale Structure. *Mon. Not. Roy. Astron. Soc.*, **251**, 600

Bonnet, H. and Mellier, Y. (1995): Statistical analysis of weak gravitational shear in the extended periphery of rich galaxy clusters. *Astron. Astrophys.*, **303**, 331

Bower, R.G. and Smail, I. (1997): A Weak Lensing Survey in the Field of $z \sim 1$ Luminous Radio Sources. *Mon. Not. Roy. Astron. Soc.*, in press

Brainerd, T.G., Blandford, R.D., and Smail, I. (1996): Weak Gravitational Lensing By Galaxies. *Astrophys. J.*, **466**, 623

Broadhurst, T.J. (1995): In: *Dark matter*, AIP Conf. Proc. 336 (eds. S.S. Holt and C.L. Bennett). AIP, New York

Broadhurst, T.J., Taylor, A.N., and Peacock, J.A. (1995): Mapping Cluster Mass Distributions via Gravitational Lensing of Background Galaxies. *Astrophys. J.*, **438**, 49

Dell'Antonio, I.P. and Tyson, J.A. (1996): Galaxy Dark Matter: Galaxy-Galaxy Lensing in the Hubble Deep Field. *Astrophys. J.*, **473**, L17

Dressler, A., Oemler, A., Butcher, H., and Gunn, J.E. (1994): The Morphology of Distant Cluster Galaxies. I. HST Observations of CL 0939+4713. *Astrophys. J.*, **430**, 107

Erben, T. (1997): Die Bestimmung von Galaxieneigenschaften durch Galaxy-Galaxy-Lensing. Diplomarbeit, TU München

Fahlman, G., Kaiser, N., Squires, G., and Woods, D. (1994): Dark Matter in MS 1224 from Distortion of Background Galaxies. *Astrophys. J.*, **437**, 56

Fort, B. and Mellier, Y. (1994): Arc(let)s in clusters of galaxies. *Astron. Astrophys. Rev.*, **5**, 239

Fort, B., Mellier, Y., Dantel-Fort, M., Bonnet, H., and Kneib, J.-P. (1996): Observations of weak lensing in the fields of luminous radio sources. *Astron. Astrophys.*, **310**, 705

Gorenstein, M.V., Falco, E.E., and Shapiro, I.I. (1988): Degeneracies in Parameter Estimates for Models of Gravitational Lens Systems. *Astrophys. J.*, **327**, 693

Griffiths, R.E., Casertano, S., Im, M., and Ratnatunga, K.U. (1996): Weak gravitational lensing around field galaxies in Hubble Space Telescope survey images. *Mon. Not. Roy. Astron. Soc.*, **282**, 1159

Jain, B. and Seljak, U. (1997): Cosmological Model Predictions for Weak Lensing: Linear and Nonlinear Regimes. *Astrophys. J.*, in press

Kaiser, N. (1992): Weak Gravitational Lensing of Distant Galaxies. *Astrophys. J.*, **388**, 272

Kaiser, N. (1995a): Nonlinear Cluster Lens Reconstruction. *Astrophys. J.*, **439**, L1

Kaiser, N. (1995b): Weak Gravitational Lensing: Current Status and Future Prospects. preprint (astro-ph/9509019)

Kaiser, N. (1996): Weak lensing and cosmology. *Astrophys. J.*, submitted

Kaiser, N. and Squires, G. (1993): Mapping the Dark Matter with Weak Gravitational Lensing. *Astrophys. J.*, **404**, 441

Kaiser, N., Squires, G., and Broadhurst, T. (1995): A Method for Weak Lensing Observations. *Astrophys. J.*, **449**, 460

Kaiser, N., Squires, G., Fahlman, G., and Woods, D. (1994): In: *Clusters of galaxies* (eds. F. Durret, A. Mazure, and J. Tran Thanh Van). Editions Frontiers, Gif-sur-Yvette

Kneib, J.-P., Ellis, R.S., Smail, I., Couch, W.J., and Sharples, R.M. (1996): Hubble Space Telescope Observations of the Lensing Cluster Abell 2218. *Astrophys. J.*, **471**, 643

Kochanek, C.S. and Hewitt, J.N. (eds.)(1996): *Astrophysical applications of gravitational lensing*, IAU Symp. 173. Kluwer, Dordrecht

Luppino, G. and Kaiser, N. (1997): Detection of Weak Lensing by a Cluster of Galaxies at z = 0.83. *Astrophys. J.*, **475**, 20

Lynds, R. and Petrosian, V. (1986): Giant luminous arcs in galaxy clusters. *Bull. Amer. Astron. Soc.*, **18**, 1014

Miralda-Escudé, J. (1991): Gravitational Lensing by Clusters of Galaxies: Constraining the Mass Distribution. *Astrophys. J.*, **370**, 1

Mould, J., Blandford, R., Villumsen, J., Brainerd, T., Smail, I., Small, T., and Kells, W. (1994): A Search for Weak Distortion of Distant Galaxy Images by Large-Scale Structure. *Mon. Not. Roy. Astron. Soc.*, **271**, 31

Narayan, R. and Bartelmann, M. (1997): *Jerusalem Winter School Lectures on gravitational lensing.*

Paczyński, B. (1986): Gravitational Microlensing by the Galactic Halo. *Astrophys. J.*, **304**, 1

Paczyński, B. (1996): Gravitational Microlensing in the Local Group. *Ann. Rev. Astron. Astrophys.*, **34**, 419

Refsdal, S. and Surdej, J. (1994): Thermometry below 1 k. *Rep. Prog. Phys.*, **56**, 117

Schindler, S. and Wambsganss, J. (1996): ROSAT/HRI study of the optically rich, lensing cluster CL0500-24. *Astron. Astrophys.*, in press

Schneider, P. (1995): Cluster lens reconstruction using only observed local data. *Astron. Astrophys.*, **302**, 639

Schneider, P. (1996): Detection of (dark) matter concentrations via weak gravitational lensing. *Mon. Not. Roy. Astron. Soc.*, **283**, 837

Schneider, P. and Bartelmann, M. (1997): Aperture multipole moments from weak gravitational lensing. *Mon. Not. Roy. Astron. Soc.*, **286**, 696

Schneider, P. and Rix, H.-W. (1997): Quantitative Analysis of Galaxy-Galaxy Lensing. *Astrophys. J.*, **474**, 25

Schneider, P. and Seitz, C. (1995): Steps towards nonlinear cluster inversion through gravitational distortions. I. Basic considerations and circular clusters. *Astron. Astrophys.*, **294**, 411

Schneider, P., Ehlers, J., and Falco, E.E. (eds.) (1992): *Gravitational lenses.* Springer, New York

Schneider, P., van Waerbeke, L., Mellier, Y., Jain, B., Seitz, S. and Fort, B. (1997): Detection of shear due to weak lensing by large-scale structure. *Astron. Astrophys.*, submitted

Seitz, C. and Schneider, P. (1995): Steps towards nonlinear cluster inversion through gravitational distortions. II Generalization of the Kaiser and Squires method. *Astron. Astrophys.*, **297**, 287

Seitz, S. and Schneider, P. (1996): Cluster lens reconstruction using only observed local data: an improved finite-field inversion technique. *Astron. Astrophys.*, **305**, 383

Seitz, C. and Schneider, P. (1997): Steps towards nonlinear cluster inversion through gravitational distortions. III. Including a redshift distribution of the sources. *Astron. Astrophys.*, **318**, 687

Seitz, S., Schneider, P., and Ehlers, J. (1994): Light propagation in arbitrary spacetimes and the gravitational lens approximation. *Class. Quan. Gravity*, **11**, 2345

Seitz, C., Kneib, J.-P., Schneider, P., and Seitz, S. (1996): The mass distribution of CL 0939+4713 obtained from a 'weak' lensing analysis of a WFPC2 image. *Astron. Astrophys.*, **314**, 707

Smail, I. and Dickinson, M. (1995): Lensing by Distant Clusters: HST Observations of Weak Shear in the Field of 3C 324. *Astrophys. J.*, **455**, L99

Smail, I., Ellis, R.S., and Fitchett, M.J. (1994): Gravitational Lensing of Distant Field Galaxies by Rich Clusters – I. Faint Galaxy Redshift Distributions. *Mon. Not. Roy. Astron. Soc.*, **270**, 245

Smail, I., Ellis, R.S., Fitchett, M.J., and Edge, A.C. (1995): Gravitational lensing of distant field galaxies by rich clusters. II. Cluster mass distributions. *Mon. Not. Roy. Astron. Soc.*, **273**, 277

Soucail, G., Fort, B., Mellier, Y., and Picat, J.P. (1987): A blue ring-like structure, in the center of the A 370 cluster of galaxies. *Astron. Astrophys.*, 172, L14

Squires, G. and Kaiser, N. (1996): Unbiased Cluster Lens Reconstruction. *Astrophys. J.*, **473**, 65

Squires, G., Kaiser, N., Babal, A., Fahlmann, G., Woods, D., Neumann, D.M., and Böhringer, H. (1996): The Dark Matter, Gas, and Galaxy Distributions in Abell 218: A Weak Gravitational Lensing and X-Ray Analysis. *Astrophys. J.*, **461**, 572

Tyson, J.A. and Fischer, P. (1995): Measurement of the Mass Profile of Abell 1869. *Astrophys. J.*, **446**, L55

Tyson, J.A., Valdes, F., Jarvis, J.F., and Mills, A.P., Jr. (1984): Galaxy Mass Distribution from Gravitational Light Deflection. *Astrophys. J.*, **281**, L59

Udalski, A. (1993): The optical gravitational lensing experiment. Discovery of the first candidate microlensing event in the direction of the galactic bulge. *Acta Astro.*, **43**, 289

Villumsen, J.V. (1996): Weak lensing by large-scale structure in open, flat and closed universes. *Mon. Not. Roy. Astron. Soc.*, **281**, 369

Walsh, D., Carswell, R.F., and Weymann, R.J. (1979): 0957 + 561 A, B – Twin quasistellar objects or gravitational lens. *Nature*, **279**, 381

Wu, X.-P. (1996): Gravitational lensing in the Universe. *Fund. Cosm. Phys.*, **17**, 1

Gravitational Microlensing:
Machos and Quasars

Joachim Wambsganss

Astrophysikalisches Institut Potsdam
An der Sternwarte 16, D-14482 Potsdam
jwambsganss@aip.de

Abstract

Gravitational microlensing deals with the deflection of light by stellar mass objects. The deflection angles are tiny (of the order of milliarcseconds or smaller), so that multiple images on this scale are not observable. However, lensing magnifies the affected sources as well, and since the lens and the source are moving relative to each other, this can be detected as a time-variable brightness.

Observations of microlensing occur in two regimes: Galactic microlensing deals with the situation where foreground stars (or other compact objects) in the disk or the halo of the Milky Way act as lenses for background stars. These can be located either in the galactic bulge or in the Large Magellanic Cloud. The microlensing signature on the light curve is a smooth, symmetric increase/decrease of the apparent magnitude of a lensed star. More than 100 cases of this type of microlensing have been observed. Quasar microlensing deals with quasars that are multiply imaged by an intervening galaxy. The stars in this lensing galaxy act as microlenses on the quasar. They are also affecting the observed brightness of the quasar, but in a more complicated way, since here the magnification is a coherent effect of many stars at the same time.

1 Introduction

Gravitational microlensing is the effect of light deflection of background sources by foreground stars. In such a scenario, the gravitational lens effect has two consequences: it produces two or more images of the background source and at least one of them is magnified, i.e. appears brighter than in the unlensed case. Since the separation of these images is only of the order of a milliarcsecond or smaller, they can not be resolved individually. However, due to the relative motion of observer, lens and source, the image configuration and hence the magnification changes with time. This can be observed by frequently measuring the apparent brightness of the source.

There are two regimes in which microlensing has been observed in recent years. One is compact objects (stars, brown dwarfs, planets, black holes ...) in the halo or the disk of our Galaxy lensing stars in the Large Magellanic Cloud (LMC) or in the central bulge of the Galaxy; this will be called *galactic* microlensing below. The other one is microlensing of a "macro-lensed" quasar by individual stars in the galaxy that acts as the lens (i.e., *extragalactic* microlensing). Both topics are treated below in some detail, with a little bit more emphasis on the former, since it has become quite popular recently. A general introduction into lensing can be found in Schneider et al. (1993).

2   Galactic Microlensing (of Stars)

This here can only be a brief summary of the basics and the most important results in galactic microlensing. For a more complete presentation, I would like to recommend a very recent and comprehensive review on all aspects of microlensing in the local group by Paczyński (1996).

2.1   Basic Idea

The idea for galactic microlensing was proposed a decade ago by Paczyński (1986). He suggested an experiment that could possibly solve a long-standing problem: the flat rotation curves of galaxies (including the Milky Way) require some kind of non-luminous or "dark matter" in the halos of galaxies. If this dark matter is made of any kind of compact astronomical object (faint star, white dwarf, neutron star, brown dwarf, planet, black hole), Paczyński calculated that every now and then one of these "MACHOs" (MAssive Compact Halo Object) – as they were later named by Griest (1991) must pass directly in front of a star in the LMC, and magnify this star in an entirely predictable way. Paczyński found that the "optical depth" to microlensing for the stars in the LMC is of the order of 10^{-6}, i.e. at any given time one out of roughly a million stars in the LMC is magnified by at least 0.3 mag (see section below) by the gravitational lens action of a "Macho", if the halo is made entirely out of them. That means, in order to observe this effect, one has to monitor of the order of 10^6 stars, ideally about once per night! What at the time sounded like science fiction, soon became feasible through big advances both in the development of CCD technology and in computer speed and disk space. Three big collaborations started to attack this problem only a few years after it was suggested.

2.2   Geometry

The setup for a gravitational lens situation in this regime of very low optical depth is given by an isolated lens L and an isolated source star S, as shown on Fig. 1. There are two images of the source, S_1 and S_2. Due to the small angular separation of the two images (of the order of a milliarcsecond or smaller) the images cannot be observed separately, but rather only the combined intensity. Due to the relative motion between observer, lens and source, the relative positions and – even more important – the relative brightnesses of the two images change with time. This is displayed in Fig. 2a for an extended source for five different instants of time, starting with a rather large projected separation between source and lens, and ending with perfect alignment. In Fig. 2b, the light curve for such a scenario is shown for a source of finite size, i.e. the combined magnifications of the two lensed images. The five instants of time corresponding to the "snapshots" displayed in Fig. 2a are marked with dotted vertical lines.

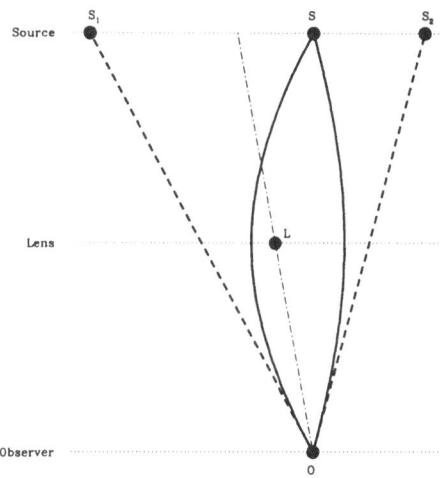

Figure 1   Setup of a gravitational lens situation: The lens L located between source S and observer O produces two images S_1 and S_2 of the background source.

2.3   Equations

The total magnification $\mu = \mu_1 + \mu_2$ of the two images entirely depends on the impact parameter $u(t) = r(t)/R_E$ between the lensed star and the lensing object (measured in the lens plane):

$$\mu(u) = \frac{u^2 + 2}{u\sqrt{u^2 + 4}} \tag{1}$$

where R_E is the Einstein radius of the lens, i.e. the radius at which a circular image appears for perfect alignment between source, lens and observer (cf. Fig. 2a, rightmost panel):

$$R_E = \left(\frac{4GM}{c^2} \frac{D_d D_{ds}}{D_s} \right)^{1/2} ; \tag{2}$$

in angular coordinates, the Einstein angle reads:

$$\theta_E = \left(\frac{4GM}{c^2} \frac{D_{ds}}{D_s D_d} \right)^{1/2} . \tag{3}$$

Here c is the velocity of light, G the Gravitational constant, and the Ds are the distances between observer and lens (D_s) and observer and source (D_s). The distance between lens and source is simply $D_{ds} = D_s - D_d$.

The time scale of such an "microlensing event" is defined as the time it takes the source to cross the Einstein radius:

$$t_0 = \frac{R_E}{v_\perp} \approx 0.21 \left(\frac{M}{M_\odot} \right)^{1/2} \left(\frac{D_d}{10\mathrm{kpc}} \right)^{1/2} \left(1 - \frac{D_d}{D_s} \right)^{1/2} \left(\frac{v_\perp}{200\mathrm{km/sec}} \right)^{-1} \mathrm{yr.} \tag{4}$$

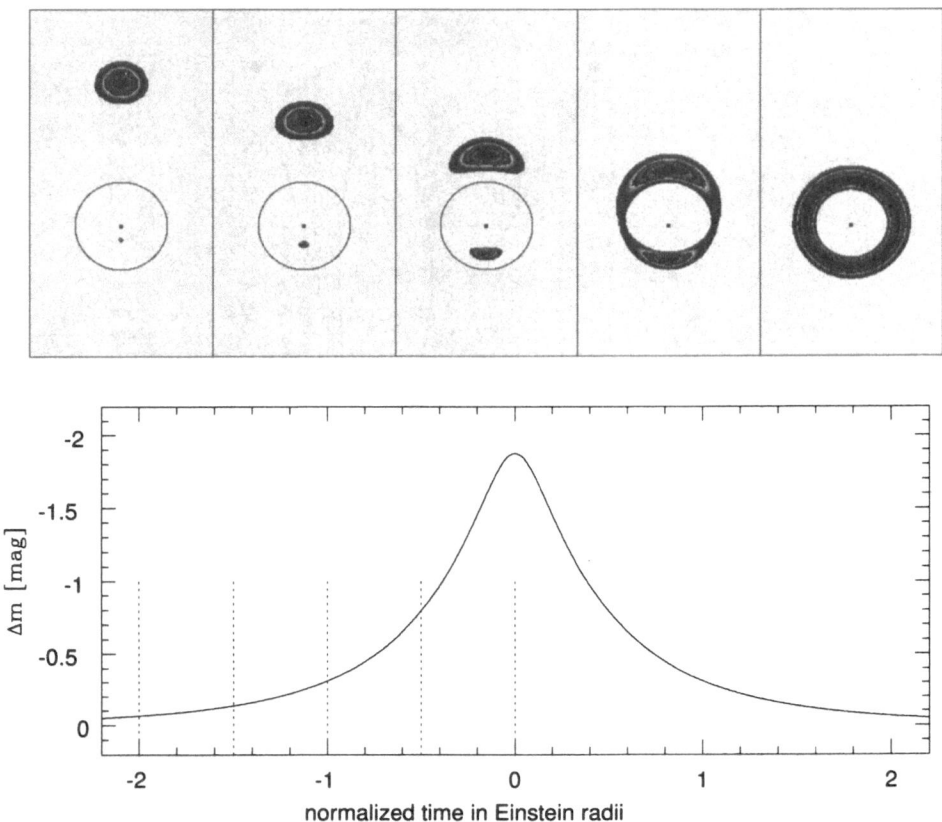

Figure 2 a) Five snapshots of a gravitational lens situation: From left to right the alignment between lens and source gets better and better, until it reaches perfect alignment in the rightmost panel, which results in an "Einstein-ring image". b) Light curve for a microlensed source. The five snapshots displayed in a) are indicated with dotted vertical lines.

Here $v_\perp$ is the (relative) transverse velocity of the lens and we parameterized by "typical" numbers for the other variables. This equation shows the important fact that it is not possible to determine the mass of the lens responsible for a certain microlensing event individually. The duration of an event is determined by three unknown parameters, namely the mass of the lens, the transverse velocity and the distance between lens and source. It is impossible to disentangle these for individual events. One has to use a model for the spatial and velocity distribution of the lensing objects, in order to be able to check "consistency" statistically.

The impact parameter as a function of time is given as

$$u(t) = \left[u_{\min}^2 + \left(\frac{t - t_{\min}}{t_0} \right)^2 \right]^{1/2} .$$

(5)

Here u_{min} is the impact parameter at minimum distance, t_{min} is the time of the closest approach. The cross section for microlensing is defined in such a way that the dimensionless impact parameter has to be smaller than unity, $u \leq 1$, i.e. the source position is inside the Einstein radius of the lens. Plugged in Eq. 1, this translates into a minimum magnification $\mu \geq 3/5^{1/2} \approx 1.3416$; this translates into $\Delta m \approx 0.32$mag. With this definition, the optical depth can be visualized as the fraction of the sky that is covered by the "Einstein circles" of the lensing objects.

2.4 Predictions

It was obvious from the beginning, that a massive monitoring campaign would produce a very large number of variable stars, and that the few microlensing events per 10 million stars would rather be like the needle in the haystack of other variables. However, microlensing light curves have a few characteristics which make it possible to distinguish them from all kinds of (known) variable stars. For an individual microlensing light curve the following statements must hold:

- the microlensing event is achromatic
- the microlensing event is symmetric and follows Equation 1
- the microlensing event does not recur
- the lensed star does not have a peculiar spectrum

For a large number of microlensing light curves, produced by "Machos", the following statistical arguments must hold:

- the positions of the lensed stars in the sky must be distributed proportional to the star density in the LMC
- the positions of the lensed stars in the color-magnitude diagram must be distributed proportional to the density of the stars there (and not accumulate in any peculiar corner)
- the distribution of maximum magnifications for events that are within the Einstein ring (i.e., $\mu_{thresh} \approx 1.34$) must follow the predetermined relation $f(> \mu) = \sqrt{2} \ [\ (1 - \mu^{-2})^{-1/2} - 1 \ ]^{1/2}$, or $f(> \mu) \propto \mu^{-1}$ for $\mu \gg 1$.

2.5 Experiments

Originally, three groups started these massive searches for microlensing events.

- MACHO collaboration (MAssive Compact Halo Objects; USA/Australia): big CCD camera, blue and red simultaneously, at Mt. Stromlo, observing LMC and galactic bulge (Alcock et al. 1993)
- EROS collaboration (Expérience de Recherche d'Objets Sombres; France) CCD and photographic, La Silla, observing LMC (Aubourg et al. 1993)
- OGLE collaboration (Optical Gravitational Lens Experiment; Poland/USA) Las Campanas, observing the galactic bulge (Udalski et al. 1993)

The various groups followed different strategies. In order to cover different mass scales for the (unknown) lensing objects, the monitoring frequency varied from about two observations per month up to about 10 times per night. The groups had also different emphases. OGLE monitored only the galactic bulge, whereas MACHO and EROS looked preferentially at the LMC.

2.6   First Results and Impact

In September 1993, almost at the same time, all three groups mentioned above reported their first events (Alcock et al. 1993, Aubourg et al. 1993, Udalski et al. 1993) the LMC and the galactic bulge, respectively. Since then many more events have been found, and new teams joint the race (Ansarie et al. 1996, Alard et al. 1995).
In Fig. 3, the light curves of OGLE events nos 1-4 are displayed, the first microlensing events towards the galactic bulge. Figure 4 shows small pieces of the mapped region in the bulge of the Galaxy, with the lensed stars at minimum and maximum magnification, respectively. These maps give a hint on how difficult it is to do good photometry in such crowded fields!
The impact of the microlensing searches can easily be seen when looking at the following numbers. According to a compilation by Surdej and Pospieszalska-Surdej (1996)a total of 262 papers were published or submitted in the field of gravitational lensing. It turns out that 84 of them, almost a third, deal with this type of galactic microlensing!

2.7   Complications

In section 2.4 the "ideal" situation was presented, as can be easily derived from gravitational lens theory. The real world, though, is not that perfect. In fact, many microlensing light curves are not perfectly achromatic, nor perfectly symmetric, nor perfectly following the light curve as described in Eq. 1. There are various reasons for this, and I can list only a few here:

- Binary Lenses: More than half of the stars exist in double or multiple star arrangements. For a certain range in separation (roughly, if the projected distance between the stars is comparable to the Einstein radius of the total mass), they act as a double lens. Such complicated lenses produce light curves that are quite different from the smooth ones shown in Fig. 2b. The reason is that with such a lens configuration "caustics" occur, locations which correspond to formally infinite magnification. These caustics can produce very steep and high amplitude features in the light curve. Figure 5 shows the light curve of the first binary event observed, OGLE no 7. So far at least three binary microlensing events have been observed.

- Repeating Events: Even if the double star acting as a lens has a wide separation, so that each lens acts as a single rather than as a binary lens, it is imaginable (though not very likely) that the separation of the two stars is just along direction of the relative motion. That means, there could be two individual microlensing events for the same source star.

- Parallax: If the duration of the microlensing event is of the order a year, the motion of the earth around the sun introduces a second (vector) component to the relative

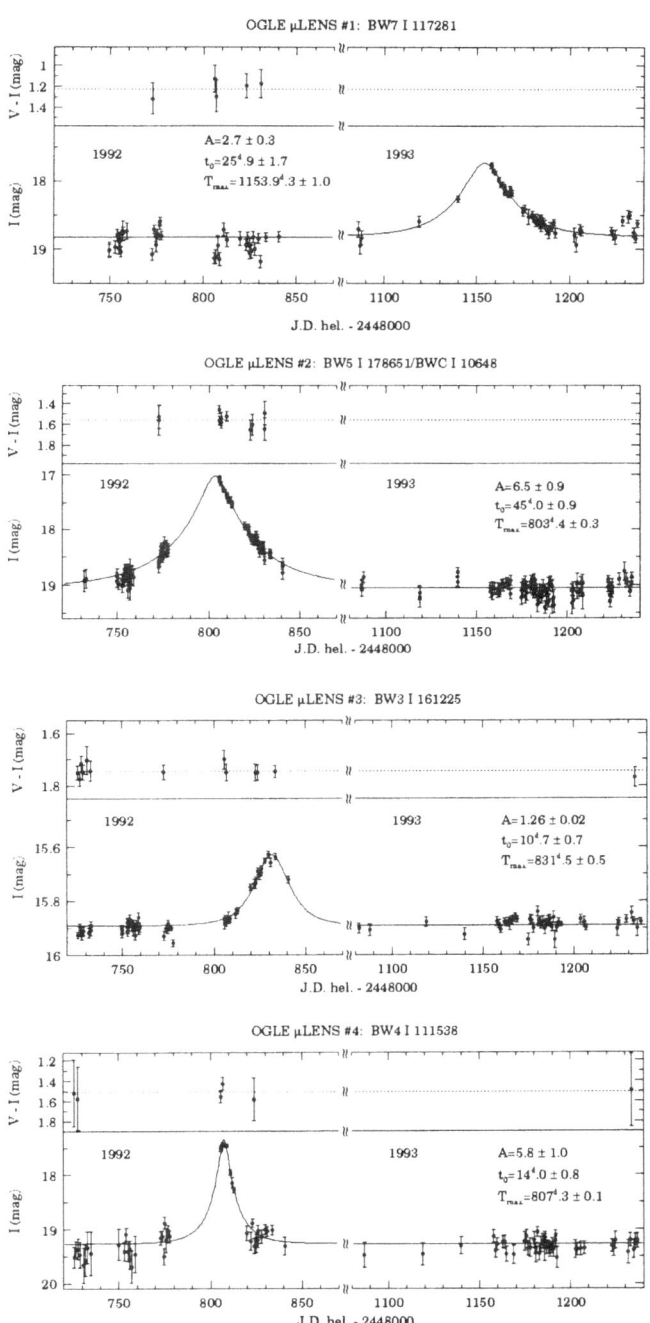

Figure 3 Light curves for OGLE events nos 1-4.

OGLE #1   OGLE #2   OGLE #3   OGLE #4

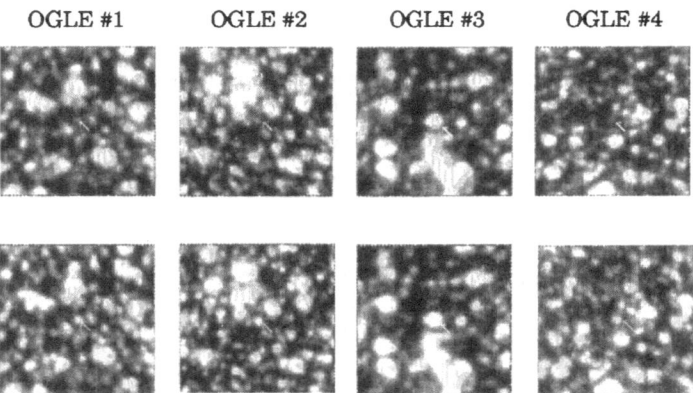

Figure 4 Crowded star fields with the locations of the stars responsible for OGLE events nos 1-4, both in "low" and "high" state.

velocity vector between source, lens and observer (Gould 1994). That means a (small) deviation from the smooth curves. At least one of such events has been observed by the MACHO team (Alcock et al. 1995).

- Extended Source: The idealized scenario for microlensed light curves mentioned above assumed point sources. Since in the real world all physical sources are finite, at some point there are deviations between ideal and real behavior. The finite size of the source can lead to higher or lower amplitude events (Witt and Mao 1994).

- Binary Source: If an unresolved double star is gravitationally lensed, only one of the two partners is magnified. If the two stars have different colors, the color of the combined light curve changes during the event.

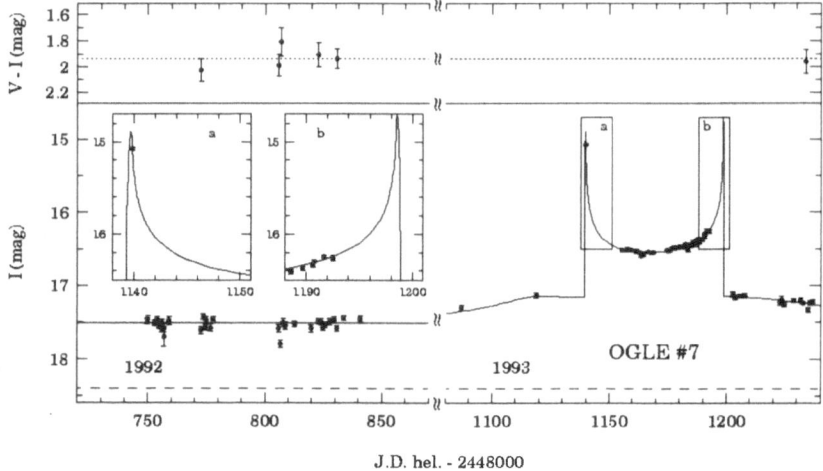

Figure 5 Light curves for OGLE event no 7, the first binary-lens event.

Another complication arouse from the possible positions of the lenses. Originally only a simple model for the galactic halo (galactic disk) was considered as the location of the possible lenses for monitoring the stars in the LMC (galactic bulge). By now there is evidence that these models were too simplistic. Various other lens distributions are discussed, among them a non-standard halo (i.e. highly flattened); foreground stars in the LMC (bulge) could act as lenses to background stars in the LMC (bulge); lenses in the galactic corona or in the galactic spheroid; population III stars; a heavy disk; lenses in the galactic bar (cf. Paczyński 1996).

2.8   The Current State

By now more than 100 microlensing events have been detected, most of them towards the bulge of the Galaxy. Recently, the MACHO collaboration published eight events towards the LMC. The duration of the events ranges from nine to 200 days (Alcock et al. 1996). The distribution of the "lensed" bulge stars in the color magnitude diagram and the distribution of their maximum magnifications is consistent with the events being microlensed, there are no peculiarities.

There is no evidence, though, that the bulge events are related to dark matter, all can be produced by ordinary low mass stars. The quantitative picture is such that the number of events towards the galactic bulge is about 3 times higher than predicted/expected. This clearly shows that the Galactic structure was not understood well enough when the experiments started. Thus, lensing is also a new tool for studying the structure of the Galaxy (Paczyński 1996).

The number of events towards the LMC is smaller than expected. It depends on the exact model of the halo to quantify this. The most recent analysis estimates that about 50% of the matter in the halo could be due to machos. However, the most likely mass of the lenses is quite high, of the order of a solar mass (Alcock et al. 1996).

The MACHO and the OGLE groups have implemented an on-line data reduction system so that microlensing events can be "caught in the act" and "early warnings" or "alerts" can be released. This does not just allow a much better time coverage of the light curve, but also taking spectra during and after the event, so that one can really see whether the spectrum changed during the event. So far about 30 (MACHO) and 6 (OGLE) of such events have been detected in real time (Alcock et al. 1996, Udalski et al. 1994).

2.9   The Future

Alerts/Follow-up Observation

The current experiments will continue. The OGLE group will soon get a dedicated telescope in Las Campanas, so that they also can observe year round. There are also attempts by at least two groups [PLANET - Probing Lens Anomalies NETwork (Albrow et al. 1996); GMAN - Global Microlensing Alert Network (Pratt et al. 1996)] to set up coordinated follow-up observing networks around the globe. This should make sure to get a real good time coverage of the microlensed light curves. In principle, time scales from about ten minutes to about 100 years are accessible with the current technology, covering 13 orders of magnitude in lens masses.

Search for Planets

Another promising aspect for the future of these microlensing searches is the possibility of detecting planets. In fact, it is just an extrapolation from the binary lens scenario to mass ratios of the order of 1:1000 or higher. It turns out that with high enough time resolution (ideally: once per hour or higher) and high enough photometric accuracy (0.01 mag or better), in a fraction of the microlensed light curves small perturbations could be detected which are the fingerprints of planetary companions to the main lensing stars (see Wambsganss 1997 and references therein).

"Pixel Lensing"

A third very interesting development is the so-called "pixel-lensing". This deals with the fact that in galaxies further away from us than the LMC, individual stars cannot be resolved any more. That means, in each pixel on a CCD the images of many stars are merged. If now one of these stars undergoes a microlensing effect, the total number of photons in this pixel increases, but the "light curve" is heavily damped due to the additional light of the other stars in this pixel. Nevertheless, it is possible to subtract CCD frames from each other, and – after a lot of technical difficulties have been overcome – filter out those pixels which correspond to light curves of variable stars and microlensing events. First attempts to implement this technique are underway and will ultimately be able to detect microlensing in stars of the Andromeda galaxy M31.

3 Extragalactic Microlensing (of Quasars)

3.1 What is Quasar Microlensing?

So far we know of about 20 quasars which are gravitationally lensed, i.e. we see two or more images of them. The lenses are galaxies or galaxy clusters which happen to lie along the line of sight between the quasar and us. The quasar images are usually seen through a certain part of a lensing galaxy. That means there are stars directly "in front of" the quasar. These stars (and possibly other compact objects, such as black holes, brown dwarfs, or planets) act as gravitational (micro-)lenses as well. They split each of the quasar (macro-)images into two or more. The separation is only of the order of microarcseconds, so it is not observable. But lensing by stars causes a magnification as well, and since this changes with time due to the relative motion, it is observable!

The surface mass density of the stars in front of a multiply imaged quasar is high, due to the fact that the quasar is "macro-lensed". It is of the order of the "critical surface mass density":

$$\Sigma_{\text{crit}} = \frac{c^2}{4\pi G} \frac{D_s}{D_d D_{ds}}; \tag{6}$$

here the Ds are the angular diameter distances between observer and lens (D_d), observer and source (D_s), and lens and source (D_{ds}). Note in particular that for these cosmological distances, the latter is not just the difference of the two former values: $D_{ds} \neq D_s - D_d$! (The value of the critical surface mass density is very roughly $\Sigma_{\text{crit}} \approx 0.5 \text{gcm}^{-2}$ for typical quasar lenses.)

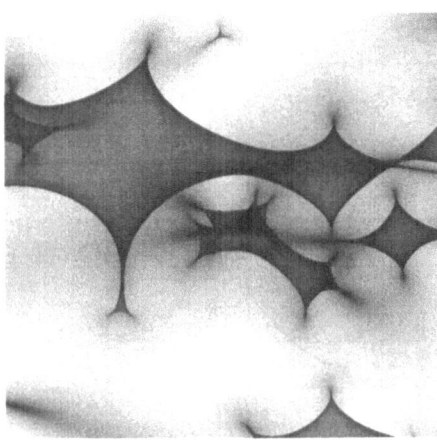

Figure 6 Magnification pattern in the quasar plane, produced by a dense field of stars in the lensing galaxy. The gray color reflects the magnification as a function of the quasar position: dark means high magnification.

For such a high surface mass density, the stars produce a quite complicated magnification pattern in the plane of the quasar. It consists of many caustics, locations that correspond to formally infinitely high magnification. An example for such a magnification pattern is shown in Fig. 6. This is the two-dimensional distribution of the microlensing magnification as a function of position in the quasar plane. The darker the gray, the higher the magnification at this location. Due to the relative motion of quasar, galaxy and observer, the quasar changes its position relative to this arrangement of caustics, i.e. at different times the quasar is differently magnified. That means the apparent brightness of the quasar image changes with time.

3.2 Why is it Interesting?

Microlens-induced fluctuations in the observed brightness of quasars contain information both about the light-emitting source (size of continuum region or broad line region of the quasar, brightness profile of quasar) and about the lensing objects (masses, density, transverse velocity). It is not trivial, though, to extract this information.

We saw already in Section 2, that the masses of the lenses responsible for individual microlensing events cannot be obtained directly from the light curves. Here the situation is even more complicated, since microlensing in this regime of high optical depth is a coherent effect of many stars. That means individual mass determinations are not just impossible from the detection of a single caustic-crossing microlensing event, but it does not even make sense to try to do so, since these events are not produced by individual lenses. Mass determinations can be done only in a statistical sense, by comparing good observations (frequently sampled, high accuracy) with simulations.

So far the best example of a microlensed quasar is the quadruple quasar Q2237+0305 (Corrigan et al. 1991). Comparing the observed fluctuations in this quasar with simulations allowed to conclude that the continuum emitting region cannot be larger than about 2×10^{15}cm (Wambsganss et al. 1990).

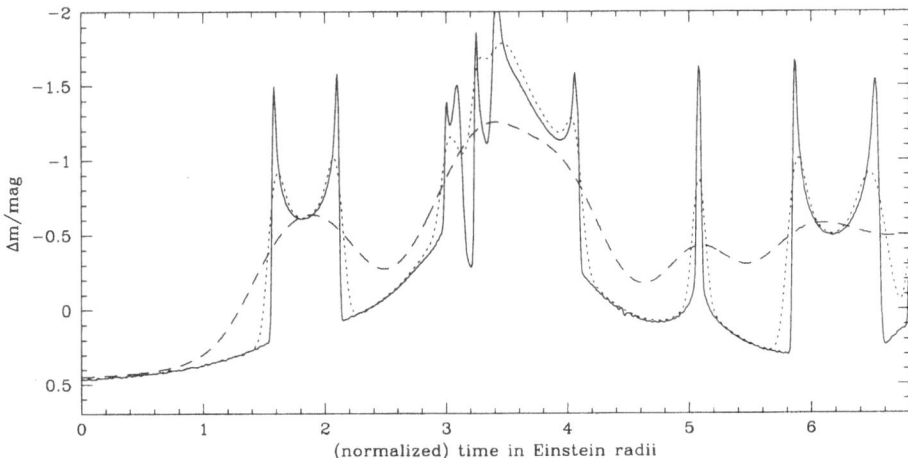

Figure 7 Microlensed light curves taken along the central horizontal track of the magnification pattern in Fig. 6, for three different source sizes R_S (Gaussian profiles); solid line: $R_S = 0.008R_E$; dotted line: $R_S = 0.040R_E$; dashed line: $R_S = 0.200R_E$.

3.3 Microlensing Simulations

There exist two complementary methods for the determination of the magnification distributions produced by microlenses. One method determines the deflection angles of light rays by all the microlenses and follows the rays to the quasar plane. The density of the rays in the quasar plane reflects the magnification at this location (see, e.g., Wambsganss et al. 1990) The other method is based on the parametric representation of the caustics, and finds the exact positions of them (Witt 1993, Lewis et al. 1993). A combination of the two methods (ray shooting and parametric representation of caustics) can be found in Wambsganss et al. (1992).

Microlensing simulations are done to compare theory with observations. The most straightforward result to extract from a simulated two-dimensional magnification pattern like the one in Fig. 6 are light curves, i.e. one-dimensional cuts through it. These simulated light curves correspond to a situation in which the relative motion between quasar, lensing galaxy and observer is a straight line (which is a pretty good approximation). One can "fold in" the brightness profile of the quasar, usually taken as a Gaussian brightness distribution. But light curves for other quasar models (like one consisting of a number of hotspots) can as easily be determined.

An example for light curves determined along the horizontal central line of Fig. 6 is shown in Fig. 7, for three different source sizes. It is obvious that the amplitude of the peaks strongly depends on the source size, being higher and more peaked for the smaller sources.

The fact that the microlensed light curves depend on the source size introduces a "chromatic" effect on quasars. It is well known that the deflection of a light ray is achromatic, gravity does not depend on the energy or the wavelength of a photon. Nevertheless, a chromatic effect can be observed in microlensed quasars. The reason is that the physical model of a quasar, namely an accretion disk around a central black hole, predicts higher

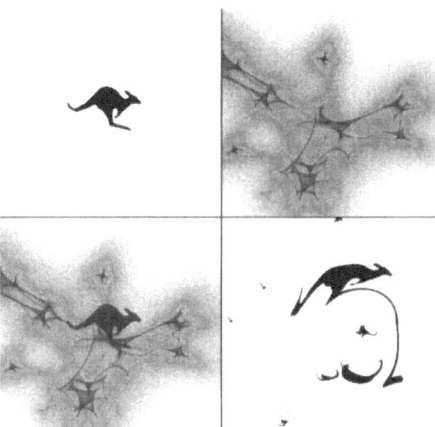

Figure 8 Visualization of the gravitational lens effect on an asymmetric extended source. The source shape is shown in the top left panel, the caustic/magnification pattern produced by a lens consisting of a few mass concentrations in the top right panel. The bottom left panel shows the source location relative to the caustics, and the bottom right panel presents the view we would have if we could see such an asymmetric source through this particular lens pattern. Note the distorted images, the mirror-inverted demagnified images, and the highly enlarged Giant Luminous Kang-Arc-roo.

temperatures for the regions of the disk closer to the center, and lower temperatures farther out the disk. This means the central part of an accretion disk should be "bluer", the outer parts "redder". In other words, the source size of a quasar is different when viewed in red light and in blue light. Therefore the amplitudes of microlensing peaks produced should be higher in a filter corresponding to a shorter wavelength, the "color" of the quasar should change during a microlensing event. This was explored quantitatively by (Wambsganss and Paczyński 1991).

In "standard" microlensing simulations, it is assumed that the relative motion between quasar, galaxy and observer is dominated by the large scale motions. This approximation is not bad, but for more quantitative analysis the motion of the individual stars in the lensing galaxy has to be considered as well, since this causes not just a lateral motion of the caustics, but rather a change of their shapes and configuration. Wambsganss and Kundíc (1995) the microlens-induced peaks in the light curve can be much narrower/sharper in this case, since the caustics themselves can move with arbitrarily high velocities. A nice visualization of the moving and changing caustics based on the analysis of Wambsganss and Kundíc (1995) can be obtained as an MPEG-movie for three different surface mass densities at `http://www.aip.de:8080/~jkw`.

4 Macropod Microlensing (of Kangaroos)

For the visualization of the multiple imaging aspects of gravitational lensing it is useful to study the effects for a source shape which is not perfectly symmetric (unlike the standard sources circle or ellipse). For asymmetric sources, it becomes obvious to the eye, that lensing does not just magnify an extended source by a constant factor, but rather that a realistic lens consisting of various components can differentially magnify parts of the source, even produce multiple images of only parts of the source (those parts that happen to lie within the caustics). This effect is illustrated in Fig. 8. The top left panel shows the shape of the unlensed source, reminiscent of a kangaroo. The top right part shows the particular magnification/caustic distribution in the source plane. The bottom left panel shows the superposition of the kangaroo-source ontop of the magnification distribution, and the bottom right panel presents the view of the distorted images of the kangaroo which one would see for the given distribution of point lenses (marked as little squares). It is quite amazing to see how different parts of the kangaroo-shaped source get distorted and magnified, producing images like the Giant Luminous Kang-Arc-roo!

References

Alard, C., Mao, S., and Guibert, J. (1995): Object DUO 2: A New Binary Lens Candidate? *Astron. Astrophys.*, **300**, L17

Albrow, M., Birch, P., Caldwell, J., Martin, R., Menzies, J., Pel, J.-W., Pollard, K., Sackett, P.D., Sahu, K., Vreeswijk, P., Williams, A., and Zwaan, M. (1996): The Contribution Of Binaries To The Observed Galactic Microlensing Events. In: *IAU Symposium*, **173** (eds. C.S. Kochanek and J.N. Hewitt). Kluwer, Dordrecht, p. 227

Alcock, C., Allsman, R.A., Alves, D., Axelrod, T.S., Bennett, D.P., Cook, K.H., Freeman, K.C., Griest, K., Guern, J., Lehner, M.J., Marshall, S.L., Peterson, B.A., Pratt, M.R., Quinn, P.J., Rodgers, A.W., Stubbs, C.W., and Sutherland, W. (The MACHO Collaboration)(1995): First Observation of Pusallax in a Gravitational Microlensing Event. *Astrophys. J.*, **454**, L125

Alcock, C., Akerlof, C.W., Allsman, R.A., Axelrod, T.S., Bennett, D.P., Chan, S., Cook, K.A., Freeman, K.C., Griest, K., Marshall, S.L., Park, H.-S., Perlmutter, S., Peterson, B.A., Pratt, M.R., Quinn, P.J., Rodgers, A.W., Stubbs, C.W., and Sutherland, W. (1993): Possible Gravitational Microlensing of a Star in the LMC. *Nature*, **365**, 621

Alcock, C., Allsman, R.A., Axelrod, T.S., Bennett, D.P., Cook, K.H., Freeman, K.C., Griest, K., Guern, J.A., Lehner, M.J., Marshall, S.L., Park, H.-S., Perlmutter, S., Peterson, B.A., Pratt, M.R., Quinn, P.J., Rodgers, A.W., Stubbs, C.W., and Sutherland, W. (The MACHO Collaboration) (1996): The MACHO Project First-Year LMC Results: The Microlensing Rate and the Nature of the Galactic Dark Halo. *Astrophys. J.*, **461**, 84

Ansari, R., Cavalier, F., Moniez, M., Aubourg, E., and Bareyre, P. (1996): Observational limits on the contribution of substellar and stellar objects to the galactic halo. *Astron. Astrophys.*, **314**, 94

Aubourg, E., Bareyre, P., Bréhin, S., Gros, M., Lachièze-Rey, M., Laurent, B., Lesquoy, E., Magneville, C., Milsztajn, A., Moscoso, L., Queinnec, F., Rich, J., Spiro, M., Vigroux, L., Zylberajch, S., Ansari, R., Cavelier, F., Moniez, M., Beaulieu, J.-P., Ferlet, R., Grisan, Ph., Vidal-Madjar, A., Guibert, J., Moreau, O., Tajahmady, F., Maurice, E., Prévôt, L., and Gry, C. (1993): Evidence for Gravitational Microlensing by Dark Objects in the Galaktic Halo. *Nature*, **365**, 623

Corrigan, R.T., Irwin, M.J., Arnaud, J., Fahlman, G.G., and Fletcher, J.M. (1991): Initial lightcurve of Q2237+0305, *Astron. J.*, **102**, 34

Gould, A. (1994): MACHO Velocities from Satellite-Based Parallaxes. *Astrophys. J.*, **421**, L75

Griest, K. (1991): Galactic Microlensing as a Method of Detecting Massive Compact Halo Objects. *Astrophys. J.*, **366**, 412

Lewis, G.F., Miralda-Escudé, J., Richardson, D.C., and Wambsganss, J. (1993): Microlensing Light Curves: A New and Efficient Numerical Method. *Mon. Not. Roy. Astron. Soc.*, **261**, 647

Paczyński, B. (1986): Gravitational Microlensing by the Galactic Halo. *Astrophys. J.*, **304**, 1

Paczyński, B. (1996): Gravitational Microlensing in the Local Group. *Ann. Rev. Astron. Astroph.*, **34**, 419-459

Pratt, M.R., Alcock, C., Allsman, R.A., Alves, D., and Axelrod, T.S. (1996): Real-Time Detection Of Gravitational Microlensing. In: *IAU Symposium*, **173** (eds. C.S. Kochanek and J.N. Hewitt). Kluwer, Dordrecht, p. 221

Schneider, P., Ehlers, J., and Falco, E.E. (1993): *Gravitational Lenses*. Springer, Berlin

Surdej, J. and Pospieszalska-Surdej, A. (1996):
   `http://www.stsci.edu/ftp/stsci/library/grav_lens/grav_lens.html`

Udalski, A., Szymanski, M., Kaluzny, J., Kubiak, M., Mateo, M., Krzeminski, W., and Paczyński, B. (1994): The Optical Gravitational Lensing Experiment. The Early Warning System: Real Time Microlensing. *Acta Astron.*, **44**, 227

Udalski, A., Szymanski, M., Kaluzny, J., Kubiak, M., Krzeminski, W., Mateo, M., Preston, G.W., and Paczyński, B. (1993): The Optical Gravitational Lensing Experiment. Discovery of the First Candidate Microlensing Event in the Direction of the Galactic Bulge. *Acta Astron.*, **43**, 289

Wambsganss, J. (1997): Discovering Galactic planets by gravitational microlensing: magnification patterns and light curves. *Mon. Not. Roy. Astron. Soc.*, **284**, 172

Wambsganss, J. and Kundíc, T. (1995): Gravitational Microlensing by Random Motion of Stars: Analysis of Light Curves. *Astrophys. J.*, **450**, 19

Wambsganss, J. and Paczyński, B. (1991): Expected color variations of the gravitationally microlensed QSO 2237+0305. *Astron. J.*, **102**, 864

Wambsganss, J., Paczyński, B., and Schneider, P. (1990): Interpretation of the Microlensing Event in QSO 2237+0305. *Astrophys. J.*, **358**, L33

Wambsganss, J., Witt, H.J., and Schneider, P. (1992): Gravitational Microlensing: Powerful Combination of Ray-Shooting and Parametric Representation of Caustics. *Astron. Astrophys.*, **258**, 591

Witt, H.J. (1993): An Efficient Method to Compute Microlensed Light Curves For Point Sources. *Astrophys. J.*, **403**, 530

Witt, H.J. and Mao, S. (1994): Can Lensed Stars be Regarded as Pointlike for Microlensing by MACHOs? *Astrophys. J.*, **430**, 505

Laser-Interferometric Gravitational Wave Detectors – on the Ground and in Deep Space

Karsten Danzmann

Institut für Atom- und Molekülphysik
Universität Hannover, Appelstr. 2, D-30167 Hannover
kvd@mpq.mpg.de

Abstract

Worldwide activities in laser-interferometric detection of gravitational waves, so far concerned with building ever more sensitive prototypes, are beginning to bear fruit: several large ground-based interferometers have received funding, and some of them are already under construction. The sensitivities achieved with prototypes so far, and the features of the large interferometer projects, in particular the German-British GEO 600 project, will be presented. The regime of extremely low frequencies is not accessible with ground-based detectors and can be covered only with space-borne detectors. The project LISA, selected by ESA as one of the cornerstone space projects under the program Horizon 2000+, will deploy six spacecraft in Earth-like orbits, forming up to three interferometers with arm lengths of 5 million km. Launch is expected around the year 2015.

1 Introduction

This article is about interferometric gravitational wave detectors located either on *Earth's surface*, or even far out in *interplanetary space*. The difficulties (and thus the great challenge) stem from the fact that gravitational waves have so little interaction with matter and the measuring apparatus. Great scientific and technological efforts, large detectors, and a good deal of patience are required to detect and measure this elusive type of radiation. And yet – just on account of their weak interaction – gravitational waves can give us knowledge about cosmic events to which the electromagnetic window will be closed forever. Gravitational wave detection can be regarded as a new window to the universe. To open this window, however, we still have to go a long way in building and perfecting our antennas, and it will only be after these large interferometers are completed that we can reap the fruits of this enormous effort: a sensitivity that will allow us to look far out in the universe, far beyond our own galaxy.

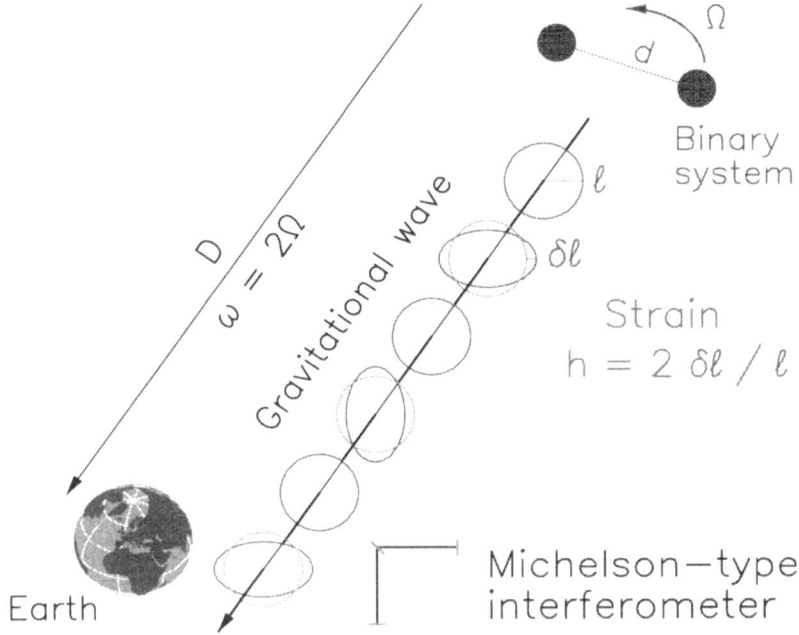

Figure 1 Generation and propagation of a gravitational wave, emitted by a binary star.

1.1 Gravitational Waves

Gravitational waves of measurable strengths are emitted only when large cosmic masses undergo strong accelerations, for instance, as shown schematically in Fig. 1, in the orbits of (close) binary stars. The strength of the gravitational wave is proportional to the second time derivative of the quadrupole moment of the constellation, its emitted frequency ω is twice the rotational frequency Ω. The effect of such a gravitational wave is an apparent strain in space, transverse to the direction of propagation, that makes distances ℓ between test bodies shrink and expand by small amounts $\delta\ell$. The strength of the gravitational wave, its "amplitude", is generally expressed by $h = 2\,\delta\ell/\ell$.

An interferometer of the Michelson type, typically consisting of two orthogonal arms, is an ideal instrument to register such strains in space. The problem lies in the magnitude, or better the smallness, of the effect. From a supernova, or from the in-spiral of a binary of two close neutron stars out at the Virgo cluster (a cluster of 2000 galaxies, 10 Mpc away), we could expect a strain of about $h \approx 10^{-22}$. So all we have to do is to measure – in a Michelson interferometer of kilometer dimensions – path changes in the order of 10^{-19} m. We shall see that this is not as unrealistic as it may appear.

Figure 2 shows some typical sources of gravitational radiation. These sources range in frequency over a vast spectrum, from the kHz region of supernovae and final mergers of binary stars down to mHz events due to formation and coalescence of supermassive black holes. Indicated are sources in two clearly separated regimes: events in the range from, say, some 10 Hz to several kHz (and only these will be detectable with terrestrial antennas) and a low-frequency regime, 10^{-4} to 10^{-1} Hz, accessible only with a space project.

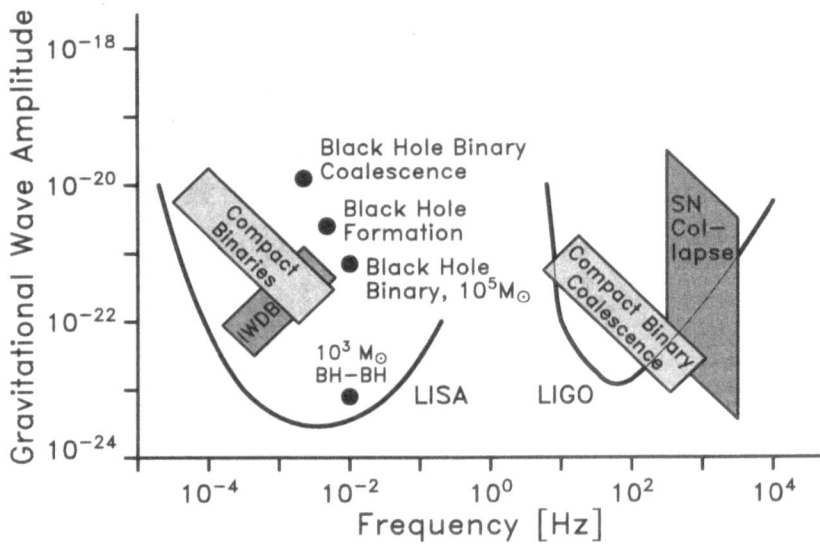

Figure 2 Some sources of gravitational waves.

2   Ground-Based Interferometers

The underlying concept of all our detectors is the Michelson interferometer in which an incoming laser beam is divided into two beams going along different arms. On their return, these two beams are recombined, and the interference between the two (measured with a photodiode PD) depends on the difference in gravitational wave effects that the two beams have experienced.

The changes δL in optical path become the greater the longer optical paths L are made, optimally about half the wavelength of the gravitational wave: e.g. to a seemingly unrealistic 150 km for a 1 kHz signal. Schemes were devised to make the optical path L significantly longer than the geometrical arm length ℓ. One scheme is to use "optical delay lines" in the arms, the beam being folded back and forth between two concave mirrors; the other is to use Fabry-Perot cavities, again with the aim of increasing the interaction time of the light beam with the gravitational wave. After pioneering work by Rai Weiss at MIT in 1972, other groups at Munich/Garching, at Glasgow, then Caltech, Paris/Orsay, Pisa, and later in Japan and Australia, also entered the scene. Their prototypes range from a few meters up to 30, 40, and even 100 m. Even though some of these prototypes have by now reached the sensitivities of the cryogenic resonant-mass antennas, they were never meant to be used as detectors, but rather as test-beds to verify new schemes and configurations devised to overcome otherwise limiting noise effects.

2.1   The Large-Scale Projects

All of the large-scale projects will use low-noise Nd:YAG lasers (wavelength 1.064 μm), pumped with laser diodes for high overall efficiency. Furthest advanced, and the most ambitious, is the US project LIGO. It comprises *two* facilities at two widely separated sites (Vogt 1989). Both will house a 4 km, Hanford an additional 2 km interferometer. At both sites construction has started. Funding of the order of 300 M\$ has been approved. These three interferometers are designed for coincidence operation, and a declared aim is to provide the flexibility to harbor experiments also of other user groups, i.e. to provide a true facility for the nation-wide or even world-wide community.

Next in size (3 km), and also securely funded, is the French-Italian project VIRGO, to be built near Pisa, Italy (Giazotto and Brillet 1989). An elaborate seismic isolation system will allow this project to measure down to a frequency of 10 Hz or even below. A British-German collaboration GEO, originally planning to build a 3 km antenna, has now reduced the project to a length of only 600 m, very scant accommodation in minimum sized houses: GEO 600. This has cut down cost, but also compels us to employ advanced optical techniques to make up for the shorter arms (Danzmann et al. 1994). Ground work at the site near Hannover is almost finished, with trenches and buildings completed, and welding of the vacuum tubes in the final stages. In Japan, by merging efforts at ISAS (where the delay-line prototypes of 10 m and 100 m were built) and at other institutions (with their Fabry-Perot prototypes of up to 20 m), plans are going ahead for a common 300 m project called TAMA 300. It will be equipped, just as LIGO and VIRGO, with standard Fabry-Perot cavities in the arms. Construction has started and will be finished around 1998.

As it happens, all of these projects are scheduled in such a way that first installation of optics and first operation can be expected before the year 2000. That the various projects are rather well in synchronism is a great stroke of luck. For the received signal to be meaningful, coincident recordings of at least two detectors at well-separated sites are essential. A minimum of three detectors is required to locate the position of the source, and there is general agreement that only with at least four detectors can we speak of a veritable gravitational wave *astronomy*.

2.2   Noise and Sensitivity

Let us look at the interferometric detectors and their projected sensitivities in some more detail. The mirrors between which the distances are to be monitored are suspended as pendulums in vacuum to isolate them from extraneous vibrations: from seismic and acoustic noise. Various schemes (pendulum suspension, lead-and-rubber stacks, even active position control) are used (often in combination) to reduce seismic noise by many powers of 10, which is relatively easy for frequencies above, say, 100 Hz.

Each optical component – and in particular the mirrors – will cause fluctuations in the optical paths also due to their thermal vibrations, i.e., their *Brownian motion*. By choice of materials (high mechanical Q) and appropriate shaping (to keep their resonant frequencies above our kHz range) the effect of these thermal motions can be reduced. But for both of the thermal noise effects, the internal vibrations of the mirrors as well as the pendulation mode, and similarly also for the seismic disturbances, the sensitivity goal can only be reached if we choose the armlength ℓ long enough. This is where our desire for kilometer dimensions comes from.

Particularly at higher frequencies, the sensitivity is limited by another fundamental source of noise, the so-called shot noise, a fluctuation in the measured interference coming from the "graininess" of the light. These statistical fluctuations fake apparent fluctuations in the optical path difference ΔL that are inversely proportional to the square root of the light power P used in the interferometer. For the very ambitious aims of the "Advanced LIGO" antenna, and also for GEO and VIRGO, about $10\,\text{kW}$ of light power, in the visible or the near infrared, would be required. This is not as unrealistic as it may sound. By applying the (now so fashionable) concept of "recycling", we can attain such enormous light powers with only moderate input light from the laser.

The laser interferometers are planned to monitor the (gravitational-wave induced) changes δL of the light path by observing the dark fringe of the interferometer in one output port. The (unused) light going out at the other port of the beam splitter can be fed back in correct phase with the incoming light via an additional mirror. This scheme was proposed by Ron Drever in 1981, at the same time as it came as a natural consequence of the Garching 30 m prototype, where the appropriate feedback had been implemented already for an efficient frequency stabilisation of the laser. It was with that Garching prototype that this recycling scheme was first demonstrated.

An additional "recycling" scheme was later proposed by Brian Meers, and it is the baseline for the GEO 600 interferometer: "signal recycling" (Meers and Strain 1991a). A further mirror is added to the interferometer, this one in the output port. This mirror can be adjusted in such a way that the signal sideband is also made to be resonant in the interferometer, providing an enhancement of the signal, at the cost, however, of reduced measuring bandwidth.

3   GEO 600

GEO 600 is a gravitational wave detector with 600 m arms, being built near Hannover, Germany, by a collaboration involving the University of Hannover, the Max-Planck-Institut für Quantenoptik (MPQ) at Garching, the University of Glasgow and the University of Wales (College of Cardiff). The civil engineering work is being undertaken under the direction of Hannover, the vacuum system constructed by Hannover with the advice of J.R.J. Bennett of Rutherford Appleton Laboratory, the technology has been developed by Hannover, MPQ and Glasgow working in close collaboration (K. Danzmann and colleagues and J. Hough and colleagues); theoretical input on astrophysical matters and the development of data analysis techniques and algorithms is being undertaken at the University of Wales and the Albert-Einstein-Institut in Potsdam (B. Schutz and colleagues). The original German-British project (Hough et al. 1989) – later known as GEO – which was proposed and agreed on in principle in the UK in 1990 and was ranking high in the list of projects to be funded in Germany, was however cancelled in 1991 due to lack of funding in both countries.

3.1   Astrophysical Aims

Long-Term Aims

The overall target of all the gravitational wave detector projects is the detection and study of gravitational wave fluxes and waveforms from various astronomical sources. The strength of a gravitational wave signal can be characterised by the strain in space at a detector, the gravitational wave amplitude h being defined as twice this strain. For example, observations over a year of coalescing binary systems of neutron stars or black holes with a detector capable of detecting bursts with an amplitude of around 10^{-22} would provide crucial information about neutron star masses and binary evolution. If three or more detectors see such an event, both the location in the sky and the distance of the binary system can be measured, allowing a high-accuracy determination of the Hubble constant (Schutz 1986), and giving definitive information about whether gamma-ray bursts originate in such binaries. If *four* detectors see an event, this would allow the spin 2 nature of the graviton to be definitively checked. Optical and gravitational observations of a supernova could allow the relative speed of gravitational waves and light to be tested to approximately 1 part in 10^9. Cross-correlations between two detectors would either detect background stochastic gravitational waves or decrease by 5 orders of magnitude the current upper limit on their energy density. This could put tight constraints on theories of galaxy formation. Pulsars could be searched for by integrating over times of a month or so, and one might expect one or more detectable pulsars in our galaxy today if 1% of all neutron stars were born with rotation periods of 2 ms or less, and with symmetries allowed by current theory and observation. The recent detection of a 2.1 ms pulsar in SN 1987A suggests that this is a relatively conservative estimate. Moreover, the pulsar PSR J0437-4715, discovered in 1993 at a distance of 100 pc and having a rotation frequency of approximately 173 Hz, could be expected to produce a signal of amplitude up to 3×10^{-26} at ~ 346 Hz. Similar techniques might detect a rapidly spinning neutron star rotating at the Chandrasekhar-Friedman-Schutz (CFS) instability point.

Some Possible Target Sources

Sources that could be targeted by the network of four detectors include:

Strong bursts from supernovae. It becomes more and more evident that there is a large variety of supernovae, however, many of those with low luminosities (such as SN 1987A) are missing in surrounding galaxies. Computer simulations are still not able to predict realistically what will happen in a gravitational collapse with high angular momentum, which is the situation likely to lead to gravitational radiation. Until recently, it was assumed that rotation is not important because all young pulsars, such as the Crab, are relatively slowly rotating. However, the unpublished observations by Middleditch et al. (private communication) of a 467 Hz optical pulsar in SN 1987A, spinning down on a time-scale less than 10^5 yr, show that, contrary to this prejudice, rapid rotation may be common or even normal (Michel 1994) in gravitational collapse. The rapid spin may be associated with the unusually low optical brightness of this supernova, and may indicate that a substantial population of supernovae with rapidly rotating cores has been missed in supernova statistics. This would greatly increase the likelihood of strong bursts of gravitational waves, detectable even if they came from the Virgo Cluster by first-stage detectors. The event rate could quite plausibly be several per year. Moreover, pulsar evidence (Lyne and Lorimer 1994) now suggests that the mean space velocity of pulsars is three times higher than it had previously been estimated: typical speeds are 450 km/s. This linear velocity must come from some non-axisymmetric asymmetry in the gravitational collapse, and this would also enhance one's expectations of gravitational radiation. If this velocity is acquired on the timescale of the bounce, 1 ms, then the *minimum* amplitude of gravitational radiation would be about 5×10^{-21} for a supernova at 1 kpc. Of course, if the collapse is messy and non-symmetric, the radiation would be expected to be much stronger than this.

Coalescing binary systems. Observations of pulsars like the Hulse-Taylor pulsar PSR
1913+16 have suggested that the nearest of such systems that will actually co-
alesce within any year will be about 100 Mpc away. This would be easily detectable
by the Stage-2 detectors, with their higher sensitivity and (more important) better performance at low frequencies; but until recently such events seemed out of
reach of the first-stage detectors. However, theoretical studies of binary evolution
(Tutukov and Yungelson 1993, Yamaoka et al. 1993) have recently suggested that
there should be a large population of very tight neutron-star binaries that have such
short gravitational-wave inspiral times. The times are so short that the chances of
seeing one at any particular time in our Galaxy are small, but the coalescence rate
integrated over time in our Galaxy could be at least 100 times larger than before.
That would move the nearest of such coalescences within one year at about the distance of the Virgo Cluster, where it might well be detectable by Stage-1 detectors,
including GEO 600.

Pulsars and accreting neutron stars. If the newly discovered nearby pulsar, PSR J0437-
4715, radiates gravitational energy at a rate comparable to the rate at which it
is losing rotational energy (as inferred from its spindown), then it would produce
a signal of an amplitude up to 3×10^{-26} at 346 Hz. This should be detectable by
GEO 600 within one year of observation. Significantly, the new pulsar in SN 1987A
is spinning down much more rapidly, and would radiate at $h \sim 1.4 \times 10^{-26}$ at 934 Hz
on the same assumption; this might also be detectable. In the case of SN 1987A,
the assumption that the radiation of gravitational waves is the dominant energy
loss is not unreasonable: the magnetic field may well be weak at present, and there
could still be significant irregularities in the shape of the star if it were formed
in a rapidly rotating collapse. Moreover, the SN 1987A pulsar also brings about
the possibility that there are nearer pulsars, formed by weak supernovae in our
Galaxy, that could still be strong radiators. These might be found by doing wide-
band gravitational-wave searches of particular regions of the galactic plane in a
year-long data set.

Stochastic background. The detection of a cosmological background of gravitational waves
would be one of the most significant events in astrophysics since the detection of
the cosmic microwave background. There have been no recent developments suggesting that the cosmological gravitational wave background at frequencies above
100 Hz should be any larger than we estimated in the original GEO proposal in
1989. However, searches will certainly be made with Stage-1 detectors, and it is
likely that GEO 600 will be able to do a better job with the VIRGO detector than
the two LIGO detectors could do at Stage-1. The reason is proximity: to get a good
correlation, detectors should be as close together as possible, so that they respond
to the same (random) gravitational waves at the same time. The separation of the
two LIGO detectors is more than 3 times greater than the separation of GEO 600
and VIRGO, leading to the loss of a factor of about 10 in sensitivity to energy density. It is unlikely that at Stage-1 the LIGO detectors could completely overcome
this disadvantage with improved sensitivity or wider bandwidth. The two European
detectors will be likely to set limits on the ratio Ω_{gw}, i.e., the energy density of the
gravitational wave background to the closure density, to around 10^{-6} at 300 Hz.
This is comparable to limits set at very low frequencies (sub-μHz) by observations
of millisecond pulsars and will allow the testing of some of the predictions of cosmic
string theory.

Tests of gravitation theory. The direct detection of gravitational waves will, of course, be a momentous event in physics, but beyond this, the observation of gravitational waves provides significant information about gravitation theory. If a supernova in VIRGO is detected, then the delay between the arrival of the gravitational waves and the light signal tests the speed of gravitational waves. The light should lag behind the gravitational radiation by no more than about 1 day, due to the propagation of the shock in the star. This uncertainty, over a travel time of about 60 Myr, tests the relative speeds to one part in 10^9 or better. Another fundamental aspect of gravity is the existence of other polarisations than those predicted by general relativity. These would indicate other spin fields, such as scalar fields, which have lately been the subject of renewed speculation from the point of view of unified field theories involving gravitation. Present limits on the couplings of other fields are about 10^{-3} of standard gravity. A gravitational wave observed by *four* detectors would have enough redundant information to provide an independent test of these couplings. If the source were strong enough, such as a gravitational collapse in our galaxy, then it would be possible to improve present limits on additional gravitational fields.

Future Developments

Exciting as some of these possibilities are, it must be stressed that, at the sensitivity levels of initial experiments, detection of gravitational wave signals cannot be guaranteed. Therefore, a further valuable aspect of the 600 m detector would be its use as a development system for future, more sensitive detectors. It is possible that the 600 m detector could have its sensitivity, in particular for narrow band sources, considerably enhanced by cooling of the detector test masses to reduce thermal noise and by optimising the design of the laser interferometry used. The initial design of the systems will allow for such future developments.

Required Detector Performance

To achieve all the detection goals mentioned above, a sensitivity of 10^{-22} over a bandwidth of approximately 1 kHz, or a sensitivity spectral density of $3 \times 10^{-24}/\sqrt{Hz}$ from about 100 Hz is likely to be required; however, the initial goals of the LIGO and VIRGO detectors are somewhat more modest than that – a few times $10^{-23}/\sqrt{Hz}$. Initial coincidence experiments, likely to last several years around the end of the century, are proposed to be at this level. Increasing the sensitivity will require the development of more advanced detectors, and at present this is planned as a second stage for both LIGO and VIRGO.

We believe that a detector with shorter arm length (600 m) using more advanced techniques than are currently proposed for LIGO or VIRGO could achieve a sensitivity close to their initial sensitivities (at least above 200 Hz and possibly within a limited bandwidth) and on a timescale similar to or somewhat earlier than these detectors. The modest scale of GEO 600 will make it easier to introduce the sophisticated technology required. The building of such an instrument could allow a more sensitive coincidence experiment to be carried out in the early stages of operation of the long detectors. Furthermore such a detector could by itself carry out a meaningful search for gravitational radiation from nearby pulsars. The interferometer design developed for GEO 600 would be a very strong candidate for installation in the LIGO and VIRGO instruments at a later stage to allow these systems to attain their advanced target performance.

3.2   Joint German/British 600 m Detector – GEO 600

Based on many years of development work and collaborative efforts at the University of Glasgow and the Max-Planck-Institut für Quantenoptik, with strong theoretical support from the University of Wales (Cardiff), the research groups in Hannover, Garching, Glasgow and Cardiff are jointly building a gravitational wave detector using laser interferometry to sense the motion of essentially free test masses which form two perpendicular arms 600 m in length. This instrument is built on farmland owned by the University of Hannover. The detector is built just below ground.

Vacuum System

The vacuum system designed by J.R.J. Bennett from RAL (Bennett et al. 1992a,b) would consist of a cluster of up to 9 stainless steel vacuum tanks, each of 1 m in diameter, at the center of the system and one tank of the same diameter at each end of the perpendicular arms. The end tanks will be joined to the cluster in the center by stainless steel vacuum pipes of 0.6 m in diameter. The system will be pumped by a combination of turbo molecular pumps and NEG pumps. The design will provide a vacuum pressure close to 10^{-8} mbar for H_2 and 10^{-9} mbar for other heavier gases, which is adequate for the design sensitivity of the instrument.

Interferometer Arrangement

The optical scheme is a somewhat simplified version of the original delay line interferometer (Hough et al. 1989) proposed for the 3 km GEO detector, a delay line interferometer with four passes in each arm. Power recycling will be implemented to allow a standing power at the beamsplitter of approximately 6 kW, a figure consistent with the need to avoid excessive distortion of the optical phase fronts due to heating effects mainly in the beamsplitter (based on Winkler et al. 1991). Signal recycling as proposed by Meers (1988) and demonstrated experimentally by Meers and Strain (1991a) will be used to increase the storage time of the system for the signal sidebands and thus increase the detected signal size in the interferometer.

The input laser power would be approximately 5 W from a stabilised all-solid-state diode-pumped Nd:YAG laser system currently being developed at the Laser Zentrum Hannover (Freitag et al. 1995). The excess noise due to relaxation oscillations in such lasers and its relevance for gravitational wave detectors has been studied by Campbell et al. (1992), and the necessary reduction in such noise by electronic feedback has been demonstrated by Rowan et al. (1994) and by Harb et al. (1994). The laser will be frequency prestabilised to a small optical cavity. The main beam will have its direction, beam diameter and convergence stabilized by passing through two in-line mode cleaning cavities of the type originally proposed and implemented by Rüdiger et al. (1981). The interferometer will be locked on a null fringe using techniques partly outlined in the original German/British proposal (Hough et al. 1989) These locking techniques are currently being implemented in the suspended-mass 30 m prototype at Garching. It should be noted that this optical system makes full use of the very low-loss mirror coatings and specially developed low absorption fused silica substrates that are now available. Achieving the required signal storage time with only a small number of bounces and a high degree of signal recycling has advantages over a conventional design, because it allows the detector to be tuned for narrow-band sources and makes it highly immune to optical aberrations (Meers and Strain 1991b).

The beamsplitter and the test masses that form the main mirrors of the interferometer consist of solid cylinders of fused silica, approximately 25 cm in diameter and 15 cm in thickness, with supersmooth and dielectrically coated surfaces. The recycling mirrors are similar. The coatings will be of laser gyro quality, resulting in optical losses of only a few parts per million.

Seismic Isolation, Position and Orientation Control of Test Mass

The test masses in the interferometer will be isolated from ground motions using a multilayer stack of heavy metal and neoprene or silicone rubber. In order to prevent contamination of the mirror surfaces by impurities it is important that only a negligible surface area of rubber is exposed to the vacuum system. The rubber will therefore be encapsulated in very soft metal bellows which would be pumped separately from the rest of the system. Each test mass, the beamsplitter, and each recycling mirror will be suspended on a double loop as the lower mass of a double-pendulum suspension. The upper mass will be suspended on a single loop to allow the orientation of the lower mass to be controlled by tilting and rotation of the upper mass. Another pendulum will be mounted with a reaction mass close to the upper mass to allow electronic damping and orientation control of the pendulum system. For some of the test masses a second reaction mass will be suspended below the first one to allow direct electronic feedback to the position of the test mass itself. In most cases, the forces required for control of position and orientation will be imposed by coil-and-magnet systems, but electrostatic systems may be used for those forces which have to be applied directly to the test masses. Experimental studies of multiple mass systems of a similar type have been carried out in Glasgow (Veitch et al. 1993), and systems of a similar type have been installed in the 30 m prototype at MPQ. Automatic alignment systems based on the technique demonstrated by Morrison et al. (1994a,b) on the 10 m prototype interferometer in Glasgow, and now being installed in the 30 m prototype at MPQ, will be implemented to maintain optimum orientation of the principal optical components.

Thermal Noise of Pendulum and Internal Modes

A dominant noise source in laser interferometric detectors is expected to be thermal noise associated with the pendulum modes of the suspended test masses, with the violin modes of the suspension wires, and with the internal modes of the test masses. To minimise the thermal noise contributions in the bandwidth of interest, very high quality factors (Q), i.e. low losses, are required for the suspension wires and the test masses. It has recently been shown by Logan et al. (1993) and independently by Gillespie and Raab (1993) and by González and Saulson (1994) that the Q factor of the simple pendulum mode is related to that of the suspension violin modes. Thus for very low suspension losses very high

Q material must be used also for the suspension wires. It has been shown originally by Martin (1978) and recently by Quinn et al. (1994) that mechanical clamping of suspension wires or membranes leads to a significant lowering of pendulum Q. Thus we intend to use a material of intrinsically high Q for the test masses and suspension wires (fused silica) and to avoid mechanical clamping wherever possible.

In our interferometer the lower fused silica test masses would be suspended from upper masses, also of fused silica, on double loops of fused silica fibre. These will be welded or optically contacted to the fused silica masses in order to obtain very high Q for both the pendulum mode and the internal modes of the masses. Test experiments by Martin (1978) and more recently by Braginsky et al. (1993) suggest that Q factors of 5×10^7 for the *pendulum mode* should be obtainable and experiments on silicon by Logan (1993) and on fused silica by Traeger et al. (1997) indicate that Q factors for the *internal modes* of the test masses of 5×10^6 should be achievable with such a system. It should be noted that as a result of the mechanical filtering action of the lower pendulum, thermal noise associated with the upper mass and its pendulum is not so important and thus a loop of steel wire may be used for its suspension.

3.3   Previous Results

Both the MPQ and Glasgow groups have carried out a lot of development work over the past 15 years. The prototype detectors at Glasgow and MPQ Garching have achieved strain sensitivities of approximately $10^{-19}/\sqrt{\text{Hz}}$ over a kilohertz bandwidth, and studies of many aspects of laser systems, suspensions, autoalignment techniques, mirror quality, and mirror heating and scattering problems have been tackled. New methods of improving the sensitivity of optical interferometers have been invented both at Glasgow and MPQ, and new laser systems of high stability and power have been developed at Hannover. For a number of years the Cardiff group has been studying the characteristics of possible sources of gravitational waves. This work led to a significant program of investigation of algorithms for searching for signals in noise, and also to the development of a powerful software environment for a wide-ranging analysis of data from detectors.

The prototype detectors at MPQ and Glasgow were run in coincidence for 100 hours in 1989 with much of the data analysis undertaken by the Cardiff group. The results have been published by Nicholson et al. (1995). A significant result of this analysis is that the presence of a low rate of spurious pulses on the detector outputs had little effect on the sensitivity of the search for gravitational waves. In a search for a reported pulsar source, Niebauer et al. (1993) developed an efficient algorithm by which the large volume of data was greatly reduced, thus opening the way for sophisticated searches even for unknown sources.

3.4   Worldwide Collaboration

The *detection* and quantative *measurement* of gravitational waves strongly depends on simultaneous monitoring by a number of antennas (at least two for detection, preferably four for measurement and source location). This need for international collaboration during development as well as later for data exchange has created a climate of amiable competition amongst the research institutes involved. GEO 600 will give the collaboration of research groups (Hannover, Glasgow, Garching, Cardiff) a realistic opportunity to be a

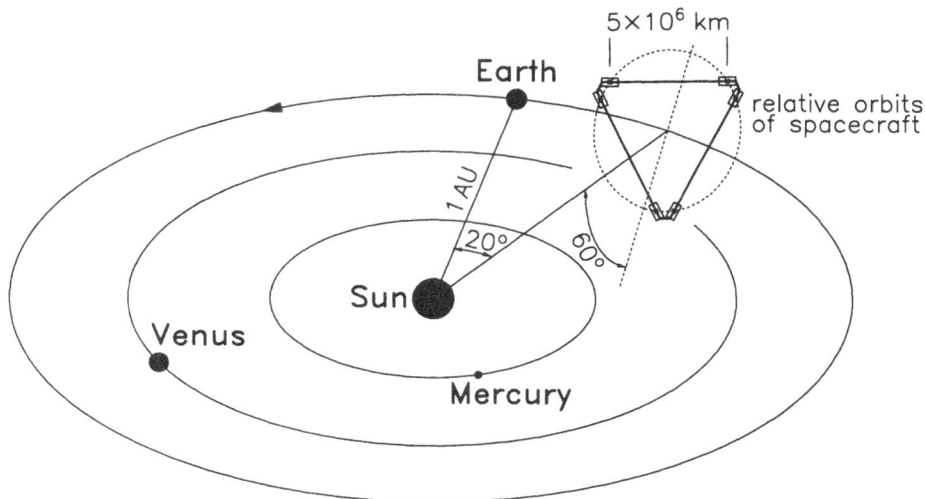

Figure 3 LISA concept. Six spacecraft in a triangle, with a pair at each vertex. Only four are required for the basic interferometry. The other two provide supplementary scientific information and some redundancy.

part of the initial coincidence experiments to search for gravitational waves at an exciting level of sensitivity and at a fraction of the cost of the original GEO experiment. It will also considerably enhance the prospects for contribution to higher sensitivity experiments in the future.

4   LISA – An Interferometer in Deep Space

4.1   Overview

Conceptually, the idea of implementing an interferometer in space is straightforward, but the practical realisation requires an intricate blend of optical technology, spacecraft engineering and control. For a start, the interferometer mirrors cannot simply float freely in space — they must be contained inside the spacecraft. Nonetheless, they can be arranged to be floating almost freely inside the spacecraft, protected from external disturbances by the spacecraft walls. Provided that the spacecraft do not disturb the mirrors, then, ideally, only gravitational waves would perturb their relative motion. "Drag-free control" can be employed to ensure that the spacecraft always remain centered on the mirrors. In principle, the Michelson interferometer could be realized using three spacecraft: one at the "corner" to house the light source, beam splitter, and detector, plus one at each "end" to house the remote mirrors. But there would be immediate practical problems with such a configuration. All three spacecraft would drift around, and the corner spacecraft would not be able to keep itself aligned with both of the end spacecraft at the same time. One way around this would be to have steerable optics inside the corner spacecraft so that alignment could be maintained with the two arms independently. To avoid this complexity, LISA uses six spacecraft, arranged in a triangular configuration with two at each vertex. With this setup, each of the corner spacecraft can dedicate itself to pointing at only one of the end spacecraft, thus eliminating the need to steer the main optics. The corner spacecraft must, nevertheless, communicate with each other using steerable optics — but the separation distance is so much less that the steerable components can be much smaller, and hence are more manoeuverable.

Each "corner" pair of spacecraft, separated by 200 km, is located at the vertex of a large triangle whose sides measure 5×10^6 km in length (Fig. 3). This arm length has been chosen to optimize the sensitivity of LISA at the frequencies of known and expected sources. An increase by a factor of two may be desirable. However, an arm-length increase beyond that would begin to compromise the high-frequency sensitivity when the light in the arms experiences more than half of the gravitational wave period. An interferometer shorter than 5×10^6 km would begin to lose the interesting low-frequency massive black-hole sources. It would give less scientific information but would not be any easier to build or operate because the spacecraft and the interferometry would be essentially the same.

Nominally in such an arrangement of spacecraft, any two sides of the triangle (i.e. four spacecraft) can be used for the main interferometry, with the third arm giving supplementary information and redundancy. With the six spacecraft configuration, up to two can be lost without jeopardising the mission (as long as the two failures are not at the same corner), since the basic group of four in an approximate "L" shape is sufficient to perform the full interferometry.

Each spacecraft is actually in its own orbit around the Sun. The six individual orbits have their inclinations and eccentricities arranged in such a way that, relative to each other, the spacecraft rotate on a circle 'drawn through' the vertices of the giant triangle which is tilted at 60° with respect to the ecliptic. With this special choice of orbits, the triangular geometry of the interferometer is largely maintained throughout the mission. The center of the triangle is located on the ecliptic — 20° behind the Earth — and follows the Earth on its orbit around the Sun. Ideally, the constellation should be as far from Earth as possible in order to minimize gravitational disturbances. The choice of 20° is a practical compromise based on launch vehicle and telemetry capabilities.

The once-per-year orbital rotation of the LISA constellation around the Sun provides the instrument with angular resolution, i.e. the ability to pin-point the particular direction of a source. An interferometer is rather omnidirectional in its response to gravitational waves. In one sense this is advantageous — it means that more sources can be detected at any given time — but it has the disadvantage that the antenna cannot be "aimed" at a particular location in space. For a given source direction, the orbital motion of the interferometer Doppler-shifts the signal, and also affects the observed amplitude. By measuring these effects the angular position can thus be determined. This is analogous to the technique used by radio astronomers to determine pulsar locations.

It is expected that the strongest LISA sources (from very distant supermassive black holes) should be resolvable to better than an arcminute, and to within one degree even for the weaker sources (galactic binaries) throughout the entire galaxy.

A LISA spacecraft is shown in Fig. 4, and a cross-section of the payload in Fig. 5. Each spacecraft has its own 1 W laser (actually two, one for redundancy), its own two-mirror telescope for sending and receiving light, and an optical bench which is a mechanically stable structure on which various sensitive optical components are mounted. The mirrors enclosed in each spacecraft are actually 40 mm gold-platinum cubes (also referred to as the 'proof masses'). Each one is located inside a titanium vacuum can at the centre of the respective optical bench. Quartz windows allow access for the laser light.

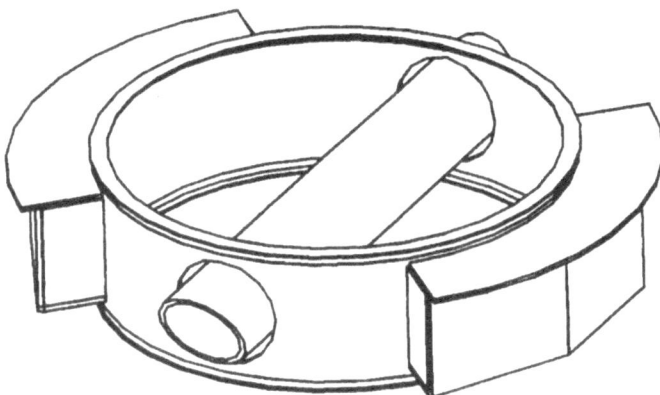

Figure 4 One of the six identical LISA spacecraft. The main structure is a ring with a diameter of 2.6 m and a height of 0.7 m, made from carbon-epoxy for low thermal expansion. The ring supports the payload cylinder, as shown. Equipment boxes are mounted on the outside of the ring to house non-precision electronics (e.g. power regulator, computer, radios). FEEP control thrusters (not shown) are mounted at various locations on the outer spacecraft structure. The tops of the equipment boxes support two annular sections of solar array for power generation. A lid on top of the spacecraft (not shown) protects the thermal shields and payload cylinder from direct sunlight.

Within the corner pair of spacecraft, one laser is the 'master', and a fraction of its light (10 mW) is bounced off the back surface of its cube, and sent to the neighbouring corner spacecraft (via the small steerable optics), where it is used as a reference to 'slave' the local laser. In this way, the main (~ 1 W) beams going out along each arm can be considered to originate from a single laser. This is vital to the function of the interferometer.

The light sent out along an arm is received by the end spacecraft telescope, bounced off its cube, then amplified using its local laser, in such a way that the phase of the incoming light is maintained. The amplified light is then sent to the corner spacecraft. Amplification at the end spacecraft is required due to divergence of the beam over the very great distances. Even though each outgoing beam is extremely narrow — a few micro radians — it is about 20 km wide when it reaches the distant spacecraft. This diffraction effect, together with unavoidable optical losses, means that only a small fraction of the

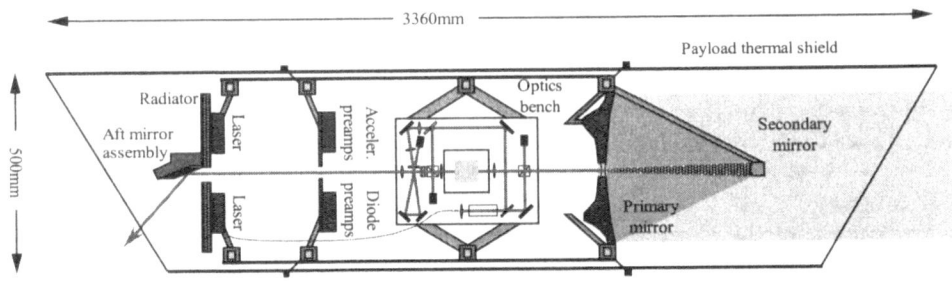

Figure 5 Cross-section of the payload on each of the six identical LISA spacecraft.

original output power ($\sim 10^{-10}$) finally reaches the end diode. If this was simply reflected and sent all the way back, only about 200 photons per hour would reach the corner diode after the round-trip. The phase-signals they carry would be swamped by shot noise, i.e., the quantum-mechanical fluctuations in the arrival times of the photons. The amplification brings the number back up to over 10^8 photons per second — which makes the signal detection straightforward using standard photodiodes. The phase precision requirement for this measurement is seven orders of magnitude less demanding than is routinely achieved (at higher frequencies) in ground-based prototype interferometers.

The resulting round-trip journey from the corner to the end and back defines one arm of the large interferometer. On its return to the corner spacecraft, the incoming light is bounced off the cube and then mixed with a fraction of the outgoing light on a sensitive photodetector, where interference is detected. The resulting brightness variations contain the phase-shift information for one arm of the interferometer. This signal is then compared (using software on the on-board computer) with the corresponding signals from the other two arms, and some preliminary data processing is done. The results are then transmitted to Earth by radio link. The LISA spacecraft must be designed to minimize the total mass and required power. Preliminary results yield a mass of 300 kg per spacecraft and an operational power requirement of 192 W per spacecraft.

4.2 Lasers

Lasers have extremely narrow beams that can survive long journeys through space. In addition, they are very stable in frequency (and phase) which is crucial to interferometry since phase "noise" just looks like gravitational waves. Furthermore, the infrared light has a frequency of 3×10^{14} Hz which renders it immune from refraction caused by the charged particles (plasma) which permeate interplanetary space.

The lasers for LISA must deliver sufficient power at high efficiency, as well as be compact, stable (in frequency and amplitude), and reliable. The plan is to use solid-state diode-pumped monolithic miniature Nd:YAG ring lasers which generate a continuous 1 W infrared beam with a wavelength of 1.064 μm.

4.3 Drag-free and Attitude Control

An essential task of the spacecraft is to protect the mirrors from any disturbances which could jostle them around and create phase-signals that appear as gravitational waves. For example, consider the momentum of the light from the Sun which amounts to an average pressure of about 5×10^{-6} N/m^2. The internal dynamics of the Sun lead to small variations — less than one percent — in this photon pressure, which occur at the low frequencies within LISA's range of interest. Although this variable photon pressure may seem rather small, if it were allowed to act on the cubical mirrors, the resulting motion would be 10^4 times larger than the tiny motions due to gravitational waves that LISA is looking for. By simply "wrapping a spacecraft around each of the cubes", they are isolated from the solar pressure — but this is not the complete picture. When the solar pressure blows on the surface of the spacecraft, it will move relative to the freely-floating cube. Left alone, this motion would build up to unacceptable levels — in the extreme case, the cube would eventually "hit the wall". To stop this from happening, the relative motion can be measured very precisely by monitoring the change in electrical capacitance between the cube and electrodes mounted on the spacecraft. This measurement is then converted into a force-command which instructs thrusters mounted on the outer structure of the spacecraft to fire against the solar pressure and keep the spacecraft centred on the cube. This concept is, for historical reasons, known as "drag-free control", since it was originally invented in the 1960s to shield Earth-orbiting satellites from the aerodynamic drag due to the residual atmospheric gases. The method was first demonstrated on the TRIAD spacecraft, flown by the US Navy in 1972, where the drag-free controller designed at Stanford University in collaboration with the Johns Hopkins Applied Physics Laboratory, was effective in reducing the effects of atmospheric drag by a factor of 10^3. Since then, the technique has undergone continued development, most notably for use on NASA's Gravity Probe B mission, which is the proposed space experiment to search for the relativistic precessions of gyroscopes orbiting the Earth.

The thrusters used on conventional spacecraft are far too powerful for LISA. The drag-free system only needs to develop a force of a few micro-Newtons. Furthermore, the delivered force must be smoothly controllable so that the varying disturbance forces can be matched without introducing a further disturbance from the thrust system itself. Surprisingly, it is not a trivial task to build a thruster which generates such a small force and yet operates smoothly and does not consume too much power. By good fortune, ESA has been developing them for years, as an alternative to hydrazine rockets for station-keeping of communication satellites. They are called FEEPs (Field Emission Electric Propulsion). They operate by accelerating ions in an electric field and ejecting them to develop the thrust.

4.4 Ultrastable Structures

The small variations in the intensity of sunlight will cause fluctuations in the heat-load applied to the spacecraft. This could lead to thermal gradients across the optical bench, which would upset the stability of the laser cavity. To obtain the required thermal stability, most structural elements are made from carbon-epoxy which has a thermal expansion coefficient of $4 \times 10^{-7}/\text{K}$ and the optical bench is made from ULE, which has a temperature coefficient at least a factor 4 lower over the possible temperature range of the LISA payload. Furthermore, low emissivity coatings are used on most surfaces inside the spacecraft and a thermal shield surrounds the payload cylinder, in order to provide isolation from the temperature variations of the spacecraft skin that is exposed to the Sun. These shields are only effective against heat fluctuations faster than a few hours to half a day. The slower variations will get through, thus causing the sensitivity of LISA to deteriorate rapidly below roughly $10^{-4}\,\text{Hz}$. The use of carbon-epoxy structures also minimizes any thermally-induced mechanical distortions which could produce physical changes in the optical path-length, as well as local gravitational disturbances on the mirror-cubes.

4.5 Data Transmission

Each spacecraft will be equipped with two (one spare) X-band transponders with steerable 30 cm high-gain antennas for communication with the Earth. On average, the transmissions will require about eight hours per day, at a data rate of roughly 600 bits per second. The entire LISA data set, after a nominal two-year mission, will be stored on about a hundred CD-ROMs.

References

Abramovici, A., Lyons, T.T., and Raab, F.J. (1995): Measured limits to contamination of optical surfaces by elastomers in vacuum. *Applied Optics*, **34**, 183–185

Bennett, J.R.J. and Elsey, R.J. (1992a): The design of the vacuum system for the joint German-British interferometric gravitational wave detector GEO. *Vacuum*, **43**, 35–39

Bennett, J.R.J., Elsey, R.J., and Malton, R.W. (1992b): Convoluted vacuum tubes for long baseline interferometric gravitational wave detectors. *Vacuum*, **43**, 531–535

Braginsky, V.B., Mitrofanov, V.P., and Okhrimenko, O.A. (1993): The isolation of test masses for gravitational wave antennae. *Phys. Lett. A*, **175**, 82–84

Campbell, A.M., Rowan, S., and Hough, J. (1992): Apparent relaxation oscillations in the frequency noise of a diode pumped miniataure Nd:YAG ring laser. *Phys. Lett. A*, **170**, 363–369

Danzmann, K., Lück, H., Rüdiger, A., Schilling, R., Schrempel, M., Winkler, W., Hough, J., Newton, G.P., Robertson, N.A., Ward, H., Campbell, A.M., Logan, J.E., Robertson, D.I., Strain, K.A., Bennett, J.R.J., Kose, V., Kühne, M., Schutz, B.F., Nicolsen, D., Shuttleworth, J., Welling, H., Aufmuth, P., Rinkleff, R., Tünnermann, A., and Willke, B. (1994): Proposal for a 600 m Laser-Interferometric Gravitational Wave Antenna, *Internal Report MPQ*, **190**

Freitag, I., Golla, D., Tünnermann, A., Welling, H., and Danzmann, K. (1995): Diode-pumped solid-state lasers as light sources Michelson-type gravitational-wave detectors. *Appl. Phys. B*, **60**, S255–S260

Giazotto, A. and Brillet, A. (1989): Proposal to CNRS and INFN.

Gillespie, A. and Raab, F. (1993): Thermal noise in the test mass suspensions of a laser interferometer gravitational-wave detector prototype. *Phys. Lett. A*, **178**, 357–363

González, G.I. and Saulson, P.R. (1994): Brownian motion of a mass suspended by an anelastic wire. *J. Acoust. Soc. Amer.*, **96**, 207–212

Harb, C.C., Gray, M.B., Bachor, H.-A., Schilling, R., Rottengatter, P., Freitag, I., and Welling, H. (1994): Suppression of the intensity noise in a diode-pumped neodymium: YAG nonplanar ring laser. *IEEE J. Quant. Electr.*, **30**, 2907-2913

Hough, J., Meers, B.J., Newton, G.P., Robertson, L.A., Ward, H., Leuchs, G., Niebauer, T.M., Rüdiger, A., Schilling, R., Schnupp, L., Walter, H., Winkler, W., Schutz, B.F., Ehlers, J., Kafka, P., Schäfer, G., Hamilton, M.W., Schütz, I., Welling, H., Bennett, J.R.J., Corbett, I.F., Edwards, B.W.H., Greenhalgh, R.J.S., and Kose, V. (1989): Proposal for a Joint German-Britisch Interferometric Gravitational Wave Detector. *Internal Report MPQ*, **147**

Logan, J.E. (1993): An investigation of some mechanical and optical properties of materials for test masses in laser interferometric graviational wave detectors. Ph.D. Thesis, University of Glasgow

Logan, J.E., Hough, J., and Robertson, N.A. (1993): Aspects of the thermal motion of a mass suspended as a pendulum by wires. *Phys. Lett. A*, **183**, 145–152

Lyne, A.G. and Lorimer, D.R. (1994): High birth velocities of radio pulsars. *Nature*, **369**, 127–129

Martin, W. (1978): Experiments and techniques for the detection of gravitational and pulsed electromagnetic radiation of astrophysical sources. Ph.D. Thesis, University of Glasgow

Meers, B.J. (1988): Recycling in laser-interferometric gravitational-wave detectors. *Phys. Rev. D*, **38**, 2317–2326

Meers, B.J. and Strain, K.A. (1991a): Experimental Demonstration of Dual Recycling Interferometric Gravitational-Wave Detectors. *Phys. Rev. Lett.*, **66**, 1391–1394

Meers, B.J. and Strain, K.A. (1991b): Wave front distortion in laser interferometric gravitational-wave detectors. *Phys. Rev. D*, **43**, 3117–3129

Michel, F.C. (1994): A fast pulsar in SN1987A? *Mon. Not. Roy. Astron. Soc.*, **267**, L4

Morrison, E., Meers, B.J., Robertson, D.I., and Ward, H. (1994a): Automatic alignment of optical interferometers. *Applied Optics*, **33**, 5041-5049

Morrison, E., Meers, B.J., Robertson, D.I., and Ward, H. (1994b): Experimental demonstration of an automatic alignment system for optical interferometers. *Applied Optics*, **33**, 5037-5040

Nicholson, D., Dickson, C.A., Watkins, W.J., Schutz, B.F., Shuttleworth, J., Jones, G.S., Robertson, D.I., Mackanzie, N.L., Strain, K.A., Meers, B.J., Newton, G.P., Ward, H., Cantley, C.A., Robertson, N.A., Hough, J., Danzmann, K., Niebauer, T.M., Rüdiger, A., Schilling, R., Schnupp, L., and Winkler, W. (1995): Results of the first coincident observations by two laser-interferometric gravitational wave detectors. *Phys. Lett. A*, **218**, 175–180

Niebauer, T.M., Rüdiger, A., Schilling, R., Schnupp, L., Winkler, W., and Danzmann, K. (1993): Pulsar search using data compression with the Garching gravitational wave detector. *Phys. Rev. D*, **47**, 3106–3123

Quinn, T.J., Speake, C.C., Tew, W., Davis, R.S., and Brown, L.M. (1994): Report presented at Workshop on Thermal Noise in Laser Interferometers, Caltech, January 1994

Rowan, S., Campbell, A.M., Skeldon, K., and Hough, J. (1994): Broadband intensity stabilization of a diode-pumped monolithic miiniature ND:YAG ring laser. *J. Mod. Optics*, **41**, 1263–1269

Rüdiger, A., Schilling, R., Schnupp, L., Winkler, W., Billing, H., and Maischberger, K. (1981): A mode selector to suppress fluctuations in laser beam geometry. *Optica Acta*, **28**, 641–658

Schutz, B.F. (1986): Determining the Hubble constant from gravitational wave observations. *Nature*, **323**, 310

Traeger, S., Willke, B., and Danzmann, K. (1997): Monolithically Suspended Fused Silica Substrates with very High Mechanical Q. *Phys. Lett. A*, **225**, 39–44

Tutukov, A.V. and Yungelson, L.R. (1993): The merger rate of neutron star and black hole binaries. *Mon. Not. Roy. Astron. Soc.*, **260**, 675–678

Veitch, P.J., Robertson, N.A., Cantley, C.A., and Hough, J. (1993): Active control of a balanced two stage pendulum vibration isolation system and its application to laser interferometric gravity wave detectors. *Rev. Sci. Instr.*, **64**, 1330–1336

Vogt, R.E. (1989): Proposal to the National Science Foundation.

Winkler, W., Danzmann, K., Rüdiger, A., and Schilling, R. (1991): Heating by optical absorption and the performance of interferometric gravitational-wave detectors. *Phys. Rev. A*, **44**, 7022–7036

Yamaoka, H., Shigeyama, T., and Nomoto, K. (1993): Formation of double neutron star systems and antisymmetric supernova explosions. *Astron. Astrophys.*, **267**, 433–438

Light Deflection Near Neutron Stars

Ute Kraus

Institut für Theoretische Astrophysik
Universität Tübingen, Auf der Morgenstelle 10, D-72076 Tübingen
kraus@tat.physik.uni-tuebingen.de

Abstract

This contribution describes and illustrates light deflection near neutron stars as an example of the significance of general relativity for astrophysics. First, a summary is given of the properties of photon orbits in the Schwarzschild metric, the Schwarzschild metric being a good approximation to the exterior metric of slowly rotating neutron stars. Secondly, it is illustrated how light deflection affects the observation of sources on the surface or close to the surface of a neutron star. Thirdly, it is illustrated that it is imperative to take light deflection into account when interpreting the pulse profiles of accreting X-ray pulsars, because the ratio of neutron star radius to Schwarzschild radius strongly affects the pulse profiles predicted from models of the pulsar's X-ray emission regions.

1   Introduction

When observed radiation is analyzed to deduce the properties of the source, there are three quantities of interest: photon energy, intensity and angular distribution. In order to relate the observed photon energy, intensity, and angular distribution to the photon energy, intensity, and angular distribution at the source one has to *i)* follow the photon orbit between source and observer, *ii)* find the variation of photon energy along the orbit, and *iii)* find the variation of intensity along the orbit.

This is straightforward in the absence of gravitational fields, where energy and intensity are constants along the photon orbits and photon orbits are straight lines. In the presence of gravitational fields one has to proceed in the following way:

1. Photon orbits are null geodesics $x^\mu(\lambda)$, i.e. the photon four-momentum $p^\mu(\lambda) = dx/d\lambda$ satisfies the geodesic equation and is null:

$$\frac{dp^\mu}{d\lambda} + \Gamma^\mu_{\nu\kappa} p^\nu p^\kappa = 0 \qquad\qquad g_{\mu\nu} p^\mu p^\nu = 0.$$

2. The energy measured by a local inertial observer with four-velocity u^μ at some location λ along the orbit is

$$E = -g_{\mu\nu} u^\mu p^\nu.$$

3. The specific intensity I_ν measured by a local inertial observer is obtained via the photon number density f which is a Lorentz invariant quantity and is constant along null geodesics in vacuum:

$$\frac{d}{d\lambda} f(x^\mu(\lambda), p^\mu(\lambda)) = 0 \qquad f = I_\nu c^2/(h^4 \nu^3).$$

2 Photons in the Schwarzschild Metric

2.1 The Geodesic Equation

As a specific example for the general equations given above we will turn to photon orbits in the Schwarzschild metric. The Schwarzschild metric is the metric in vacuum outside a spherical mass distribution. It depends on a single parameter, the Schwarzschild radius of the central mass M, defined by $r_s = 2GM/c^2$ (G the gravitational constant, c the speed of light). In the usual Schwarzschild coordinates t, r, θ, and ϕ the Schwarzschild metric is diagonal and reads

$$g_{\mu\nu} = \text{diag}(-(1 - r_s/r), 1/(1 - r_s/r), r^2, r^2 \sin^2 \theta).$$

Because of the spherical symmetry, a photon orbit is always confined to a single plane. If one chooses the polar coordinates θ and ϕ such that this plane is the equatorial plane, then $\theta = \pi/2$ is constant along the orbit and $p^\theta = d\theta/d\lambda = 0$.

The geodesic equations for the momentum components $p^t = dt/d\lambda$, $p^r = dr/d\lambda$, and $p^\phi = d\phi/d\lambda$ are:

$$\frac{dp^t}{d\lambda} + \frac{r_s}{r^2(1 - r_s/r)} p^t p^r = 0$$

$$\frac{dp^r}{d\lambda} + \frac{r_s(1 - r_s/r)}{2r^2}(p^t)^2 - \frac{r_s}{2r^2(1 - r_s/r)}(p^r)^2$$

$$- r(1 - r_s/r)(p^\phi)^2 = 0$$

$$\frac{dp^\phi}{d\lambda} + \frac{2}{r} p^\phi p^r = 0.$$

These three equations can be integrated analytically once and give

$$p^t = k^t/(1 - r_s/r) \tag{1}$$

$$p^r = \pm\sqrt{(k^t)^2 - (k^\phi)^2(1 - r_s/r)/r^2 + k^r(1 - r_s/r)} \tag{2}$$

$$p^\phi = k^\phi/r^2 \tag{3}$$

where k^t, k^r, k^ϕ are constants of integration. The constants of integration parameterize the photon orbits. The next step, therefore, is to clarify their meaning.

2.2   The Constants of Integration

The condition that p^μ be a null vector gives $0 = g_{\mu\nu}p^\mu p^\nu = k^r \Rightarrow k^r = 0$. In order to see the meaning of k^t, consider a measurement of photon energy by a local inertial observer momentarily at rest. The four-velocity of such an observer is $u^\mu = (c/\sqrt{1 - r_s/r}, 0, 0, 0)$ and with the expressions (1) to (3) for the photon momentum, the energy measured turns out to be $E = -g_{\mu\nu}u^\mu p^\nu = k^t c/\sqrt{1 - r_s/r}$. At large r the measured energy approaches $E_\infty = k^t c$, so that the constant of integration k^t can be identified as photon energy at infinity over c.

Concerning the third constant of integration, k^ϕ, it is instructive to look at the ratio $b \equiv k^\phi/k^t$ which has the simple geometric meaning of impact parameter of the photon trajectory. This can most easily be seen in the limit of vanishing central mass $r_s \to 0$ as follows. The photon orbit is completely described by three functions $t(\lambda)$, $r(\lambda)$, and $\phi(\lambda)$. If one is not interested in the lapse of coordinate time along the orbit, it is sufficient to find the trajectory $\phi(r)$, i.e., the set of all points (r, ϕ) that the photon passes through. The equation for the trajectory is

$$\frac{d\phi}{dr} = \frac{d\phi/d\lambda}{dr/d\lambda} = \frac{p^\phi}{p^r}$$

Inserting the expressions (2) and (3) for p^ϕ and p^r and setting $r_s = 0$ one obtains

$$d\phi/dr = \pm b/(r^2\sqrt{1 - b^2/r^2})$$

which can be integrated to give

$$\sin(\phi - \phi_0) = \pm b/r \ . \tag{4}$$

Equation (4) is the equation of a straight line expressed in polar coordinates. The meaning of b is illustrated in Fig. 1: When a straight line is drawn parallel to the trajectory and such that it passes through the center of coordinates, then the photon trajectory is a distance b away from this straight line. Therefore, b is the impact parameter of the $(r_s = 0)$-trajectory. Since the photon trajectory for nonvanishing mass $r_s > 0$ approaches the zero mass trajectory at large values of r, b is also the impact parameter of the $(r_s > 0)$-trajectory.

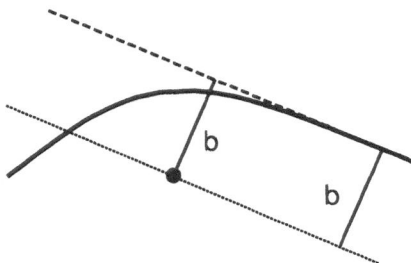

Figure 1   For $r_s = 0$, the photon trajectory with impact parameter b (dashed line) is a distance b away from a parallel through the center of coordinates (dotted line). The photon trajectory for $r_s > 0$ (solid line) approaches the $(r_s = 0)$-trajectory for $r \gg r_s$.

In summary: Photon orbits in the Schwarzschild metric are parameterized by the photon energy at infinity E_∞ and the impact parameter b. The photon four-momentum $p^\mu = dx^\mu/d\lambda$ is given by

$$p^\mu = \frac{E_\infty}{c}\left(\frac{1}{1 - r_s/r}, \pm\sqrt{1 - \frac{b^2}{r^2}(1 - r_s/r)}, 0, \frac{b}{r^2}\right) \tag{5}$$

and the photon trajectory $\phi(r)$ is defined by

$$\frac{d\phi}{dr} = \frac{d\phi/d\lambda}{dr/d\lambda} = \frac{p^\phi}{p^r} = \pm\frac{b}{r^2\sqrt{1 - \frac{b^2}{r^2}(1 - \frac{r_s}{r})}}. \tag{6}$$

2.3 The Trajectory

The photon trajectory is given by

$$\phi(r) = \phi_0 \pm \int_{r_0}^{r} \frac{b\,dr}{r^2\sqrt{1 - \frac{b^2}{r^2}(1 - \frac{r_s}{r})}}. \tag{7}$$

This integral is an elliptic integral of the first kind. There is no analytic solution, but numerical integration using standard routines is straightforward (Press et al. 1985). There are two distinct types of photon orbits, depending on whether the square root in the denominator of equation (7) has zeroes or not.

1. If $b < b_c = 1.5\sqrt{3}r_s$, then $\sqrt{1 - \frac{b^2}{r^2}(1 - \frac{r_s}{r})} > 0$, always. These photon orbits are defined for $0 < r < \infty$. Examples are shown in Fig. 2: the closer b is to the critical value b_c, the stronger the deflection of the orbit from a straight line. When b gets arbitrarily close to b_c, then the orbit circles the center of coordinates arbitrarily many times. That part of the orbit where the spiralling about the center of coordinates occurs and which is nearly circular, is in the immediate vicinity of $r = 1.5r_s$. The sphere with $r = 1.5r_s$ is called the photon sphere.

2. If $b > b_c$, then $\sqrt{1 - \frac{b^2}{r^2}(1 - \frac{r_s}{r})}$ has two zeroes, r_1 and r_2, say, one inside and one outside the photon sphere: $r_1 < 1.5r_s < r_2$. For each b there are two photon orbits, one defined for $0 < r < r_1$ and one defined for $r_2 < r < \infty$, i.e., photon orbits are either completely inside or completely outside the photon sphere.

   Fig. 3 (left) shows photon orbits with $r > r_2$. When b approaches b_c from above, the orbit spirals more and more often around the center of coordinates. The nearly circular part of the orbit is very close to and just outside the photon sphere. Orbits with $r < r_1$ are plotted in Fig. 3 (right). For b approaching b_c, the orbit is very close to and just inside the photon sphere.

If the central mass is a star, then only those parts of the orbits exist that are outside the star. Since most of the deflection occurs close to $r = 1.5r_s$, light deflection is important only if the radius of the star is not too large compared to its Schwarzschild radius.

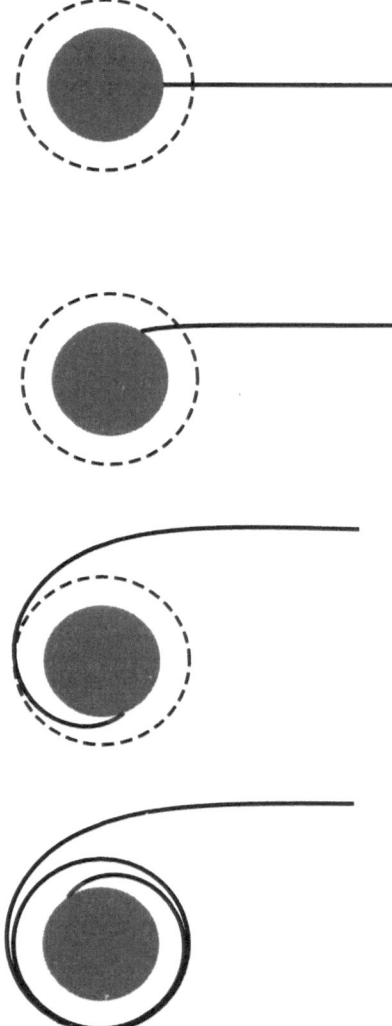

Figure 2   Photon orbits for impact parameters $b < b_c$. The shaded area is the region $r < r_s$, the dashed line marks the photon sphere at $r = 1.5\,r_s$. Orbits are shown for the impact parameters 0, $0.37\,b_c$, $0.98\,b_c$, and $(1 - 2 \cdot 10^{-6})\,b_c$ (top to bottom).

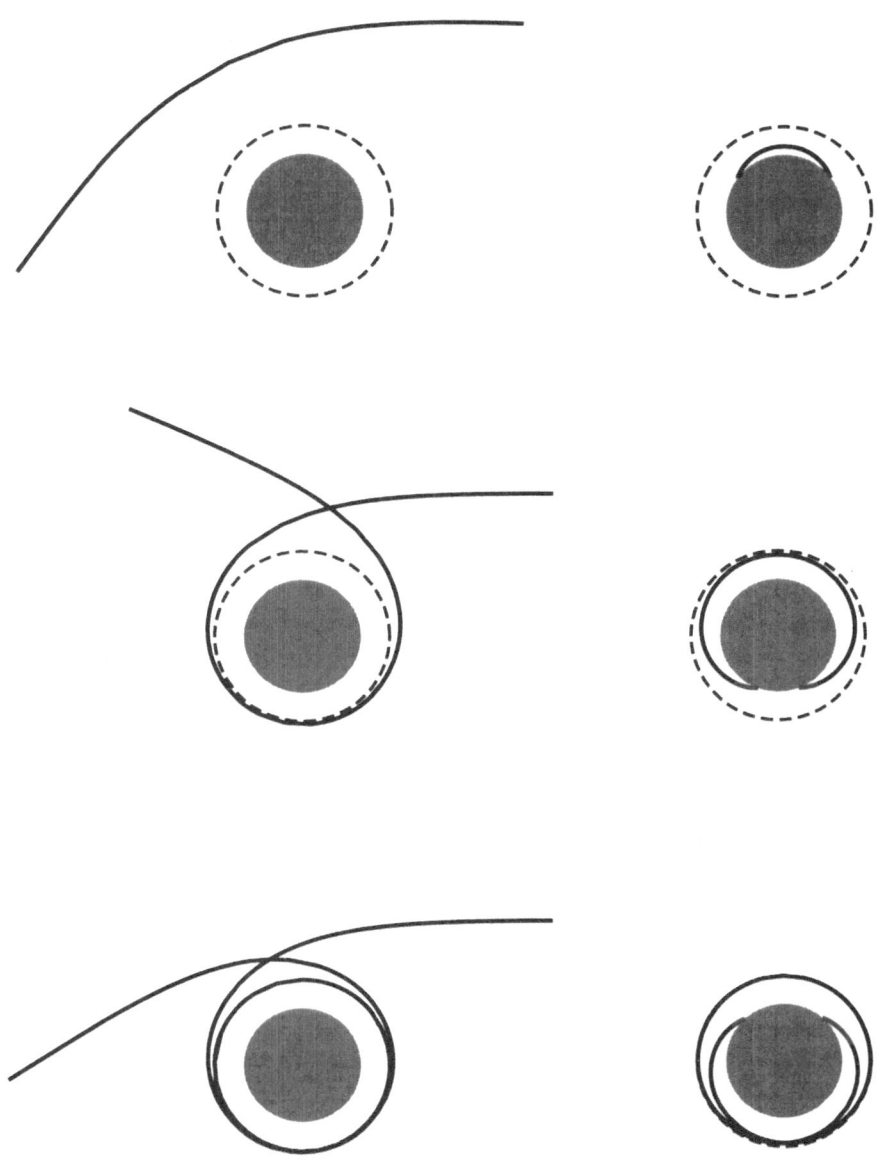

Figure 3 Photon orbits for impact parameters $b > b_c$. Orbits outside the photon sphere are shown on the left hand side for the impact parameters $1.38\,b_c$, $1.002\,b_c$, and $(1 + 1.3 \cdot 10^{-6})\,b_c$ (top to bottom). Orbits inside the photon sphere are shown on the right hand side for the impact parameters $1.23\,b_c$, $1.0043\,b_c$, and $(1 + 1.7 \cdot 10^{-5})\,b_c$ (top to bottom).

3   Light Deflection near Neutron Stars

3.1   Masses and Radii of Neutron Stars

As we have seen, the decisive quantity regarding light deflection near a neutron star is the ratio of its radius R and its Schwarzschild radius r_s.

a) Observational Evidence

There are several mass determinations from the observation of neutron stars in binary systems (Shapiro and Teukolsky 1983). The most accurately known neutron star masses are those of the Hulse-Taylor binary pulsar PSR 1913+16 and its companion which are $1.442 \pm 0.003 M_\odot$ and $1.386 \pm 0.003 M_\odot$, respectively (Taylor and Weisberg 1989).

The less precisely determined masses of six eclipsing X-ray pulsars seem to be consistent with the "canonical" neutron star mass of $1.4 M_\odot$. Observational determinations of neutron star radii are nonexistent.

b) Neutron Star Models

Models of nonrotating neutron stars are solutions to the Oppenheimer-Volkoff equations of hydrostatic equilibrium together with an equation of state (Shapiro and Teukolsky 1983). The key uncertainty here is the equation of state, especially at nuclear matter density and beyond. For a given equation of state the neutron star model depends on a single parameter, the central density, and there is a range of central densities which produces a series of stable neutron stars. The most massive neutron star in this series is the one with the lowest value of R/r_s. For different equations of state that are considered realistic the minimum values of R/r_s lie between 1.52 and 2.3. On the other hand one may be interested in the value of R/r_s for the $1.4 M_\odot$ neutron stars that seem to be favored by observations. For this case, different equations of state predict values of R/r_s between 2 and 3.8. According to all of these models, a neutron star is larger than its photon sphere, but not necessarily by very much.

c) Fundamental Limits

Because of the uncertainty of the equation of state it is interesting to note that there are lower limits on R/r_s based only on the conditions of stability and causality (Shapiro and Teukolsky 1983, Weinberg 1972). According to Buchdahl's theorem, any stable star must have $R/r_s > 9/8 = 1.125$. With some additional assumptions, mainly the requirement that the speed of sound be less than the speed of light, Rhoades and Ruffini found a lower limit of $R/r_s > 1.235$. In principle, therefore, one cannot rule out the existence of ultracompact neutron stars that are smaller than their photon spheres.

3.2   Radiation from the Neutron Star Surface

Orbits of photons starting at the neutron star surface and reaching an observer some distance away are shown in Fig. 4. The impact parameters of these orbits lie between $b = 0$ (photon emitted radially) and some maximum value $b_{\max}$. For a neutron star larger than its photon sphere as depicted in Fig. 4 the impact parameter is maximum

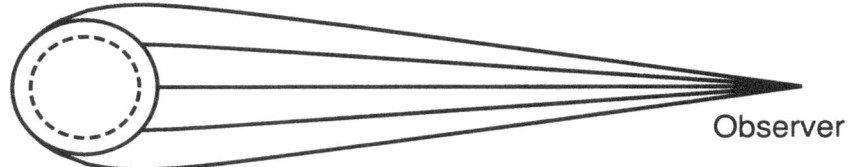

Figure 4 Orbits of photons reaching an observer from the neutron star surface. The dashed line marks the photon sphere.

for a photon emitted in tangential direction. From the condition that $p^r = 0$ at $r = R$ then follows with equation (5) that $b_{\max} = R/\sqrt{1 - r_s/R}$. For a neutron star smaller than its photon sphere, a photon that starts tangentially to the stellar surface is on an orbit confined within the photon sphere and does not reach the observer. In order for the photon to leave the photon sphere its impact parameter must be smaller than the critical impact parameter. Therefore, in this case, $b_{\max} = b_c = 1.5\sqrt{3}r_s$.

Two consequences of light deflection are immediately apparent from Fig. 4: enhanced surface visibility and increased angular size. Consider an observer that is close enough to resolve the neutron star but at the same time many neutron star radii away: $r_0 \gg R$. Without light deflection the near side of the neutron star is visible, the far side hidden from view. According to Fig. 4, light deflection means that photons emitted on the far side may also reach the observer, so that a larger part of the surface is visible. Here are some figures that illustrate the enhanced surface visibility:

R/r_s	∞	3	2	1.75	1.509	1.5
visible part of the surface	50%	74%	94%	100%	200%	∞

For the same observer at an intermediate distance, the angular size of the star as determined by the outermost photon orbit is $\alpha = b_{\max}/r_0$ (cf. Fig. 5). As summarized in this table

R/r_s	∞	> 1.5	< 1.5
$b_{\max}$	R	$R/\sqrt{1 - r_s/R}$	$b_c = 1.5\sqrt{3}r_s$
$\alpha = b_{\max}/r_0$	$\alpha = \alpha(R)$	$\alpha = \alpha(M,R)$	$\alpha = \alpha(M)$

the angular size of a star larger

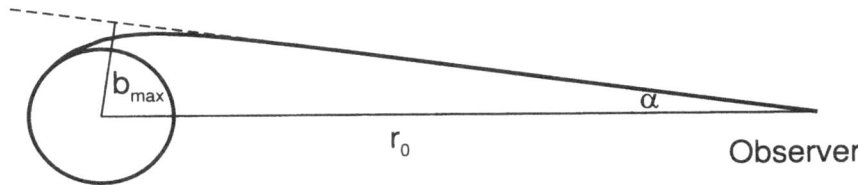

Figure 5 The angular size of the neutron star, α, is determined by the impact parameter $b_{\max}$ of the outermost photon orbit between the stellar surface and the observer.

than its photon sphere depends on both its mass and its radius. For a star smaller than its photon sphere, the angular size is a function of mass only and completely independent of the geometric size.

Surface visibility and angular size are illustrated in Fig. 6 which shows five images of "neutron stars" with identical radii R and different masses so that $R/r_s = \infty$, 3, 2, 1.7 and 1.52 (after Zahn 1991). In the last case, the complete surface is visible and part of the surface is imaged a second time in a thin circular strip at the border of the image.

For a source on the neutron star surface, the photon energies measured by local inertial observers momentarily at rest at the neutron star surface and at $r \gg r_s$, respectively, are related by

$$E_\infty = E_{em}\sqrt{1 - r_s/R}$$

(cf. Sect. 2.2). The intensity change between source and observer is then given by

$$I_\infty = I_{em}(E_\infty/E_{em})^3 = I_{em}\sqrt{1 - r_s/R}^3$$

(cf. Sect. 1). Since the factor $\sqrt{1 - r_s/R}^3$ has the same value for all points on the neutron star, a uniformly bright star will produce a uniformly bright image.

For a given neutron star radius R, an increase in r_s makes the image both larger and fainter. These two changes compensate in the sense that the observed photon flux remains constant: The neutron star subtends a solid angle

$$\Delta\Omega = \pi\alpha^2 = \pi\frac{R^2}{r_0^2(1 - r_s/R)}.$$

The observed photon flux is then

$$N = \frac{I_\infty\Delta\Omega}{E_\infty} = \frac{I_{em}}{E_{em}}\pi\frac{R^2}{r_0^2}$$

independent of r_s.

3.3 Radiation from above the Neutron Star Surface

Since redshift and intensity change between source and observer depend on the radial coordinate of the source, the total spectrum observed from a source extended in height is a superposition of spectra with different redshifts and intensity changes.

For a source above the neutron star surface, the two major consequences of light deflection are enhanced visibility (as before) and focussing.

As regards enhanced visibility, there is a minimum height h_{min} so that a source at a height $h \geq h_{min}$ above the neutron star can never be eclipsed by the star. This is illustrated in Fig. 7. For an observer far away from the neutron star ($r_0 \gg R$), here are some figures for h_{min}:

R/r_s	∞	10	5	3	2.5	2	1.75
h_{min}	∞	3.8R	1.32R	0.39R	0.19R	0.03R	0

Neutron stars with $R/r_s \leq 1.75$ cannot eclipse anything at all!

Figure 6   Five images of "neutron stars" with identical geometrical radii R and different masses: $R/r_\mathrm{s} = \infty$, 3, 2, 1.7, and 1.52 (top to bottom) (after Zahn 1991). Note the enhanced surface visibility and the larger angular size with decreasing R/r_s.

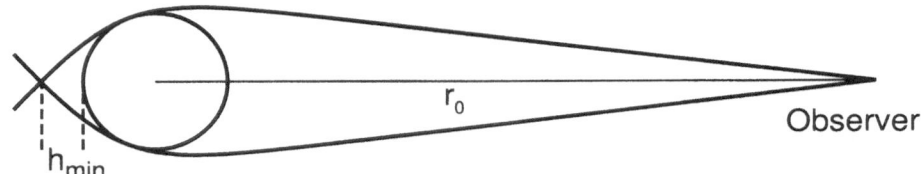

Figure 7   A source at a height $h \geq h_{min}$ can never be eclipsed by the neutron star.

Focussing is illustrated in Fig. 8: A small source with isotropic emission is placed at $r = 1.25R$ (Fig. 8a). The flux measured by a distant observer with viewing angle θ is plotted in polar diagrams for $R/r_s = \infty$ (Fig. 8b) and $R/r_s = 2.5$ (Fig. 8c). In the former case, the source is visible for $\theta \leq \theta_{max} = 126°$ (with the flux independent of θ) and hidden behind the star for $\theta > \theta_{max}$. In the latter case, since $h = 0.025R > h_{min} = 0.19R$, the source is always visible. At $\theta \approx 180°$ it is not only visible but also exceedingly bright!

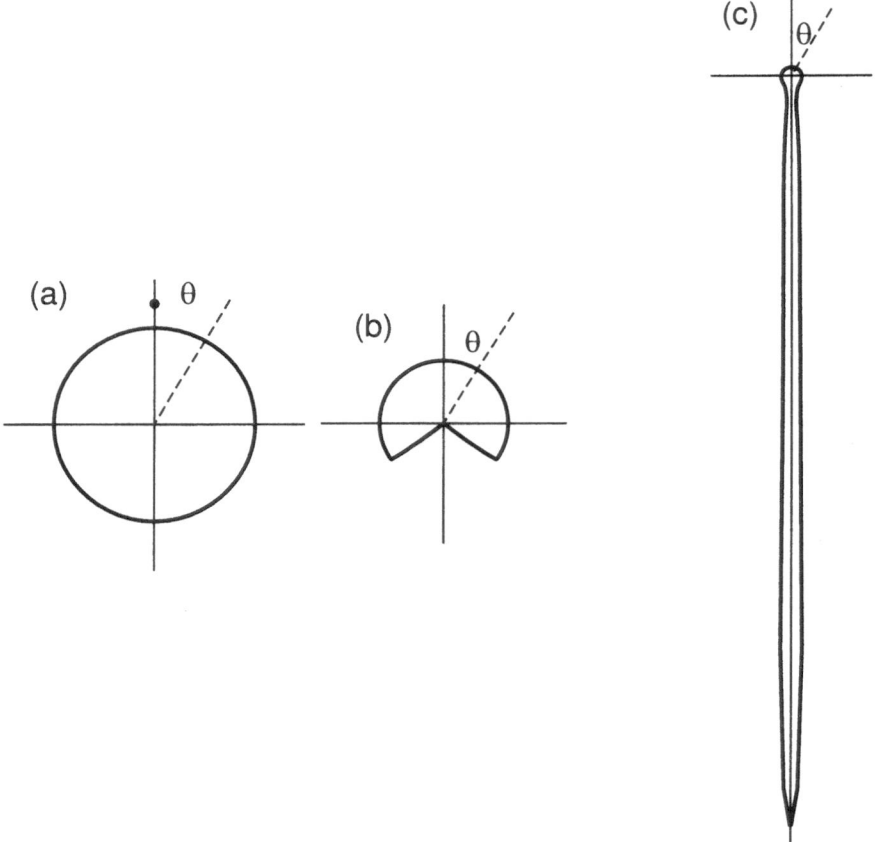

Figure 8   A small source above the neutron star (a) and the observed flux as a function of viewing direction for $R/r_s = \infty$ (b) and $R/r_s = 2.5$ (c). [Note that (b) and (c) are not to scale].

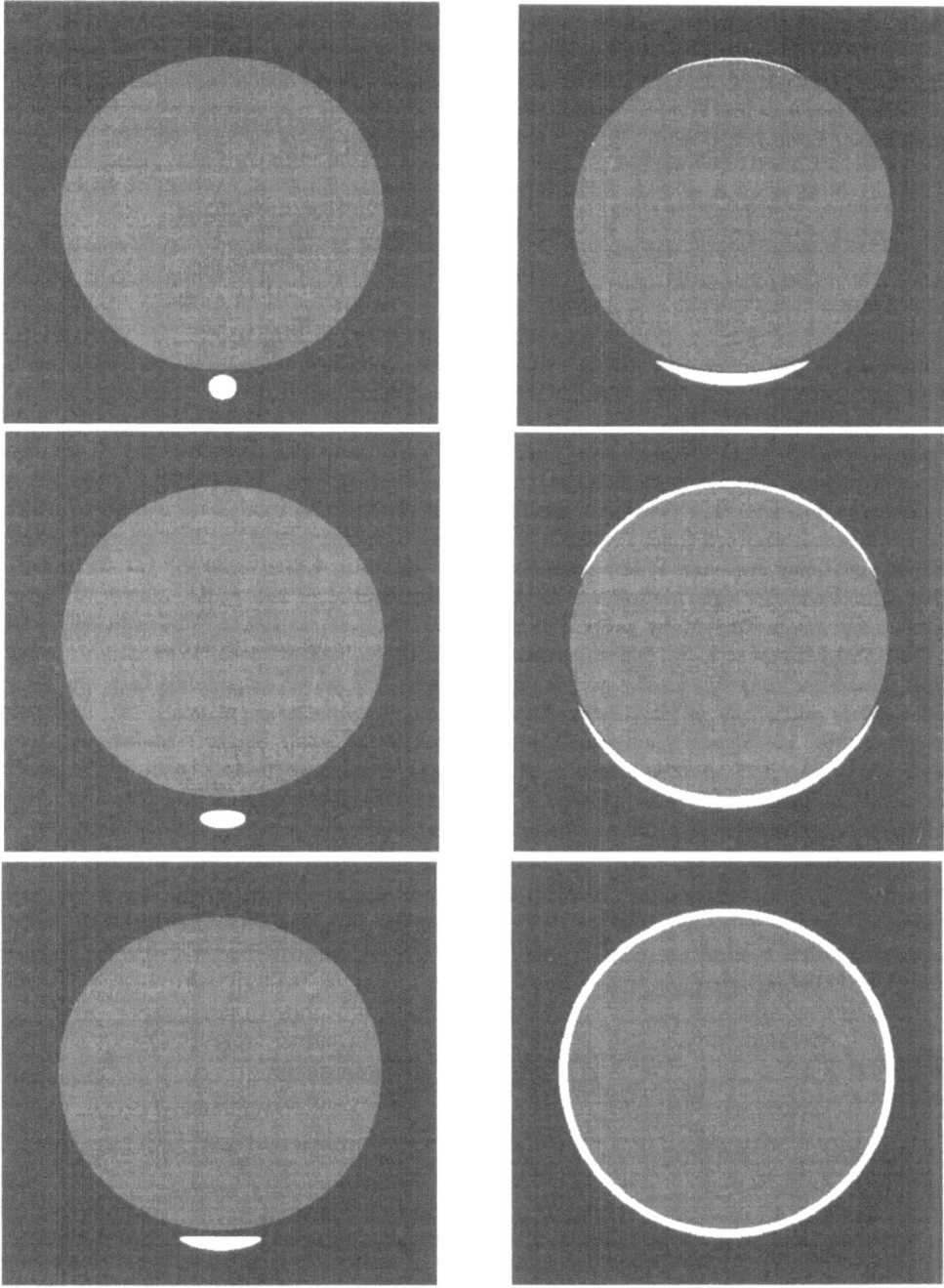

Figure 9 Visual appearance of a small source above the neutron star as seen from different directions. The viewing angle (defined to be zero when the source is in between the observer and the center of the neutron star is $90°$, $140°$, $160°$ (left, top to bottom), $170°$, $175°$, $180°$ (right, top to bottom). The images have been computed for a neutron star with $R/r_s = 2.5$ and a source that is $h = 0.25R$ above the stellar surface.

Fig. 9 illustrates what the source looks like as seen from various directions θ. Note that at $\theta = 140°$ the source would be eclipsed if $R/r_s = \infty$. With $R/r_s = 2.5$ it is instead slightly elongated. At $\theta = 170°$ there are two images of the source, produced by photons passing above and below the neutron star, respectively. As θ increases, these two images grow more elongated and finally at $\theta = 180°$ merge into a ring.

4   Light Deflection in Accreting X-Ray Pulsars

An accreting X-ray pulsar is a strongly magnetized neutron star in a binary system that accretes matter from its non-degenerate companion. According to the standard model, the strong magnetic field (a typical surface field strength is $10^8\,T$) channels the matter along the magnetic field lines onto the magnetic poles where it is stopped and its kinetic energy converted to X-radiation. When the neutron star rotates, the two polar X-ray emission regions pass through the observer's field of view and therefore the X-ray flux appears pulsed with the period of rotation of the neutron star. The models of the X-ray emission region are often classified into two types: *i)* polar cap models according to which the radiation is emitted from the surface of the neutron star and *ii)* column models where the site of X-ray emission is the accretion funnel just above the neutron star surface.

The significance of light deflection for the understanding of the pulse shapes of X-ray pulsars has been studied by several authors (Pechenik et al. 1983, Ftaclas et al. 1986, Riffert and Mészáros 1988, Mészáros and Riffert 1988, Nollert et al. 1989, Riffert et al. 1993, Bulik et al. 1995, Leahy and Li 1995). Here, the most simplified phenomenological models will serve as illustrative examples. A phenomenological polar cap model is shown in Fig. 10. There is isotropic emission from a uniformly bright circular polar cap (Fig. 10a). The half-opening angle α of the cap is usually estimated to be quite small; we adopt a "standard" value of $\alpha = 5°$ (Shapiro and Teukolsky 1983). The flux from this cap as measured by a distant observer depends on the viewing angle θ as shown in polar diagrams for $R/r_s = \infty$, i.e., without light deflection (Fig. 10b) and $R/r_s = 2.4$ (Fig. 10c). In the former case the flux is maximum for $\theta = 0°$ when the observer looks directly onto the polar cap and drops to zero shortly after $\theta = 90°$ when the cap disappears from view behind the neutron star. In the latter case, the flux is also maximum at $\theta = 0°$ but the cap remains visible up to $\theta = 135°$.

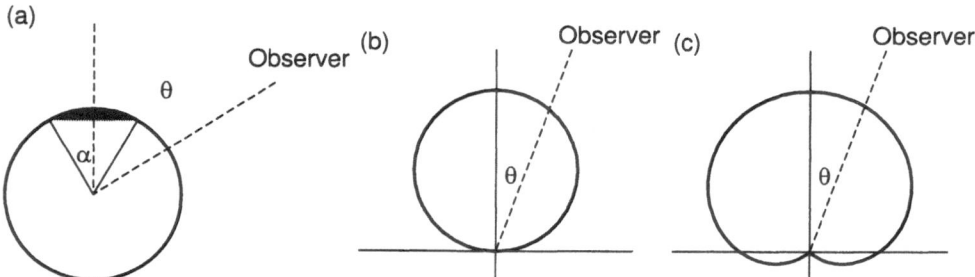

Figure 10   A phenomenological polar cap model (a) and the observed flux as a function of direction for $R/r_s = \infty$ (b) and $R/r_s = 2.4$ (c). The polar cap has a half-opening angle $\alpha = 5°$, is uniformly bright, and emits radiation isotropically. [Note that (b) and (c) are not to scale].

The changes in the angular flux distribution are much more dramatic for column models as illustrated in Fig. 11. Here, the emission comes from the side of a truncated radial cone (Fig. 11a). In the simplest case, the surface of the cone is uniformly bright and emits isotropically. The dependence of observed flux on viewing angle θ for a radial cone with half-opening angle $\alpha = 5°$, truncated at $r = 1.05\,R$ is shown in polar diagrams for $R/r_\mathrm{s} = \infty$ (Fig. 11b), $R/r_\mathrm{s} = 2.5$ (Fig. 11c), and $R/r_\mathrm{s} = 2$ (Fig. 11d). Seen from above at $\theta = 0°$, the flux is zero, because the top of the truncated cone does not radiate. In Fig. 11b and 11c the maximum flux is seen sideways and at high θ the cone disappears behind the neutron star. In the case of Fig. 11d, however, the height of the cone is $h > h_\mathrm{min}$, so that the source is never eclipsed by the neutron star and focussing produces a sharp increase in flux towards $\theta = 180°$.

Since the pulse profiles directly reflect the angular flux distribution, it is clear that for a given model of the emission region the parameter R/r_s plays a decisive role in determining the pulse shapes. This is illustrated in Fig. 12 for the cap model. A neutron star with

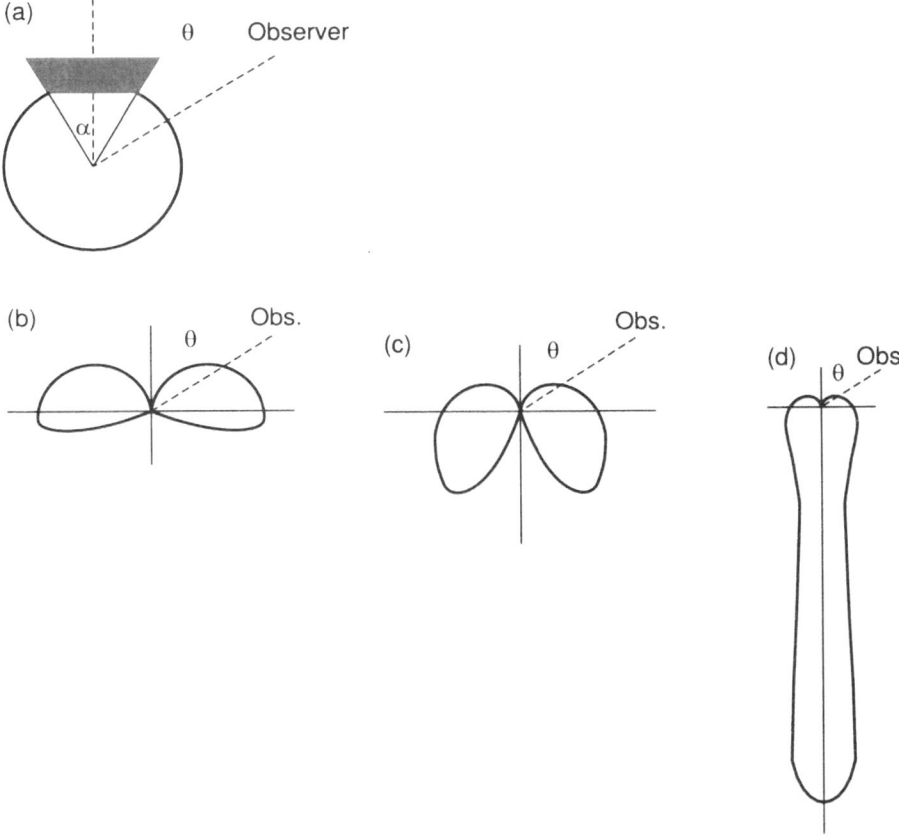

Figure 11   A phenomenological column model (a) and the observed flux as a function of direction for $R/r_\mathrm{s} = \infty$ (b), $R/r_\mathrm{s} = 2.5$ (c), and $R/r_\mathrm{s} = 2$ (d). The column is a truncated radial cone of half-opening angle $\alpha = 5°$ with uniform and isotropic emission from the side between radial coordinates R and $1.05\,R$. [Note that (b), (c) and (d) are not to scale].

two polar caps ($\alpha = 5°$) that are $\theta_m = 20°$ away from the rotation axis is observed at
$\theta_o = 80°$ (Fig. 12, top). Without light deflection ($R/r_s = \infty$, Fig. 12, left) both polar
caps are visible only part of the time. Their contributions to the pulse profile, labeled F_1
and F_2, therefore vanish during part of the pulse period. The total pulse profile, labeled
$F_1 + F_2$, can at nearly all phases be attributed to either one or the other polar cap.
At $R/r_s = 2.4$ (Fig. 12, right), both polar caps are always visible so that F_1 and F_2
never vanish. The total pulse profile is at all phases due to contributions of both polar
caps. The most conspicuous change is the dramatic reduction in modulation of the pulse
profile. This can also be understood in terms of the enhanced surface visibility: 84%

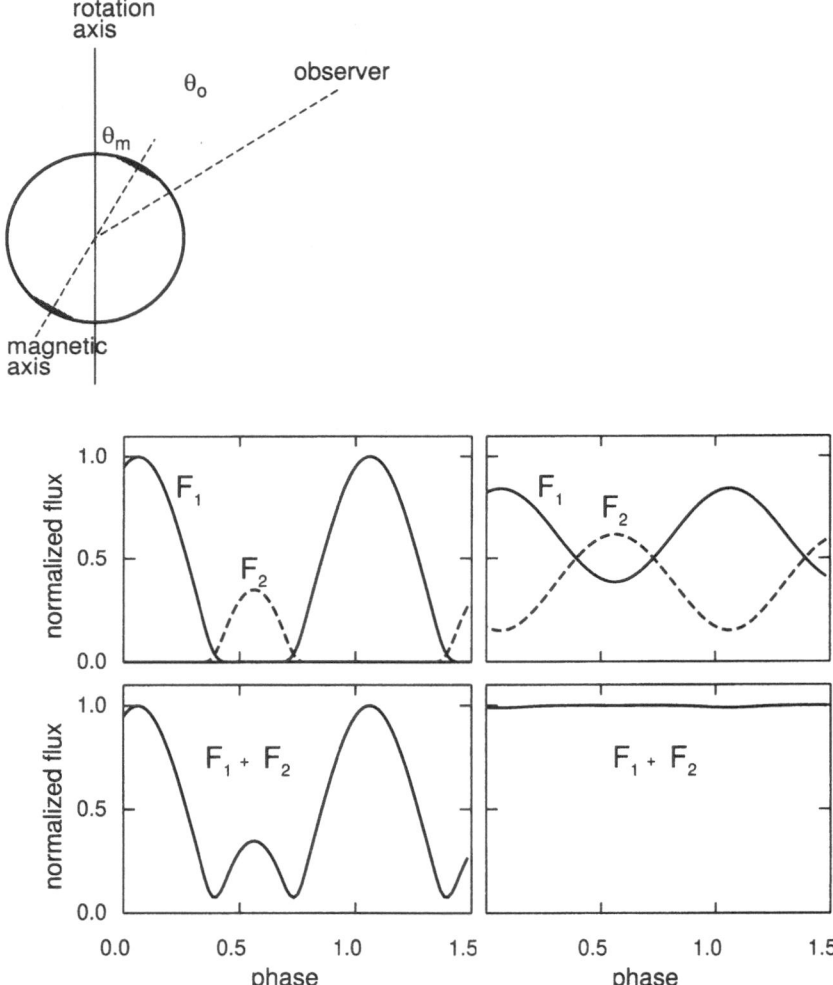

Figure 12 Pulse profiles predicted for two polar caps (top) with the same parameters as in
Fig. 10 for $R/r_s = \infty$ (left) and $R/r_s = 2.4$ (right). The magnetic axis is $\theta_m = 20°$ away from
the rotation axis and the direction of observation is $\theta_o = 80°$.

of the neutron star surface are visible to the observer for $R/r_s = 2.4$. When basically everything is visible all the time, then rotation of the star cannot produce a substantial amount of modulation of the flux.

More detailed modelling of polar caps and of columns will in general predict non-uniform and non-isotropic emission. In this case, light deflection will not necessarily suppress the modulation, but it will certainly affect the predicted pulse shape in a significant way.

References

Bulik, T., Riffert, H., Mészáros, P., Makishima, K., Mihara, T., and Thomas, B. (1995): Geometry and pulse profiles of X-ray pulsars: Asymmetric relativistic fits to 4U 1538-52 and Vela X-1. *Astrophys. J.*, **444**, 405-414

Ftaclas, C., Kearney, M.W., and Pechenick, K. (1986): Hot spots on neutron stars. II. The observer's sky. *Astrophys. J.*, **300**, 203-208

Leahy, D.A. and Li, L. (1995): Including the effect of gravitational light bending in X-ray profile modelling. *Mon. Not. Roy. Astron. Soc.*, **277**, 1177-1184

Mészáros, P. and Riffert, H. (1988): Gravitational light bending near neutron stars. II. Accreting pulsar spectra as a function of phase. *Astrophys. J.*, **327**, 712-722

Nollert, H.-P., Ruder, H., Herold, H., and Kraus, U. (1989): The relativistic "looks" of a neutron star. *Astron. Astrophys.*, **208**, 153-156

Pechenick, K.R., Ftaclas, C., and Cohen, J.M. (1983): Hot spots on neutron stars: the near-field gravitational lens. *Astrophys. J.*, **274**, 846-857

Press, W., Flannery, B., Teukolsky, S., and Vetterling, W. (1985): Numerical Recipes. Cambridge University Press, Cambridge

Riffert, H. and Mészáros, P. (1988): Gravitational light bending near neutron stars. I. Emission from columns and hot spots. *Astrophys. J.*, **325**, 207-217

Riffert, H., Nollert, H.-P., Kraus, U., and Ruder, H. (1993): Fitting pulse profiles of X-ray pulsars: the effects of relativistic light deflection. *Astrophys. J.*, **406**, 195-200

Shapiro, S. and Teukolsky, S. (1983): Black Holes, White Dwarfs and Neutron Stars. John Wiley & Sons, New York

Taylor, J.H. and Weisberg, J.M. (1989): Further experimental tests of relativistic gravity using the binary pulsar PSR 1913+16. *Astrophys. J.*, **345**, 434-450

Weinberg, S. (1972): Gravitation and Cosmology. John Wiley & Sons, New York

Zahn, C. (1991): Vierdimensionales Ray-Tracing in einer gekrümmten Raumzeit. Diplomarbeit, Universität Stuttgart

Magnetohydrodynamics of Rotating Black Holes

Max Camenzind

Landessternwarte Königstuhl
D-69117 Heidelberg
`M.Camenzind@lsh.uni-heidelberg.de`

Abstract

Magnetic fields are important for the physics of accretion disks around black holes, accretion flows should be formulated in terms of magnetohydrodynamics. Such magnetohydrodynamic flows near rapidly rotating compact objects must include the relativistic effects of the underlying metric. Of special importance are supermassive black holes in the centers of galaxies. The evidence for the existence of these objects is shortly discussed. The basic features of resistive magnetohydrodynamics (RMHD) are reviewed for plasma flows on the background of rotating black holes. We give a comprehensive derivation of the basic equations, with particular emphasis on axisymmetric configurations. We derive the relevant time–dependent equations for the poloidal magnetic flux and the poloidal current function for accretion disks around rapidly rotating black holes. In particular, we discuss the role of the gravitomagnetic force for the evolution of magnetic fields around black holes.

Magnetic fields in accretion disks also influence the accretion process onto black holes. These processes are discussed within a stationary approach using infinite conductivity near black holes. The theory of stationary MHD flows on rotating backgrounds is now complete and can be applied to various situations, outflows, and inflows. We discuss in particular magnetic accretion onto a rotating black hole and demonstrate the Blandford–Znajek process; for sufficiently rapidly rotating black holes, the accretion can carry a negative total angular momentum inwards, spinning down in this way the black hole. We also shortly discuss magnetized disk outflows near black holes, as well as the question of magnetic jet collimation.

1 Introduction

Magnetic fields are essential ingredients in the physics of accretion disks around rapidly rotating black holes. Seed fields advected inwards from the ambient medium will be sheared into helical fields. Once these fields are advected into the neighborhood of the horizon, additional forces appear which are related to the spin of the black hole. These effects could amplify the fields and produce a dynamo action in a boundary layer between the horizon and the disk (Khanna and Camenzind 1994, 1996). In this way, the black hole could be immersed into a rotating magnetosphere.

The physics of black hole magnetospheres goes back to the pioneering work of Bland-ford and Znajek (1977). These authors investigated the interaction of a horizon with an external force–free magnetosphere and came to the conclusion that by means of this interaction rotational energy could be extracted out of the immense energy reservoir of a supermassive black hole. In the last twenty years, many authors worked on this problem which could be important for the launch of energetic jets in quasars and ra-dio galaxies (Blandford 1992, Camenzind 1995). Most of these investigations were based on the force–free approximation, where plasma inertia is neglected in the Lorentz force (Okamoto 1992, Thorne et al. 1986). This force–free approximation is, however, not suit-able for the discussion of plasma processes that are relevant in accretion and outflows. For this reason, we will present an introduction into the complete magnetohydrodynam-ics of rotating black holes including currents and plasma motion. This theory is quite complicated, when working however in the 3+1 split of Kerr space the equations can be formulated in a way familiar to astrophysicists. This derivation is presented in Sect. 3.

In the next section we outline some aspects of supermassive black holes residing in the center of galaxies. Besides the stellar black holes, there is growing evidence for the exis-tence of such objects in many nearby luminous galaxies. This is a crucial question, since the explanation of the quasar activity at high redshifts in terms of black hole accretion requires the existence of very massive remnants in at least some type of galaxies. The core ellipticals recently found in HST observations are in fact very good candidates. The Virgo galaxy M 87 is probably a prototype of such a galaxy.

In Sect. 4 we discuss the general time–evolution of axisymmetric magnetic fields near ro-tating black holes. This theory is important for the investigation of the magnetic structure in accretion disks. The coupling between the spin of the black hole and electromagnetic fields leads to the interesting property of a dynamo action near the horizon. It turns out that Cowling's theorem is no longer valid near a black hole. The theory of stationary MHD flows on Kerr space is given in Sect. 5. This theory is now quite complete, appli-cations are discussed for disk in– and outflows. The question of jet formation is shortly addressed at the end.

2   Black Holes in the Centers of Galaxies

Most models of the energy production in quasars and active galactic nuclei invoke the presence of a nuclear black hole with a mass M_H in the range of 10^6 to $10^{10} M_\odot$ (Camen-zind 1997). The number of quasars at high redshift indicates that many normal galaxies we observe today must have gone through an active phase, if the lifetime of the activity is short in comparison with the Hubble time. This would indicate that supermassive black holes could be common in normal galactic nuclei (Rees 1984). The high luminosity of quasars at high redshifts also requires that those objects populate the more massive end of the mass distribution. This excludes normal spiral galaxies as hosts of high redshift quasars.

A massive central black hole significantly influences the motion of the surrounding stars out to a radius $R_{BH} = GM_H/\sigma_*^2$, where σ_* denotes the characteristic velocity dispersion of stars in the center of the host galaxy. σ_* is observed to be $\simeq 100$ to 300 km s^{-1} for elliptical and S0 galaxies. This amounts to radii of about 100 parsecs in giant ellipticals such as M87 and therefore to a scale of about one arcsecond in nearby galaxies. This scale can be resolved with modern HST observations. In this inner part, the velocity dispersion and circular velocities should diverge as $1/\sqrt{r}$ when $r \to 0$. As a result, high resolution photometric and kinematic observations are required to find direct dynamical evidence for the existence of massive central black holes.

2.1 Energetic Considerations

Since the pioneering works by Salpeter (1964), Lynden–Bell (1969) and Blandford and Rees (1974) (see Rees 1984), the standard model for the nuclear activity in galaxies has envisaged a supermassive compact object, usually a black hole with mass $10^6 -$ $10^{10}\, M_\odot$, which acts as a central engine in the production of radio emission (Blandford et al. 1993). In this picture, the radio power is derived from twin collimated beams of relativistic material which are ejected along the black hole spin axis. The total energy output is related to both the mass M_H of the central object and the accretion rate $\dot{M}$. The maximum source luminosity is set by the Eddington limit,

$$L_{\mathrm{Ed}} = 1.3 \times 10^{47}\, \mathrm{erg\, s^{-1}}\, \frac{M_H}{10^9\, M_\odot} \tag{1}$$

and the required accretion rate

$$\dot{M}_{\mathrm{Ed}} \simeq 20\, M_\odot\, \mathrm{yr}^{-1}\, \frac{0.1}{\epsilon_H}\, \frac{L}{10^{47}\, \mathrm{erg\, s^{-1}}}\; . \tag{2}$$

ϵ_H is the efficiency of mass–energy conversion. Producing powerful 3C radio galaxies and quasars with luminosities $L \simeq 10^{46} - 10^{47}$ erg s^{-1} requires minimum black hole masses of $10^9\, M_\odot$ and accretion rates of a few $M_\odot$ yr^{-1} of infalling gas. The total amount of gas accreted in the lifetime of the quasar is similar to the black hole mass itself. This amount of gas is indeed available on the parsec scale of the host galaxies. The total amount of gas assembled in the core of these giant elliptical galaxies can be an appreciable fraction of the core mass itself.

The mass of a dead quasar can also be estimated from the total dissipation

$$M_H = \frac{L_Q \tau}{\epsilon c^2} = 7 \times 10^8\, M_\odot\, \frac{L_Q}{10^{12}\, L_\odot}\, \frac{\tau}{10^9\, yr}\, \frac{0.1}{\epsilon}\, , \tag{3}$$

where L_Q is the quasar luminosity and τ its lifetime. 10^9 years is an upper limit to the lifetime. The density of such remnants follows from the integrated comoving energy density in quasar light (Chokshi and Turner 1992)

$$u \simeq 1.3 \times 10^{-15}\, \mathrm{erg\, cm^{-3}}\, . \tag{4}$$

The corresponding mass density is then a question of efficiency

$$\rho_H = \frac{u}{\epsilon c^2} = 2.2 \times 10^5\, M_\odot\, \mathrm{Mpc}^{-3}\, \frac{0.1}{\epsilon}\, . \tag{5}$$

On the other hand, the luminosity density of galaxies is $j \simeq 1.5 \times 10^8\, L_\odot$ Mpc^{-3}, and a typical bright hot component contributes $8.5 \times 10^9\, L_\odot$, depending somewhat on the Hubble constant. This provides us with a number density of 2×10^{-2} hot galaxies Mpc^{-3}. The mean black hole mass required per hot galaxy is only $10^7\, M_\odot$, which is in accordance with the observations.

Rotating objects have in general an additional source of energy. The rotational energy of neutron stars is behind the powerful pair winds which are injected into surrounding nebulae (the Crab nebula e.g.). For black holes, energies add quadratically

$$M_H = \sqrt{M_{\mathrm{irr}}^2 + \left(\frac{J_H}{2GM_{\mathrm{irr}}/c}\right)^2}. \tag{6}$$

The irreducible mass M_{irr} is a kind of rest mass for a rotating black hole that cannot be further reduced by any physical processes, except Hawking radiation – which is however unimportant for macroscopic black holes. The second term is due to the rotational energy of black holes and is given in terms of the angular momentum J_H of the black hole. Therefore, each black hole contains a rotational energy

$$E_{\mathrm{rot}} = (M_H - M_{\mathrm{irr}})c^2 = M_H c^2 \left(1 - \sqrt{\frac{1}{2}\left(1 + \sqrt{1 - (a_H/M_H)^2}\right)}\right), \tag{7}$$

which could be extracted by means of some electrodynamic processes. a_H denotes the specific angular momentum (Kerr parameter). Since $E_{\mathrm{rot}} \leq 0.29\,M_H c^2$, the rotational energy is a considerable amount of energy

$$E_{\mathrm{rot}} < 5 \times 10^{55}\,\mathrm{Watt\,s}\,\frac{M_H}{10^9\,M_\odot}, \tag{8}$$

and no other object can approach this upper limit for the rotational energy. The mass M_H of black holes residing in giant elliptical galaxies can easily exceed $10^9\,M_\odot$. If this rotational energy could be dissipated in a kind of pulsar process, this would represent a considerable luminosity

$$L_{\mathrm{rot}} \simeq \frac{E_{\mathrm{rot}}}{t_{\mathrm{diss}}} \simeq 1.6 \times 10^{40}\,\mathrm{Watt}\,\frac{M_H}{10^9\,M_\odot}\,\frac{10^9\,\mathrm{yr}}{t_{\mathrm{diss}}}, \tag{9}$$

essentially comparable to the mean luminosity of radio quasars at a redshift of 2 – provided the system is able to dissipate the energy on a time–scale of a few billion years. This process could stand behind the energetization of bright quasars, such as 3C 273 and 3C 345. In fact, Rees et al. have suggested in 1982 that the non–thermal power of radio–loud objects (quasars and radio galaxies) could be accounted for by this rotational energy.

This energy loss rate is an interesting number which can be compared to the bulk kinetic power Q_j in jets estimated from the by–products of the jets, the large–scale lobes (Rawlings and Saunders 1991)

$$Q_j = \frac{k\,U}{\tau_j}. \tag{10}$$

Here U is the energy stored in the lobes, taken from the equipartition energy U_{eq}; $k \simeq 2$ allows for $P\,dV$ work expended by the jet on pushing back and warming up the extended medium. Here τ_j is the age of the jet, estimated e.g. from the spectral age of the radio lobes. This amounts to a maximal jet power $\simeq 10^{40}$ Watt for bright radio quasars and narrow–line radio galaxies for $0.5 < z < 1.0$. On the other hand, material radiatively excited by an AGN cools by line emission. Radio galaxies are usually only narrow–line emitters, and one can therefore easily estimate the total narrow–line luminosity L_{NLR} in all narrow lines. Rawlings and Saunders (1991) obtained for an unbiased sample of FR II radio galaxies and low–power FR Is a correlation between the jet power and L_{NLR}

- $Q_j \propto L_{\mathrm{NLR}}$ for FR I and FR II radio galaxies, as well as for radio quasars;

- $Q_j \simeq 100\, L_{\mathrm{NLR}} \simeq 10^{36} - 10^{40}$ Watt;

- Radio–quiet quasars do not lie on the $Q_j - L_{\mathrm{NLR}}$ correlation.

The last point strongly indicates that the jet power is some extra power provided, e.g., by the rotational energy of the central source. But also radio–loud objects have photoionizers which, by virtue of $Q_j \propto L_{\mathrm{NLR}}$, are controlled by the jet driving mechanism. It is also interesting that jet sources with given Q_j have a higher low–frequency luminosity when the sources are in a dense cluster environment. This probably indicates that the narrow line emission is due to the interaction of the jets with the ambient medium. Since the above correlation extends over more than 4 orders of magnitude, this could reflect the scaling of the central mass from $\simeq 10^6\, M_\odot$ in faint ellipticals to $\simeq 10^{10}\, M_\odot$ in giant ellipticals. If jet power were related to rotational energy of the black hole, then it is essentially only the mass and the angular momentum that dictate the jet power.

2.2   Evidence for Black Holes in the Cores of Galaxies

If black holes are the drivers of the activity seen at high redshifts, these objects should be still present at low redshifts, though the fuel has declined by many orders of magnitude. The search for these remnants in the centers of nearby galaxies is a key project in astronomy and can be done in various ways.

Surface Brightness Profiles

Surface brightness cusps have been observed in many elliptical galaxies. The ground–based measurements have now been superseded by HST data [for a recent review of these data, see Kormendy and Richstone (1995) and Faber et al. (1997)]. These cusps are consistent with the presence of central black holes, but cannot prove their existence. HST observations show that the observed surface brightness $I(R)$ can be well fitted with a two–power law model (for the centers of ellipticals and spiral bulges that are called *hot galaxies*)

$$I(R) = 2^{(\beta-\gamma)/\alpha}\, I_b \left(\frac{R_b}{r}\right)^\gamma \left[1 + \left(\frac{r}{R_b}\right)^\alpha\right]^{(\gamma-\beta)/\alpha}. \tag{11}$$

The radius R_b represents a kind of core radius, inside R_b the asymptotic logarithmic slope is $-\gamma$, the asymptotic outer slope is $-\beta$, and the parameter α parameterizes the sharpness of the break. I_b is the surface brightness at the radius R_b. Two types of hot galaxies are found (Faber et al. 1997):

- *Core galaxies* have a broken power–law surface brightness profile that changes the slope considerably at the break radius R_b. The slope in the inner region (the core) is found in the region $0 \leq \gamma < 0.3$. This indicates the existence of some cusp in the density distribution. Core galaxies are luminous objects and are often found in clusters. Black holes are found in this class of galaxies.

- *Power–law galaxies* have still fairly steep surface–brightness profiles in the center and show no significant break. The slope γ is typically 0.8 and these galaxies are generally fainter than core galaxies. The small Virgo ellipticals and the bulges of disk galaxies are of this type.

Stellar Spectroscopic Observations

Spectroscopic observations have also provided hints for a central dark mass in the nuclei of half a dozen nearby galaxies (Tab. 1). The evidence rests on an inwards increase of the line–of–sight velocity dispersion σ and on a steep central gradient in the mean light–of–sight velocity $< v_{\rm los} >$. Such a gradient is expected near a black hole, but it could also be the signature of a rapidly rotating nuclear stellar disk.

The compact E3 galaxy M32 has been carefully analyzed for the presence of a nuclear black hole. Though it shows no sign of nuclear activity or emission–line gas, it harbors a black hole of $3 \times 10^6 \, M_\odot$ (Kormendy and Richstone 1995).

M87 (E0 galaxy) is probably the archetypical object for an old quasar residing in a big cluster of galaxies. The stellar velocity dispersion increases from 270 km/s at 15" to 400 km/s at 0.5" from the nucleus. The emission–line gas in the nucleus of M87 shows evidence for bulk motion and for a strong inwards increase of the gas velocity. HST measurements now have shown that this gas resides in a small nuclear disk rotating at a speed of 550 km/s at a distance of 0.26" from the nucleus (corresponding to about 20 pc from the nucleus). This makes the case for a $3 \times 10^9 M_\odot$ black hole in M87 very compelling.

The best black hole candidate is the S0 galaxy NGC 3115 that shows evidence for a mass of $2 \times 10^9 \, M_\odot$ (Kormendy et al. 1996). Again, there is no obvious activity associated with this mass.

Gas Disks and Jets on the Parsec–Scale

The case of M87 already demonstrates the close connection between gaseous disks on the parsec–scale and the launch of jets in AGN. The hunt for such disks has been opened by HST observers.

Dark Matter in Ellipticals and the Formation of Black Holes

Modern galaxy formation assumes that galaxies are formed in the halo of dark matter. Even ellipticals should be formed in this way. A strong hint for the presence of dark matter in ellipticals comes from the fundamental plane which implies that the global M/L of elliptical galaxies varies systematically with total luminosity L and effective radius R_e. This would indicate that dark matter contributes significantly to M/L in the luminous region, $M/L \propto L^{0.25}$. Recent work has provided strong evidence for dark matter around the cD galaxy NGC 1399 in the Fornax cluster. The central velocity dispersion is 400 km/s corresponding to $M/L \simeq 10$ for a distance of 15 Mpc. It falls by a factor of two at 1' (4.5 kpc) and rises again to 400 km/s at 5.5' (25 kpc).

Massive ellipticals are probably embedded into a dark halo. This implies that dissipative infall of baryonic matter must have played a major role in the formation of the galaxy and its central black hole. The core luminosity $L_c = \pi R_c^2 I_b$ is typically found to be $0.01 \, L_G$

Galaxy	Constellation	Type	Distance	m_V	$M_H/M_\odot$	M/L
Galactic Centre	Sagittarius	Sbc	8.2 kpc		2×10^6	1.0
NGC 221, M32	Andromeda	E2	750 kpc	8.1	3×10^6	> 10
NGC 224, M31	Andromeda	Sb	750 kpc	3.3	3×10^7	350
NGC 4258, M106	Canes Venatici	Sbc	8 Mpc	8.4	4×10^7	
NGC 3115	Sextans	S0	8.5 Mpc	8.8	2×10^9	85
NGC 4594	Virgo	Sa	10 Mpc	7.7	5×10^8	55
NGC 3377	Leo	E5	10 Mpc	10.4	8×10^7	3
NGC 4486, M87	Virgo	E0	16 Mpc	8.6	3×10^9	500
NGC 4261, 3C 270	Virgo	E2	16 Mpc	10.4	4×10^8	20

Table 1 Black Hole candidates as known in 1996. The last column gives the mass–to–light ratio in solar units as an indicator for dark mass hidden in the center.

(Faber et al. 1997) indicating $M_c = 0.01\,M_G$. This means that for the black hole masses we have $M_H \simeq 0.2 M_c$, or $M_H \simeq 0.002 M_G$. The very origin of this concentration of a small mass in the center of a galaxy is still an unsolved problem of galaxy formation.

3   The Magnetohydrodynamic Equations on Kerr Space

The metric of a rotating axisymmetric compact object can generally be expressed as (neutron stars and black holes)

$$ds^2 = -\alpha^2\,dt^2 + \varpi^2(d\phi - \omega\,dt)^2 + g_{rr}\,dr^2 + g_{\theta\theta}\,d\theta^2 \,. \tag{12}$$

If the background is the Kerr metric in Boyer–Lindquist coordinates, then we have for the metric functions

$$\alpha^2 = \frac{\Sigma\Delta}{A} \quad,\quad \Delta = r^2 - 2Mr + a^2 \,, \tag{13}$$

$$\varpi^2 = \frac{A\sin^2\theta}{\Sigma} \quad,\quad \Sigma = r^2 + a^2\cos^2\theta \,, \tag{14}$$

$$\omega = \frac{2Mar}{A} \quad,\quad A = (r^2 + a^2)^2 - \Delta a\sin^2\theta \,, \tag{15}$$

$$g_{rr} = \frac{\Sigma}{\Delta} \quad,\quad g_{\theta\theta} = \Sigma \,. \tag{16}$$

We use units with $c = 1 = G$. Under astrophysical circumstances, the gravitational field of a black hole is uniquely determined once its mass M and its angular momentum J_H are given. Useful expressions are $M_H = GM/c^2 = 10$ AU for a mass of $10^9 M_\odot$ and the specific angular momentum $a_H = J_H/M$. In all these expressions, the mass is only a scaling parameter for spatial dimensions. The appearance of the angular momentum has a deeper meaning; it generates additional force fields which are not present in a Newtonian description.

Electromagnetic fields are treated as test fields on the spacetime. They obey Maxwell's equations for the Faraday antisymmetric tensor F

$$\nabla_\nu F^{\mu\nu} = 4\pi j^\mu \tag{17}$$

$$\partial_\lambda F_{\mu\nu} + \partial_\mu F_{\nu\lambda} + \partial_\nu F_{\lambda\mu} = 0 . \tag{18}$$

In addition we have Ohm's law of the form

$$h^{\mu\nu} j_\nu = \sigma F^{\mu\rho} u_\rho , \tag{19}$$

where $h^{\mu\nu} = g^{\mu\nu} + u^\mu u^\nu$ is the projection into 3–space of the rest system of the plasma with velocity u. F is the electromagnetic field tensor, j the current density and σ the conductivity of the plasma. In a disk σ must be much smaller than any classical conductivity to be astrophysically interesting.

3.1 Standard Electrodynamics

Early investigations on electromagnetic fields were based on a coordinate description, based on the electromagnetic potential A_μ by means of

$$F_{\mu\nu} = \partial_\mu A_\nu - \partial_\nu A_\mu . \tag{20}$$

The electromagnetic field is entirely determined in terms of the two potentials $\Phi = A_t$ and $\Psi = A_\phi$, as well as in terms of the current function B_T. Due to symmetries, one also finds the relation $dA_t = -\Omega_F\, dA_\phi$. Since F is an antisymmetric tensor, the inhomogeneous Maxwell's equations can be written as

$$\frac{1}{\sqrt{-g}} \partial_\nu (\sqrt{-g} F^{\mu\nu}) = 4\pi j^\mu . \tag{21}$$

Wald (1974) derived an exact analytical vacuum solution for a Kerr black hole immersed in a uniform external magnetic field

$$A_\phi^W = \frac{1}{2} g_{\phi\phi} B_0 (1 - 2a_H\omega) . \tag{22}$$

The rotation produces an electric field which has a quadrupolar structure near the horizon and is radial far away from the horizon. The polar regions carry a negative surface charge density and the equatorial region a positive one. The total magnetic flux through the horizon is $4\pi B_0 M_H(r_H - M_H)$, which vanishes in the extreme Kerr case. As the rotation of the hole increases, magnetic flux is expelled from it. Since Boyer–Lindquist observers are not suitably defined within the ergosphere, magnetic and electric fields defined with respect to such observers are not meaningful. For this reason, physical observers must be introduced. In addition, when plasma is present, vacuum fields do not represent physically interesting magnetic structures. We have to skip to a magnetohydrodynamic approach (i.e. Maxwell's equations including a one–component plasma). Kudoh and Kaburaki (1996) have recently worked out these equations for geometrically thin accretion disks.

The electrodynamic equations in a 4–dimensional form, based on the Faraday tensor F can hardly illuminate the physical meaning. Thorne and Macdonald (1982) have rewritten these equations using the 3+1–split for a black hole background. This formalism operates with concepts familiar from Special Relativity, such as fields, charge density, and electric current density. In this way, Maxwell's equations appear in a form similar to the flat spacetime formulation. This formalism also makes use of a "membrane interpretation of black holes" (Thorne et al. 1986): the boundary of a black hole appears to distant observers as a thin membrane characterized by special thermodynamic, mechanical and electromagnetic properties (Carter 1979). In the next section we give a short modern derivation of Maxwell's equations based on the exterior algebra of forms (Durrer and Straumann 1988).

3.2   The 3+1 Formalism for Maxwell's Equations

The 3+1 formalism is of special importance for the discussion of electromagnetic fields in General Relativity. This technique is based on a slicing of the spacetime in such a way that the slices are spacelike [for a recent differential geometric representation, see Durrer and Straumann (1988)]. This defines a vector field ∂_t which can be decomposed into normal and parallel components relative to the slicing

$$\partial_t = \alpha \vec{n} + \vec{\beta} \,. \tag{23}$$

Here $\vec{n}$ is the unit normal field and $\vec{\beta}$ is tangent to the slices. α is called the lapse function (or redshift factor) and $\vec{\beta}$ the shift vector field. The line element of the Kerr geometry (12) is a special decomposition of the general expression

$$ds^2 = -\alpha^2 \, dt^2 + g_{ij}(dx^i + \beta^i dt)(dx^j + \beta^j dt) \,. \tag{24}$$

The four one–forms $\Theta^0 = \alpha dt$ and $\Theta^i = \theta^i + \beta^i dt$ $(i = 1,2,3)$ are therefore orthonormal and form a natural basis for one–forms. θ^i are normalized one–forms in the spatial slice. In Kerr space, α is explicitly known, and due to axisymmetry the shift vector has only one nonvanishing component $\beta^\phi = -\omega$. In the case of rapidly rotating neutron stars, these two functions are only given numerically.

The derivation of Maxwell's equations adapted to the slicing is quite complicated when usual techniques are involved (Thorne and Macdonald 1982). There is a very elegant derivation based on the technique of differential forms (Durrer and Straumann 1988), the electromagnetic potential A is essentially a one–form and the Faraday tensor F a two–form. Instead of using the coordinate basis (as above), the Faraday tensor is now decomposed with respect to this natural basis of one–forms

$$F = E \wedge \Theta^0 + B \tag{25}$$

with the definition of the two fields

$$E = E_i \, \Theta^i \quad , \quad B = \frac{1}{2} B_{ij} \, \Theta^i \wedge \Theta^j \,. \tag{26}$$

On the other hand, it is useful to decompose these quantities also into components adapted to the slicing

$$E = \mathcal{E} + (E_i \beta^i) \, dt = \mathcal{E} + i_\beta \mathcal{E} \, dt \,. \tag{27}$$

$\mathcal{E} = E_i \, \theta^i$ is now the *horizontal electric field*. Similarly, we have for the magnetic 2–form

$$\mathcal{B} = \frac{1}{2} \, B_{ij} \, \theta^i \wedge \theta^j \, . \tag{28}$$

We obtain therefore the following decomposition of the Faraday 2–form

$$F = \mathcal{B} + (\alpha \mathcal{E} - i_\beta \mathcal{B}) \wedge dt \, . \tag{29}$$

Since the homogeneous Maxwell's equations are given as $dF = 0$, this means

$$d\mathcal{B} + dt \wedge \partial_t \mathcal{B} + d(\alpha \mathcal{E}) \wedge dt - d(i_\beta \mathcal{B}) \wedge dt = 0 \, . \tag{30}$$

This splits into the two equations

$$d\mathcal{B} = 0 \quad , \quad d(\alpha \mathcal{E}) + \partial_t \mathcal{B} = d(i_\beta \mathcal{B}) \, . \tag{31}$$

Using the Cartan formula for the Lie derivative, $\mathcal{L}_\beta = di_\beta + i_\beta d$, the homogeneous Maxwell's equations can also be written as

$$\boxed{d\mathcal{B} = 0 \quad , \quad d(\alpha \mathcal{E}) + \partial_t \mathcal{B} = \mathcal{L}_\beta \mathcal{B} \, .} \tag{32}$$

The inhomogeneous equations follow from the Hodge dual one–forms

$$\mathcal{H} = *\mathcal{B} \quad , \quad \mathcal{E} = *\mathcal{D} \, . \tag{33}$$

($*$ denotes the 3–dimensional Hodge dual). The dual of the Faraday 2–form is therefore

$$*F = \mathcal{D} - (\alpha \mathcal{H} + i_\beta \mathcal{D}) \wedge dt \, . \tag{34}$$

In the same sense we can also decompose the current density

$$J = \rho_e \, \theta^0 + j_k \, \theta^k \, , \tag{35}$$

where ρ_e is the electric charge density with respect to this basis. We also introduce the horizontal forms

$$\vec{\rho} = \rho_e \, \theta^1 \wedge \theta^2 \wedge \theta^3 \quad , \quad \vec{j} = j_k \, \theta^k \quad , \quad \mathcal{J} = *\vec{j} \, . \tag{36}$$

From this we find for the current 3–form

$$*J = \vec{\rho} + (i_\beta \vec{\rho} - \alpha \mathcal{J}) \wedge dt \, . \tag{37}$$

With this, the inhomogeneous Maxwell's equations can be written

$$\begin{aligned} d * F &= d\mathcal{D} + dt \wedge \partial_t \mathcal{D} - d(\alpha \mathcal{H}) \wedge dt - d(i_\beta \mathcal{D}) \wedge dt \\ &= 4\pi \vec{\rho} + 4\pi (i_\beta \vec{\rho} - \alpha \mathcal{J}) \wedge dt \, . \end{aligned} \tag{38}$$

This once again splits into two equations

$$\boxed{d\mathcal{D} = 4\pi \vec{\rho} \quad , \quad d(\alpha \mathcal{H}) = (\partial_t - \mathcal{L}_\beta)\mathcal{D} + 4\pi \alpha \mathcal{J} \, .} \tag{39}$$

The first equation is Gauss's law, and the second one Ampère's law modified by a Lie derivative along the shift vector field. These are Maxwell's equations adapted to the slicing of any geometry.

This equation also implies charge conservation in the form of

$$(\partial_t - \mathcal{L}_\beta)\bar{\rho} + d(\alpha\mathcal{J}) = 0\,.\tag{40}$$

The Hodge dual of this form is equivalent to the usual form of charge conservation

$$(\partial_t - \mathcal{L}_\beta)\rho_e + \nabla \cdot (\alpha * \mathcal{J}) - \alpha\rho_e\, tr(\vec{K}) = 0\,.\tag{41}$$

$\vec{K}$ denotes the second fundamental form of the slices defined as

$$K_{ij} = \frac{1}{2\alpha}\left(\beta_{i|j} + \beta_{j|i}\right)\tag{42}$$

which is known as the gravitomagnetic field of a rotating object. Due to axisymmetry, we have in Kerr space $tr(\vec{K}) = 0$. This corresponds in fact to the charge conservation law discussed in (Thorne et al. 1986)

$$\alpha\frac{d\rho_e}{d\tau} = \left(\partial_t - \vec{\beta} \cdot \nabla\right)\rho_e = -\nabla \cdot (\alpha\vec{j})\,.\tag{43}$$

3.3   Maxwell's Equations under Axisymmetry

In the following, we will not work with forms, but with the corresponding vector fields. $\vec{E}$ and $\vec{B}$ denote the electric and magnetic fields, respectively, as measured by ZAMOS, which are special observers having a 4–velocity U perpendicular to the absolute space

$$U = \frac{1}{\alpha}\left(\partial_t - \beta^i\,\vec{e}_i\right)\,.\tag{44}$$

Together with the orthonormal tetrad in Boyer–Lindquist coordinates $\vec{e}_r$, $\vec{e}_\theta$ and $\vec{e}_\phi = (1/\varpi)\,\partial_\phi$ this forms a physical basis in the tangential space. This is the basis that is dual to the above natural one–form basis. The circumference of a circle around the rotational axis is measured by ϖ. The 4–velocity of the plasma u is then expressed as

$$u = \gamma(U + \vec{v})\,,\tag{45}$$

where $\vec{v}$ is now the 3–velocity of the plasma with respect to ZAMOS.
Instead of using the forms, Maxwell's equations (32) and (39) are expressed for the vector fields $\vec{E}$ and $\vec{B}$

$$\nabla \cdot \vec{E} = 4\pi\rho_e \quad,\quad \nabla \cdot \vec{B} = 0\tag{46}$$

$$\nabla \times (\alpha\vec{E}) = -(\partial_t - \mathcal{L}_\beta)\vec{B}\tag{47}$$

$$\nabla \times (\alpha\vec{B}) = (\partial_t - \mathcal{L}_\beta)\vec{E} + 4\pi\alpha\,\vec{j}\,.\tag{48}$$

The current density $\vec{j}$ is given by Ohm's law,

$$\vec{j} = \sigma\gamma\,(\vec{E} + \vec{v} \times \vec{B}) + \rho_e'\,\gamma\vec{v}\,,\tag{49}$$

where σ denotes the conductivity, $\vec{v}$ the bulk velocity of the plasma, γ the corresponding Lorentz factor, and ρ_e' the charge density in the rest frame of the plasma. As compared to flat spacetimes, there are two important additional terms related to the shift vector of the slicing (or the frame–dragging effect in Kerr space).
For axisymmetric fields, Maxwell's equations assume the simple form

$$\nabla \times (\alpha \vec{E}) = -\frac{\partial \vec{B}}{\partial t} + (\vec{B} \cdot \nabla \omega) \, \varpi \vec{e}_\phi \tag{50}$$

$$\nabla \times (\alpha \vec{B}) = \frac{\partial \vec{E}}{\partial t} - (\vec{E} \cdot \nabla \omega) \, \varpi \vec{e}_\phi + 4\pi \alpha \vec{j} . \tag{51}$$

On an axisymmetric spacetime, it is now useful to split all the vector fields into poloidal and toroidal components, $\vec{B} = \vec{B}_p + \vec{B}_T$ with $\vec{B}_T = B^{\hat{\phi}} \vec{e}_\phi$ and $\vec{B}_p = B^{\hat{r}} \vec{e}_r + B^{\hat{\theta}} \vec{e}_\theta$. $\vec{B}_p = \nabla \times (A^{\hat{\phi}} \vec{e}_\phi)$. The induction and Ampère's equations give the following relations, when decomposed into poloidal and toroidal components

$$\frac{\partial A^{\hat{\phi}}}{\partial t} = -\alpha \, E^{\hat{\phi}} \tag{52}$$

$$\frac{\partial B^{\hat{\phi}}}{\partial t} = \varpi \, \vec{B}_p \cdot \nabla \omega - \vec{e}_\phi \cdot (\nabla \times \alpha \vec{E}_p) \tag{53}$$

$$\frac{\partial E^{\hat{\phi}}}{\partial t} = \varpi \vec{E}_p \cdot \nabla \omega - \mathcal{G}_2[A^{\hat{\phi}}] - 4\pi \alpha j^{\hat{\phi}} \tag{54}$$

$$\frac{\partial \vec{E}_p}{\partial t} = \nabla \times (\alpha B^{\hat{\phi}}) - 4\pi \alpha \vec{j}_p . \tag{55}$$

$\mathcal{G}_2[A] \equiv -\vec{e}_\phi \cdot [\nabla \times \alpha \vec{B}_p]$ is the Grad–Shafranov operator for the poloidal flux function $\Psi \equiv \varpi A^{\hat{\phi}}$

$$\mathcal{G}_2[\Psi] \equiv \varpi \nabla \cdot \left[\frac{\alpha}{\varpi^2} \, \nabla \Psi \right] . \tag{56}$$

In terms of this flux function Ψ, the poloidal magnetic field is given in the standard form

$$\vec{B}_p = \frac{1}{\varpi} \, \nabla \Psi \times \vec{e}_\phi . \tag{57}$$

In addition, Ohm's law has the two components

$$\vec{j}_p = \sigma \gamma \left[\vec{E}_p + \vec{v}_T \times \vec{B}_p + \vec{v}_p \times \vec{B}_T \right] \tag{58}$$

$$j^{\hat{\phi}} = \sigma \gamma \left[E^{\hat{\phi}} + \vec{e}_\phi \cdot (\vec{v}_p \times \vec{B}_p) \right] , \tag{59}$$

provided the charge density in the plasma frame vanishes. This formulation of Maxwell's equations is suitable for implementation into a code. Due to the potential formulation, $\nabla \cdot \vec{B} = 0$ is automatically satisfied. A real 3D implementation does not yet exist. In Sect. 4 we will discuss a 2D implementation of the above equations.

3.4 Boundary Conditions at the Event Horizon

Similar to neutron stars, boundary conditions must also be specified at the event horizon. In contrast to a neutron star, a black hole has no material surface. One can think of the black hole as having a fictitious surface charge density σ^H that compensates for the flux of electric field across the surface, and a fictitious surface current density $\vec{j}^H$ that closes tangential components of the magnetic fields. This interpretation is the basis of the membrane formalism (Thorne et al. 1986).

Gauss' law implies for the electric field perpendicular to the horizon

$$E_\perp^H \rightarrow 4\pi\sigma^H \,, \tag{60}$$

and Ampère's law for the field parallel to the horizon

$$\alpha\vec{B}_\parallel \rightarrow \vec{B}_\parallel^H = 4\pi\vec{j}^H \times \vec{n}. \tag{61}$$

In addition, Ohm's law implies

$$\alpha\vec{E}_\parallel \rightarrow \vec{E}_\parallel^H = \mathcal{R}^H\vec{j}^H \tag{62}$$

with $\mathcal{R}^H = 4\pi/c = 377$ Ohm as the effective surface resistance of the horizon. The fields $\vec{E}_\perp$ and $\vec{B}_\perp$ are finite at the horizon, $\vec{E}_\parallel$ and $\vec{B}_\parallel$ generally diverge as $1/\alpha$, but

$$|\vec{E}_\parallel - \vec{n} \times \vec{B}_\parallel| \propto \alpha \rightarrow 0 \tag{63}$$

at the horizon. This signifies that for a locally nonrotating observer the electromagnetic field at the horizon looks like a wave sinking into the black hole (ingoing wave conditions).

3.5   The Hydrodynamic Equations in the 3+1 Split

Besides the time evolution of magnetic fields we also need a description of the time evolution of the plasma part. We decompose the energy–momentum tensor T into horizontal and vertical components

$$T = \epsilon U \otimes U + U \otimes \vec{S} + \overset{\leftrightarrow}{t} \,. \tag{64}$$

ϵ is the energy–density with respect to ZAMOs, $\vec{S}$ the momentum flux and $\overset{\leftrightarrow}{t}$ the stress tensor. For a one–component non–viscous plasma the energy–momentum tensor has the form

$$T = (\rho + P)\, u \otimes u + Pg + T_{\text{em}} \tag{65}$$

With the decomposition of $u = \gamma(U + \vec{v})$ we find the following components with respect to ZAMOs

$$\epsilon = \gamma^2\left(\rho + P\vec{v}^2\right) + \epsilon_{\text{em}} \tag{66}$$

$$\vec{S} = (\rho + P)\,\gamma^2\vec{v} + \vec{S}_{\text{em}} = \rho_0 h\gamma^2\,\vec{v} + \vec{S}_{\text{em}} \tag{67}$$

$$\overset{\leftrightarrow}{t} = (\rho + P)\,\gamma^2\,\vec{v} \otimes \vec{v} + P\overset{\leftrightarrow}{g} + \overset{\leftrightarrow}{t}_{\text{em}} = \vec{S} \otimes \vec{v} + P\overset{\leftrightarrow}{g} + \overset{\leftrightarrow}{t}_{\text{em}} \,. \tag{68}$$

ρ_0 is the rest mass density and $h = (\rho + P)/n$ the relativistic specific enthalpy. $\vec{S}_{em}$ represents the Poynting flux measured by ZAMOs and $\vec{t}_{em}$ the Maxwell stresses

$$\epsilon_{\text{em}} = \frac{1}{8\pi}\left(\vec{E}^2 + \vec{B}^2\right) \tag{69}$$

$$\vec{S}_{\text{em}} = \frac{1}{4\pi}\vec{E} \times \vec{B} \tag{70}$$

$$\overset{\leftrightarrow}{t}_{\text{em}} = \frac{1}{4\pi}\left(-\vec{E} \otimes \vec{E} - \vec{B} \otimes \vec{B} + \frac{1}{2}\overset{\leftrightarrow}{g}\left(\vec{E}^2 + \vec{B}^2\right)\right). \tag{71}$$

Using the 3+1 split of the connection of Kerr space, one can now derive the 3+1–split of the hydrodynamic equations $\nabla \cdot T = 0$ [for more details, see Durrer and Straumann (1988)]. Energy conservation, given by $U \cdot (\nabla \cdot T) = 0$, can be written as

$$\frac{d\epsilon}{d\tau} = \frac{1}{\alpha}(\partial_t - \mathcal{L}_\beta)\epsilon = -\frac{1}{\alpha^2}\nabla \cdot (\alpha^2 \vec{S}) + \epsilon\, tr(\overset{\leftrightarrow}{K}) + tr(\overset{\leftrightarrow}{K} \cdot \overset{\leftrightarrow}{t})\,. \tag{72}$$

The first term in the energy conservation is the familiar divergence of the energy flux, with one factor of α inside the divergence to account for the gravitational redshift of the energy and the other to convert proper time in the definition of the flux into universal time. The last term is interesting and can be written in terms of the shear $\sigma_{ik}^K \equiv -K_{ik}$ of adjacent observers

$$\frac{1}{\alpha}(\partial_t - \vec{\beta} \cdot \nabla)\epsilon = -\frac{1}{\alpha^2}\nabla \cdot (\alpha^2 \vec{S}) - \epsilon\, tr(\overset{\leftrightarrow}{\sigma}) - \sigma_{ik}^K t^{ik}\,. \tag{73}$$

Similarly, Euler's equations, given by $h \cdot (\nabla \cdot T) = 0$, assume the form

$$\frac{d\vec{S}}{d\tau} = \frac{1}{\alpha}(\partial_t - \mathcal{L}_\beta)\vec{S} = -\epsilon\nabla(\ln\alpha) - \frac{1}{\alpha}\nabla \cdot (\alpha\overset{\leftrightarrow}{t}) + tr(\overset{\leftrightarrow}{K})\vec{S} + 2\overset{\leftrightarrow}{K} \cdot \vec{S}\,. \tag{74}$$

This can be brought into the form

$$\frac{1}{\alpha}(\partial_t - \mathcal{L}_\beta)\vec{S} = -\epsilon\nabla(\ln\alpha) - \frac{1}{\alpha}\nabla \cdot (\alpha\overset{\leftrightarrow}{t}) + tr(\overset{\leftrightarrow}{K})\vec{S} - \overset{\leftrightarrow}{\sigma} \cdot \vec{S} - \frac{1}{2}\vec{H} \times \vec{S}\,. \tag{75}$$

$-\nabla \ln\alpha$ represents the local gravitational force measured by ZAMOs. For slicings with nonvanishing fundamental form, additional couplings occur between the curvature of the absolute space and the momentum flux. The term $-(1/2)\vec{H} \times \vec{S}$ is familiar from the precession law for gyroscopes; the momentum density $\vec{S}$ will precess relative to absolute space.

For stationary flows on Kerr space this implies the two equations

$$\nabla \cdot (\alpha^2 \vec{S}) = \alpha^2\, \sigma_{ik}^K t^{ik} \tag{76}$$

$$\frac{1}{\alpha}\nabla_k(\alpha t_i^k) = -\epsilon\nabla_i(\ln\alpha) - \frac{1}{\alpha}S_\phi\nabla_i\omega\,. \tag{77}$$

In contrast to flat space, the energy flow is no longer conserved; the gravitomagnetic field, represented by the shear σ_{ik}^K, changes energy conservation. Even in the case of pure hydrodynamic disk accretion, this coupling between the plasma stress tensor, $t_{r\phi}$ (angular momentum flow vector), and the gravitomagnetic field introduces work done on the disk plasma, which is very similar to viscous dissipation, $\sigma_{r\phi}^K t^{r\phi}$, except that here we have a coupling between the shear of absolute space and the momentum flux. This term can be very important if strong magnetic fields occur near the horizon of a rapidly rotating black hole.

The time–dependent hydrodynamic equations including Lorentz forces have not yet been implemented in a suitable code. We will therefore skip the full equations in this context. The above equations will be used in Sect. 4 for the derivation of the stationary equations.

3.6   On Viscous Accretion Disks

Viscous accretion disks play a fundamental role in the understanding of the emission from active galactic nuclei. For this discussion, the energy–momentum tensor T has to include the effects of the heat vector q, as well as the viscous stress tensor $\overset{\leftrightarrow}{t} = -2\eta\,\overset{\leftrightarrow}{\sigma}$

$$T = (\rho + P)\,u \otimes u + Pg + u \otimes q + q \otimes u + t, \tag{78}$$

where $\overset{\leftrightarrow}{\sigma}$ denotes the shear of the flow field u. In a discussion of standard accretion disks, magnetic fields are usually neglected. As for standard Newtonian disks, we have mass conservation

$$\dot{M} = -2\pi r u \Sigma \tag{79}$$

with the radial component $u = u^r$ and the surface density Σ. Conservation of angular momentum leads to the angular momentum vector

$$J^r = (\rho + P)\,u u_\phi + q^r u_\phi + q_\phi u + t_\phi^r. \tag{80}$$

Neglecting the second and third term, this leads to the basic angular momentum conservation

$$\frac{\dot{M}}{2\pi}\,(L - L_0) = 2\pi H r\, t_\phi^r = -\nu \Sigma r t_\phi^r. \tag{81}$$

$L = \mu u_\phi$, where $\mu = (\rho + P)/n$ denotes the specific enthalpy, is the the total angular momentum and L_0 an integration constant. The third equation follows from the radial Euler equation for the accretion velocity u

$$u' = \frac{u\mathcal{N}}{\mathcal{D}} \tag{82}$$

with some suitable functions $\mathcal{N}$ and $\mathcal{D}$ (Peitz and Appl 1997). The general form of the shear tensor component σ_ϕ^r is also discussed in Peitz and Appl (1997), where solutions are presented for an adiabatic equation of state of the form $P = K\Sigma^\Gamma$.

Accretion onto black holes is different from accretion onto stellar objects. Physical solutions must cross the horizon *supersonically* for causality reasons. Therefore, a sonic point occurs at some radius r_S defined by the critical point of the Euler equation (82), $\mathcal{N}(r_S) = 0 = \mathcal{D}(r_S)$. Two types of solutions are found, characterized by a transonic transition close to or far from the marginally stable orbit. These solutions are generally sub-Keplerian and reach maximum temperatures in the range of $10^{10} - 10^{12}$ K. These hot solutions are optically thin, and cooling has to be included into the energy equation (Narayan and Yi 1995). In optically thin disks, advective cooling is also important (Narayan and Yi 1995), which can considerably lower the accretion efficiency. This topic is under great debate at the moment, but not yet fully understood.

4   Time Evolution of Magnetic and Current Flux in Turbulent Disks

The above formulation can now be used to investigate the time evolution of magnetic flux in accretion disks around rapidly rotating objects.

4.1 The Grad–Shafranov Equation

When we combine Eq. (54) with the first equation (52) of Maxwell's equations, this provides us a kind of wave equation for the poloidal flux Ψ

$$\frac{\partial^2 \Psi}{\partial t^2} - \alpha\varpi\,\mathcal{G}_2[\Psi] = -\alpha\varpi^2 \vec{E}_p \cdot \nabla\omega + 4\pi\varpi j^{\hat{\phi}}\,. \tag{83}$$

This shows explicitly that the equation is hyperbolic, as required by Maxwell's theory. Using Ohm's law, this can be rewritten as

$$\frac{\partial^2 \Psi}{\partial t^2} + 4\pi\gamma\sigma\,\alpha\frac{\partial \Psi}{\partial t} \quad - \quad \alpha\varpi\,\mathcal{G}_2[\Psi] =$$
$$-\varpi^2\alpha\vec{E}_p \cdot \nabla\omega \quad + \quad 4\pi\alpha^2\gamma\sigma\varpi\vec{e}_\phi \cdot (\vec{v}_p \times \vec{B}_p)\,. \tag{84}$$

Using the expression for the poloidal magnetic field and the definition of the magnetic diffusivity $\eta = c^2/4\pi\sigma$, we obtain the equation

$$\eta\frac{\partial^2 \Psi}{c^2\partial t^2} + \alpha\gamma\,\frac{\partial \Psi}{\partial t} \quad + \quad \alpha\gamma\,(\alpha\vec{v}_p \cdot \nabla)\Psi$$
$$- \quad \eta\,\alpha\varpi\,\mathcal{G}_2[\Psi] = -\eta\,\varpi^2\,\alpha\vec{E}_p \cdot \nabla\omega\,. \tag{85}$$

The relevance of the first two terms in a time–dependent evolution of magnetic flux can be estimated by the ansatz $\Psi(t,r,\theta) = \exp(\Gamma t)\,\tilde{\Psi}(r,\theta)$. This yields the eigenvalue equation

$$\left(\frac{\Gamma^2}{4\pi\sigma} + \alpha\gamma\Gamma\right)\tilde{\Psi} \quad + \quad \alpha\gamma\,(\alpha\vec{v}_p \cdot \nabla)\tilde{\Psi}$$
$$-\eta\,\alpha\varpi\,\mathcal{G}_2[\tilde{\Psi}] \quad = \quad -\eta\,\varpi^2\,\exp(-\Gamma t)\,\alpha\vec{E}_p \cdot \nabla\omega\,. \tag{86}$$

The second derivative in time is in general $4\pi\Gamma\sigma$ times the second term, which is first order in the time–derivative. We can neglect second order time–derivatives, whenever the growth rates Γ are much less than the microscopic scales given by the conductivity $\sigma = 3.2$ MHz $T_e^{3/2}$. In a turbulent plasma, the effective conductivity is, however, much less, and the corresponding growth times $1/\Gamma$ are of the order of the Alfvén transit time R/v_A at the radius R in a disk.

Causality requires that the second derivative is present in this transport equation for the poloidal magnetic flux. This term regulates the relaxation. For long term evolution we may neglect this part and end up with a parabolic diffusion type equation used in some simulations of the time evolution of magnetic fields (Khanna and Camenzind 1996)

$$\gamma\,\frac{\partial \Psi}{\partial t} + \gamma\,(\alpha\vec{v}_p \cdot \nabla)\Psi - \eta\,\varpi\,\mathcal{G}_2[\Psi] = -\eta\,\varpi^2\,\vec{E}_p \cdot \nabla\omega\,. \tag{87}$$

4.2 The Current Flux Equation

The second equation determines the time–evolution of the current function $T(t,r,\theta) = 2\int \alpha\vec{j}_p \cdot d\vec{A} = \alpha\varpi B^{\hat{\phi}}(t,r,\theta)$. From Equ. (53) we get

$$\frac{\partial T}{\partial t} = \alpha\varpi^2\,\vec{B}_p \cdot \nabla\omega - \alpha\varpi\vec{e}_\phi \cdot \left[\nabla \times \alpha\left(\frac{1}{\sigma\gamma}\vec{j}_p - \vec{v}_T \times \vec{B}_p - \vec{v}_p \times \vec{B}_T\right)\right]\,, \tag{88}$$

when Ohm's law is used. Together with the poloidal component of Ampère's law this can be written as

$$\frac{\partial T}{\partial t} \quad - \quad \alpha\varpi\vec{e}_\phi \cdot \left[\nabla \times \frac{\eta}{\gamma}(\nabla \times \alpha\vec{B}_T)\right] = \alpha\varpi^2\,\vec{B}_p \cdot \nabla\omega +$$

$$\alpha\varpi\vec{e}_\phi \cdot \nabla \times \left[\frac{\eta}{\gamma}\frac{\partial\vec{E}_p}{\partial t} - \vec{e}_\phi \cdot \left(\vec{v}_T \times \vec{B}_p - \vec{v}_p \times \vec{B}_T\right)\right]. \tag{89}$$

This finally leads to a diffusion type equation for the current function

$$\frac{\partial T}{\partial t} + \alpha(\vec{v}_p \cdot \nabla)T \quad - \quad \alpha\varpi^2\,\nabla \cdot \left(\frac{T}{\varpi^2}\vec{v}_p\right) - \alpha\varpi^2\,\nabla \cdot \left(\frac{\eta}{\gamma\varpi^2}\nabla T\right)$$

$$= \quad \alpha\varpi^2\,\vec{B}_p \cdot \nabla\Omega + \alpha\varpi\,\vec{e}_\phi \cdot \nabla \times \left(\frac{\eta}{\gamma}\frac{\partial\vec{E}_p}{\partial t}\right). \tag{90}$$

Since the hyperbolic nature of this equation is not obvious, it can be written in an alternative way by starting with the time–derivative of the toroidal component B_T of the induction equation (54)

$$\frac{\partial^2 B_T}{\partial t^2} - \alpha\,\mathcal{G}_2[B_T] = \varpi\,\frac{\partial\vec{B}_p}{\partial t} \cdot \nabla\omega + 4\pi\vec{e}_\phi \cdot \left[\nabla \times (\alpha^2\vec{j}_p)\right]. \tag{91}$$

This shows explicitly the hyperbolic nature of the equation. Similar to the discussion of the transport of poloidal flux we may neglect the current displacement term in equation (90) which then yields a parabolic equation for the time evolution of the current function (Khanna and Camenzind 1996).

4.3   On the Validity of Cowling's Theorem

Cowling's theorem states that stationary axisymmetric magnetic fields cannot be maintained by purely axisymmetric plasma motions. The gravitomagnetic effect in the induction equation changes this conclusion of classical electrodynamics (Khanna and Camenzind 1994, 1996). The reason is that in Kerr space the coupling between the gravitomagnetic shear $\nabla\omega$ and the poloidal electric field $\vec{E}_p$ is a potential source for the poloidal magnetic flux Ψ. This term is absent in nonrotating space and leads to the usual formulation of Cowling's theorem. In order to see this effect, one has to numerically solve the two equations, there are no analytical results known.

4.4   Time Evolution of Magnetic Fields in Disks

As discussed above, one can neglect for the long–term time–evolution of magnetic fields in accretion disks the influence of the displacement current in the two equations (87) and (90). This leads to a coupled parabolic system of equations for the poloidal magnetic flux $\Psi(t,r,\theta)$ and the poloidal current function $T(t,r,\theta)$ (Khanna and Camenzind 1994, 1996). The accretion profile with plasma rotation $\Omega(r)$ and accretion drift $v_r(r)$ must be given by some reasonable approximation, following from the accretion disk solutions. Similarly, the magnetic diffusivity η is given by some characteristic turbulent diffusivity η_0.

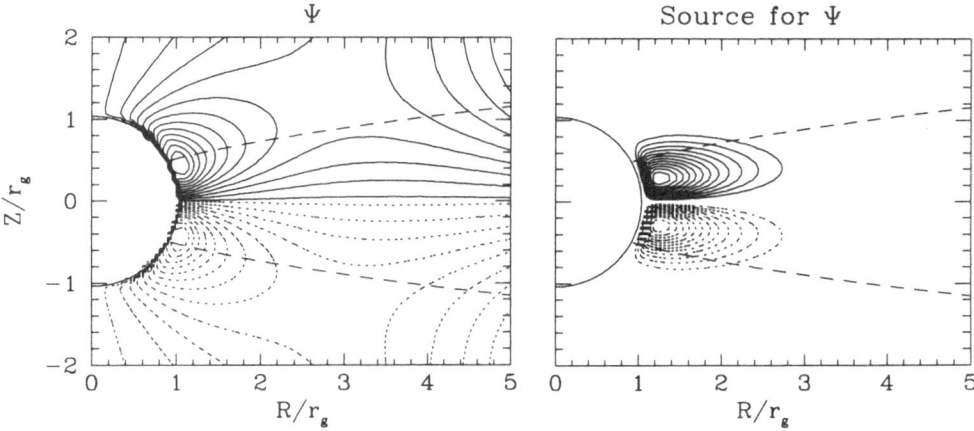

Figure 1 Quadrupolar field topology around a rapidly rotating black hole for $a_H = 0.998 M_H$ and turbulence parameter $\alpha_T = 0.25$ (R. Khanna, private communication). The right panel shows the distribution of poloidal magnetic fields advected inwards. Near the horizon, a closed quadrupolar structure emerges which is due to the gravitomagnetic source term (shown in the left panel). The dashed lines indicate the scale height of the disk.

The two equations for Ψ and T are brought into dimensionless form by scaling radii in units of M_H, velocities in units of the speed of light, and time in units of the diffusion time $t_{\mathrm{diff}} = M_H^2/\eta_0$. The advective terms are then scaled by a magnetic Reynolds number defined as

$$\mathcal{R}_m = \frac{M_H c}{\eta_0} = \frac{t_{\mathrm{diff}} c}{M_H} \simeq \frac{M_H^2}{\alpha_T H^2} \gg 1,\qquad(92)$$

where H is the disk scale–height at the horizon and $\alpha_T \leq 1$ the standard turbulence parameter of accretion disks. This Reynolds number is the only free parameter involved in the equations. The solutions shown in Fig. 1 ($a_H = 0.998$ and $\mathcal{R}_m = 25$) were calculated from an initial quadrupolar disk configuration based on advection and diffusion alone. After about 40 diffusion times, a closed quadrupole evolves around the horizon, which after 200 diffusion times extends beyond the ergosphere. The external disk field is pushed outwards and forms an X–point. The growth of this inner quadrupole is due to the right hand side in the dynamo equation (85). This effect would be completely absent for a nonrotating black hole.

We now think that this quadrupolar magnetic configuration is the natural magnetic structure realized in radio–quiet sources (Seyfert galaxies and QSOs). Disk outflows driven by quadrupolar structures cannot be efficiently accelerated, leading probably only to moderate disk winds. For efficient plasma acceleration, dipolar structures are needed, which, for some reason, are preferably built up in elliptical galaxies.

4.5 Magnetic Spin–Down of Rotating Black Holes

One consequence of this gravitomagnetic coupling, which has been discussed at various places in the literature, is the fact that a rapidly rotating black hole generates a voltage drop near the horizon which then drives a huge current system through a magnetosphere.

For this reason, we now integrate the induction law along a closed curve connecting a load region with the horizon through the disk

$$\oint_C \alpha \, \vec{E}_p \cdot d\vec{s} = -\frac{d}{dt} \int_A \vec{B} \cdot d\vec{S} - \oint_C (\vec{\beta} \times \vec{B}) \cdot d\vec{s}. \tag{93}$$

The last term acts as an additional term to Faraday's induction law for moving conductors. In a stationary situation we find the EMF

$$\mathrm{EMF}(\mathcal{C}) \equiv \oint_C \alpha \vec{E} \cdot d\vec{s} = -\oint_C (\vec{\beta} \times \vec{B}) \cdot d\vec{s}. \tag{94}$$

The integral along the flux surface outside the disk gives no contribution, and far away the gravitomagnetic effect drops out rapidly. There remains therefore only the contribution near the horizon. The main driving force is the integral near the horizon

$$\mathrm{EMF} = -\int (\vec{\beta}_H \times \vec{B}_\perp) \cdot d\vec{s} \simeq \frac{1}{2\pi} \Omega_H \, \Delta\Psi_H \simeq 10^{20} \, \mathrm{Volt} \, \frac{a_H}{M_H} \, \frac{M_H}{10^9 \, M_\odot} \, \frac{B_H}{10 \, \mathrm{kG}}, \tag{95}$$

where $\Delta\Psi_H$ denotes the magnetic flux that covers the horizon. This EMF which drives the current system near the horizon depends strongly on the rotational state Ω_H of the black hole. It vanishes for nonrotating holes. This huge voltage drop between the axis of rotation and the equatorial plane of the disk does not yet guarantee a huge power output. For this we calculate the integral of the electric field along the closed path. This gives various contributions. In the asymptotic load region we get (Okamoto 1992)

$$\Delta V_L \simeq \frac{1}{2\pi} \Omega_F \, \Delta\Psi_L \tag{96}$$

The drop near the horizon can be estimated as

$$\Delta V_H \simeq \frac{1}{2\pi} (\Omega_H - \Omega_F) \, \Delta\Psi_H. \tag{97}$$

There is also some voltage drop ΔV_D along the equatorial plane before the field leaves the disk. In total, we get the relation

$$\mathrm{EMF} = \Delta V_H + \Delta V_D + \Delta V_L. \tag{98}$$

This can also be written in terms of currents

$$\mathrm{EMF} = I(R_\perp + R_D + R_L) \tag{99}$$

with disk resistivity R_D, load resistivity R_L and some resistivity $R_\perp$ near the horizon. The total current I is therefore determined by the gravitomagnetic battery voltage EMF and the corresponding resistivities. The total power that can be dissipated in the load region

$$P_L = I \, \Delta V_L = (\mathrm{EMF})^2 \, \frac{R_L}{(R_\perp + R_D + R_L)^2} \tag{100}$$

strongly depends on the various resistivities. If impedance matching would occur, $R_L \simeq R_\perp + R_D$, the highest power could be extracted

$$P_L \simeq \frac{1}{4} \frac{(\mathrm{EMF})^2}{R_\perp}. \tag{101}$$

The maximum power which can be extracted from the rotational energy of the hole depends quadratically on the mass

$$P_{H,\text{max}} \simeq 10^{45} \,\text{erg s}^{-1} \left(\frac{a_H}{M_H}\right)^2 \left(\frac{B_H}{10^4\,G}\right)^2 \left(\frac{M_H}{10^9\,M_\odot}\right)^2. \tag{102}$$

Such a system drives a current I_H that is also given by the rotation of the field lines

$$I_H \simeq \frac{1}{2}\left(\Omega_H - \Omega^F\right) B_H r_H^2 \simeq 10^{18}\,\text{Amp}. \tag{103}$$

As in the case of Jupiter's and of stellar magnetospheres, this current–flow leads to a braking of the rotating hole (for $\Omega_H > \Omega^F$) on a typical time–scale given by

$$t_{\text{QSR}} = \frac{J_H}{dJ_H/dt} \simeq 1\,\text{Gyr} \left(\frac{B_H}{10\,kG}\right)^{-2} \left(\frac{M_H}{10^9\,M_\odot}\right)^{-1}. \tag{104}$$

Since the maximum field strength B_H also scales as $B_H \propto 1/\sqrt{M_H}$ (Eddington field strength), the braking time is practically independent of the mass M_H! This is an interesting result. When supermassive black holes are born as rapidly rotating objects at redshift $z \simeq 2-5$, they could be still moderately rotating nowadays.

The toroidal velocity $\vec{v}_T$ with respect to ZAMOs is

$$\vec{v}_T = \frac{\Omega - \omega}{\alpha}\,\vec{e}_\phi \tag{105}$$

and, instead of using the toroidal magnetic field, it is more convenient to introduce the current stream function $I = \alpha\bar{\omega}B^{\hat{\phi}}$. In terms of these quantities, the poloidal electric field for a stationary configuration can be written as

$$\alpha\vec{E}_p = \frac{\eta}{\bar{\omega}}\,\nabla I \times \vec{e}_\phi - \frac{I}{\bar{\omega}}\,\vec{v}_p \times \vec{e}_\phi - (\Omega - \omega)\,\nabla\Psi. \tag{106}$$

η is the plasma diffusivity. The poloidal electric field is the result of currents and poloidal as well as toroidal plasma motions.

Let us consider a situation where magnetic fields are advected in the disk and superposed with some external field. A dipolar field is advected inwards and closes perpendicular to the horizon. In this case, the last term in the expression for the poloidal field vanishes at the horizon, where $\Omega(r_+) = \Omega_H = \omega(r_+)$, however in such a way that $(\Omega - \omega)/\alpha$ remains finite at the horizon. This term only creates a θ–component in the electric field. The rapid radial inflow also generates only a component of this type. The only radial component of the electric field is produced by the first term due to magnetic diffusivity. For a diffusivity vanishing near the horizon, those components will also disappear near the horizon so that currents that are flowing mostly radially in the disk will be deflected away from the disk near the horizon. Outside the disk, η vanishes so that the last term is the dominant contribution to the electric field (which is always perpendicular to the flux surfaces).

5   Stationary MHD Flows on the Kerr Background

The theory of stationary and axisymmetric MHD flows is now complete and can be used for any investigation of such flows in the gravitational field of rotating compact objects. The basic features have been derived by the present author (Camenzind 1986, 1987). A recent review can be found in (Camenzind 1996).

5.1   Basic Conservation Laws

In the stationary and axisymmetric case, the poloidal magnetic field is given, as in the Newtonian case, by means of the magnetic flux function Ψ

$$\Psi = \frac{1}{2\pi} \int \vec{B}_p \cdot d\vec{S} = \varpi A^{\hat{\phi}}, \quad \vec{B}_p = \frac{1}{\varpi} \left(\nabla \Psi \wedge \vec{e}_\phi \right). \tag{107}$$

$\nabla \cdot \vec{B}_p = 0$ is therefore identically satisfied. It is also useful to introduce the current function I for the specification of the toroidal field

$$I = - \int \alpha \vec{j}_p \cdot d\vec{S} = -\frac{\alpha}{2} \varpi B^{\hat{\phi}}, \tag{108}$$

which is the current flowing inside the surface $\Psi = \text{const}$. We also assume that there are no parallel electric fields, $\vec{E} \cdot \vec{B} = 0$, and therefore that $\vec{E} \parallel \nabla \Psi$

$$\vec{E}_\perp = -\frac{\Omega^F - \omega}{\alpha} \nabla \Psi. \tag{109}$$

These are three expressions for the electromagnetic fields.
The plasma flow is given by the continuity equation

$$\nabla \cdot (n \alpha \gamma \vec{v}) = 0. \tag{110}$$

Similar to the Newtonian theory one also gets

$$\vec{u}_p = \gamma \vec{v}_p = \frac{\eta}{\alpha n} \vec{B}_p, \tag{111}$$

and the frozen–in condition requires that plasma is guided by the structure of the axisymmetric flux surfaces $\Psi = \text{const}$

$$\boxed{\vec{u} = \frac{\eta(\Psi)}{\alpha n} \vec{B} + \gamma (\Omega^F - \omega) \frac{\varpi}{\alpha} \vec{e}_\phi,} \tag{112}$$

and, as a consequence of Faraday's induction law, the rotation of the field lines remains constant at $\Omega^F(\Psi)$. For stationary and axisymmetric flows the energy equation (76) is now simply given as

$$-\frac{1}{\alpha^2} \nabla \cdot (\alpha^2 \vec{S}) + K_{ik} t^{ik} = 0 \tag{113}$$

and similarly for the momentum equation (77)

$$\frac{1}{\alpha} \nabla \cdot (\alpha \mathbf{t}) + \epsilon \nabla \ln \alpha - (\nabla \omega) S_\phi = 0. \tag{114}$$

By means of the MHD relations the momentum flux can be written as

$$\vec{S} = \mu \gamma \frac{\eta}{\alpha} \vec{B}_p + \frac{I(\Omega^F - \omega)}{2\pi \alpha^2} \vec{B}_p. \tag{115}$$

In the stationary case, this provides the following expression for the energy equation

$$\nabla \cdot \left(\mu \gamma \alpha \eta \, \vec{B}_p + \frac{I \Omega^F}{2\pi} \vec{B}_p \right) - \nabla \cdot \left(\frac{\omega I}{2\pi} \vec{B}_p \right) + (\vec{B}_p \cdot \nabla \omega) \, \mu \eta u_\phi = 0. \tag{116}$$

The angular momentum equation is simply

$$\vec{e}_\phi \cdot \nabla \cdot (\alpha \mathbf{t}) = 0 \,. \tag{117}$$

Due to axisymmetry, all other terms drop out. Since

$$\vec{e}_\phi \cdot \mathbf{t} = t_{p\phi} = \mu \vec{u}_p u_\phi - \frac{1}{4\pi} \vec{B}_p B_\phi = \frac{\mu \eta}{\alpha} \vec{B}_p u_\phi + \frac{I}{2\pi} \vec{B}_p \tag{118}$$

we get

$$\nabla \cdot \left(\mu \eta u_\phi \vec{B}_p + \frac{I}{2\pi} \vec{B}_p \right) = 0 \,. \tag{119}$$

Combining these two equations and using flux conservation $\nabla \cdot \vec{B}_p = 0$ we find for the total energy

$$\boxed{ E(\Psi) = \mu(\alpha\gamma + \omega u_\phi) + \frac{\Omega^F I}{2\pi\eta} } \tag{120}$$

and for the total angular momentum in the flux

$$\boxed{ L(\Psi) = \mu u_\phi + \frac{I}{2\pi\eta} } \,. \tag{121}$$

In the total energy, the kinetic part is redshifted and the second term is due to the *gravitomagnetic spin–spin interaction.* Together with the above relation for the toroidal motion

$$u_\phi = \frac{\eta}{\alpha n} B_\phi + \gamma(\Omega^F - \omega)\frac{\varpi}{\alpha} \,. \tag{122}$$

This represents three equations for the unknowns γ, u_ϕ and I.

In addition to these relations, an adequate equation of state is needed. For $n = n(P,s)$ and $T = T(P,s)$ the first law of thermodynamics implies

$$d\mu = \frac{1}{n} dP + T \, ds \,. \tag{123}$$

In particular, for $P = K_0(s)\, n^\Gamma$ with a polytropic index Γ, one finds

$$\mu = mc^2 + \frac{\Gamma}{\Gamma - 1} K_0(s)\, n^{\Gamma - 1} \,. \tag{124}$$

With this expression, we can derive the Newtonian limit for the total energy, $\alpha = 1 + \Phi$, $\gamma = 1 + v^2/2$ and $\omega \equiv 0$

$$E = mc^2 + \frac{1}{2} v^2 + \frac{\Gamma}{\Gamma - 1} \frac{P}{\rho_0} + \Phi + \frac{I\Omega^F}{2\pi\eta} \,. \tag{125}$$

5.2   Lorentz Factor, Angular Momentum and Poloidal Current

For given integrals of motion $E(\Psi)$, $L(\Psi)$, $\Omega^F(\Psi)$, $\eta(\Psi)$ and $K(\Psi)$ the conservation laws for stationary and axisymmetric plasma flows can be reduced to give relations for the Lorentz factor, the angular momentum and the poloidal current flux function $I(R,\Psi)$

$$\alpha\gamma = \alpha\gamma(\Psi, R) \;=\; \frac{E}{\mu}\, \frac{\alpha^2(1-\epsilon) - M^2(1-\omega L/E)}{\alpha^2 - (\Omega^F - \omega)^2\varpi^2 - M^2}\,, \tag{126}$$

$$u_\phi = u_\phi(\Psi, R) \;=\; \frac{E}{\mu}\, \frac{(1-\epsilon)(\Omega^F - \omega)\varpi^2 - M^2 L/E}{\alpha^2 - (\Omega^F - \omega)^2\varpi^2 - M^2}\,, \tag{127}$$

$$I = I(\Psi, R) \;=\; 2\pi\eta\, \frac{\alpha^2 L - (\Omega^F - \omega)\varpi^2(E - \omega L)}{\alpha^2 - (\Omega^F - \omega)^2\varpi^2 - M^2}\,, \tag{128}$$

where the quantity M, defined as

$$\boxed{\; M^2 = \frac{4\pi\mu\eta^2}{n} = \frac{\alpha^2\, u_p^2}{u_A^2}\,, \;} \tag{129}$$

represents the Alfvén Mach number. The redshift factor α corrects for the singular behavior of the poloidal velocity at the horizon of the black hole.[1] The Alfvén Mach number M is a quantity well-behaved at the horizon. The quantity ϵ defined by

$$\epsilon \equiv \frac{\Omega^F L}{E} \tag{130}$$

is a measure for the amount of energy carried by the electromagnetic fields. In special relativity, $\epsilon \leq 1$, and $\epsilon \equiv R_A^2/R_L^2$ is a direct measure for the position of the Alfvén surface in terms of the light cylinder radius R_L. For relativistic flows, $R_A \to R_L$. In the curved background of a black hole, this quantity is only a parameter that, however, still determines the position of the *Alfvén surfaces* (positions where the denominators in the above equations vanish)

$$\alpha^2(R_A) - (\Omega^F - \omega(R_A))^2 R_A^2 = M_A^2 \quad,\quad R_A^2 \equiv \varpi_A^2\,. \tag{131}$$

This equation has in general, for a given rotation Ω^F, two solutions, the outer Alfvén surface corresponding to the special relativistic one and an inner Alfvén surface near the horizon of a black hole. The light cylinder surfaces are special solutions of this equation for $M_A^2 = 1$. A rotating magnetosphere of a black hole has two light surfaces, the outer one slightly deformed by the underlying metric and an inner one due to the existence of the frame–dragging effect. For slow rotation, the outer light surface moves to infinity, and the inner one towards the static limit. In the special case, $\Omega^F = \Omega_H$, the inner light surface coincides with the horizon. Near a black hole, the quantity ϵ can exceed unity and even become negative, depending on the signs of L and E.

The regularity condition at the Alfvén point also determines the total angular momentum L as a function of the position of the Alfvén point

$$M_A^2\, \frac{L}{E} = R_A^2(\Omega^F - \omega(R_A))\,(1 - \epsilon)\,, \tag{132}$$

which simply reads in the Newtonian case as $L = R_A^2\Omega^F$.

---

[1] These relations have been derived for the first time in Camenzind (1986) without using a 3+1 split.

5.3   The Non–Relativistic Limit

The last two equations are well–known in Newtonian MHD ($\alpha = 1$, $\omega = 0$, $\epsilon \ll 1$ and $R_L \to \infty$), where they form the basis for a treatment of axisymmetric MHD winds in stellar systems (Heyvaerts and Norman 1996)

$$j = R^2\Omega \;\; = \;\; \frac{R^2\Omega^F - M^2 L}{1 - M^2}, \tag{133}$$

$$RB_\phi \;\; = \;\; -4\pi\eta \, \frac{L - R^2\Omega^F}{1 - M^2}\,. \tag{134}$$

In Newtonian MHD, the Mach number is simply $M^2 = v_p^2/v_A^2$. The first equation indicates that the plasma is corotating with the field lines for low Mach numbers, $M^2 \ll 1$, and that the specific angular momentum j is equal to the total angular momentum L for high Mach numbers, $M^2 \gg 1$. These equations tell us that Newtonian MHD is only valid inside the light cylinder $R_L = c/\Omega^F$. This approximation is therefore not justified for rapidly rotating magnetic surfaces generated by accretion disks around black holes. It is not even justified for magnetic surfaces generated by rapidly rotating young stars.

5.4   The Relativistic Wind Equation

In ideal MHD, plasma flows along the magnetic surfaces $\Psi = $ const. Since the Lorentz factor, the angular momentum (or angular velocity Ω) and the poloidal current function I are essentially only functions of the radius along the flux surface and the Mach number, the normalisation of the plasma 4–velocity, $u_\alpha u^\alpha = -1$,

$$\gamma^2 - \vec{u}^2 = 1 \quad , \quad \vec{u} = \gamma\vec{v} \tag{135}$$

leads to the equation

$$\boxed{u_p^2 + 1 = \gamma^2 - \frac{u_\phi^2}{\varpi^2}}\,. \tag{136}$$

u_ϕ is a specific angular momentum. Using the solutions for γ and u_ϕ, this can be arranged into the form

$$\alpha^2 u_p^2 + \alpha^2 = \left(\frac{E}{\mu}\right)^2 U_{\mathrm{Kerr}}(\varpi; M^2,\epsilon) \tag{137}$$

with

$$U_{\mathrm{Kerr}}(R; M^2,\epsilon) = \frac{F_{\mathrm{K}}(R; M^2,\epsilon)}{\varpi^2 D_K^2} \tag{138}$$

and the definitions

$$D_K \;\; = \;\; \alpha^2 - \varpi^2(\Omega^F - \omega)^2 - M^2 \tag{139}$$

$$F_K \;\; = \;\; \varpi^2[\alpha^2(1 - \epsilon) - M^2(1 - \omega L/E)]^2$$
$$- \;\; \alpha^2[(1 - \epsilon)(\Omega^F - \omega)\varpi^2 - M^2 L/E]^2\,. \tag{140}$$

With the relation between u_p and B_p

$$(\alpha u_p)^2 = \frac{\eta^2 B_p^2}{n^2} = \frac{B_p^2}{16\pi^2 \mu^2 \eta^2} M^4 \tag{141}$$

the above equation represents a quartic relation in M for each position along a flux surface.

Not all of these parameters are independent. For this we scale radii in units of light cylinder radii $R_L = c/\Omega^F$ (assuming $\Omega^F = $ const), $x = R/R_L$, $\bar{F}_K = F_K/R_L^2$, and multiplying the equation by R^4 we obtain

$$\alpha^2 x^2 \frac{\bar{F}_K(x; M^2, \epsilon)}{D_K^2(x; M^2)} \left(\frac{E}{\mu}\right)^2 = \alpha^2 x^4 + \frac{B_p^2 x^4}{16\pi^2 \mu^2 \eta^2} M^4 . \tag{142}$$

The last term can be written in terms of the flux function

$$\Phi_\Psi^{-1}(x) \equiv \frac{B_p \varpi^2}{B_{p*} \varpi_*^2} , \tag{143}$$

which is normalized by the footpoint R_* and its magnetic field strength B_{p*}. In addition, we introduce the *dimensionless magnetization parameter*

$$\boxed{\sigma_*(\Psi) \equiv \frac{(B_{p*} R_*^2)(\Psi)c}{4\pi \mu \eta(\Psi) R_L^2(\Psi)} ,} \tag{144}$$

which is the analogue of the magnetization parameter introduced in the Newtonian discussion. This altogether leads to the equation

$$\boxed{x^2 \frac{\bar{F}_K(x; M^2, \epsilon)}{D_K^2(x; M^2)} \left(\frac{E}{\mu}\right)^2 = \alpha^2 x^4 + \sigma_*^2 \Phi^{-2} M^4 .} \tag{145}$$

This fundamental equation clearly shows that the solutions of the wind equation, when formulated for the Mach number, depends on the following parameters

- the parameter ϵ, which defines the position of the Alfvén point

- the dimensionless energy $\bar{E} = E/mc^2$

- the magnetization parameter σ_*

- the flux tube function Φ_Ψ

- one additional parameter hidden in the specific enthalpy

$$\mu = m_p c^2 \left[1 + \frac{c_{S*}^2}{\Gamma - 1} \left(\frac{M_*^2}{M^2}\right)^{\Gamma - 1}\right] . \tag{146}$$

These are essentially 4 parameters for each flux surface $\Psi = $ const. It can be shown that the relativistic wind equation has the same critical points as the non–relativistic one: the slow magnetosonic point, the Alfvén point $D_K(R_A) = 0$ and the fast magnetosonic point. It is very important in this respect that the light cylinder is not a critical point. *The requirement that a wind solution passes through all three critical points fixes therefore three of the four parameters. We may consider σ_* as a free parameter which is fixed by the disk physics.*

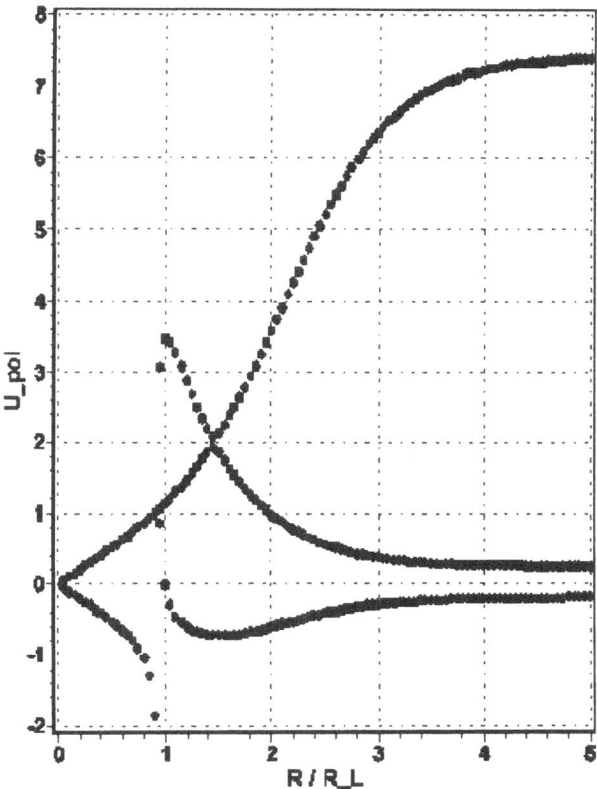

Figure 2 The various solution branches of the cold wind equation starting with low velocity near a black hole for polytropic index 5/3 and a magnetization $\sigma_* = 10$. The wind is assumed to be collimated into a cylindrical shape by the magnetic flux surface within a few light cylinder radii. The radius R along the flux surface is given in units of the light cylinder radius R_L. The upper branch is the positive wind branch which starts at low velocity, crosses the Alfvén point near the light cylinder and the fast magnetosonic point shortly after the light cylinder. Beyond 4 light cylinder radii, the wind reaches a constant poloidal velocity $u_p = 7.3c$. The acceleration occurs mostly outside the light cylinder.

5.5 Magnetically Driven Winds for Accretion Disks

As a simple application we consider the acceleration of disk winds by magnetic processes (Camenzind 1986). Winds from the corona of an accretion disk around a black hole escape along magnetic surfaces which are collimated at some distance into a cylindrical shape. The exact form of these surfaces is not important for the mechanism and will be discussed later on. We also neglect the influence of gravity, $\alpha = 1$ and $\omega = 0$. It is no major problem to include also these form functions into the wind equation.

Solutions of the wind equation can easily be obtained by using the expression for the Mach number, $\alpha = 1$ and $\omega = 0$,

$$M^2 = \frac{\mu}{mc^2}\, u_p x^2\, \frac{\Phi_\Psi}{\sigma_*}\,. \tag{147}$$

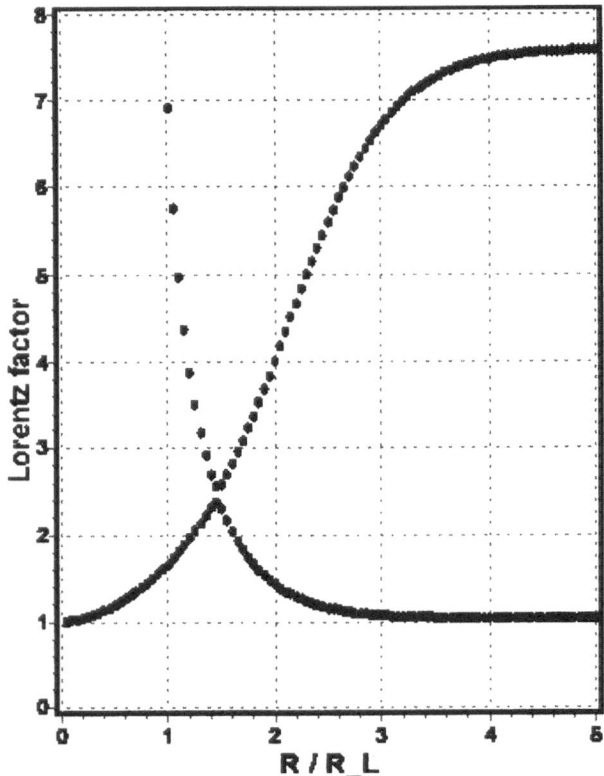

Figure 3 The Lorentz factor for the cold wind solution as a function of the radius along the collimated flux surface. The fast magnetosonic point is clearly visible as a crossing point between two solution branches. The physical wind branch starts with $\gamma = 1$ and shows successive acceleration along the collimated flux tube. About 80% of the total energy is converted into kinetic energy, the rest stays in the Poynting flux.

For normal plasma outflows, the sound velocity is not important so that flows can be approximated by cold winds, and therefore $\mu = mc^2$. For pair plasmas injected into the flux tubes, the sound speed is however relevant. From Newtonian magnetized flows (the solar wind e.g.) it is known that the wind equation will have three critical points: the slow magnetosonic point, the Alfvén point and the fast magnetosonic point (Camenzind 1986). These points occur due to the existence of three characteristic waves in magnetized plasmas: the two magnetosonic waves and the Alfvén waves. The details of these aspects, which are theoretically interesting, are not discussed here.

Neglecting the sound speed is equivalent to saying that the slow magnetosonic point is identical with the injection point. In this case, the wind equation reduces to a fourth–order polynomial in the poloidal velocity u_p

$$A_4 u_p^4 + A_3 u_p^3 + A_2 u_p^2 + A_1 u_p + A_0 = 0 \tag{148}$$

with coefficients given in Camenzind (1986) [see also Fendt et al. (1995)]

$$A_0 = x^2 (1 - x^2)^2 - E^2 x^2 (1 - x^2)(1 - \epsilon)^2, \tag{149}$$

$$A_1 = -2 g x^4 (1 - x^2 - E^2 (1 - \epsilon)), \tag{150}$$

$$A_2 = g^2 x^4 (x^2 - E^2 (x^2 - \epsilon^2)) + x^2 (1 - x^2)^2, \tag{151}$$

$$A_3 = -2 g x^4 (1 - x^2), \tag{152}$$

$$A_4 = g^2 x^6. \tag{153}$$

Here we use the quantity $g \equiv (\Phi_\Psi / \sigma_*)$.

This form is extremely useful, since it immediately produces all branches of solutions for a given flux tube function $\Phi_\Psi(x)$ and given magnetization σ_* (Fig. 2). In Fig. 2 we show all the solution branches of the wind equation (148) for plasma injected into a flux tube rooted near a black hole and collimated into a cylindrical shape beyond a few light cylinder radii. The magnetization of the plasma injected into the flux tube is a question of the physics of the corona of accretion disks. The magnetization σ_* can easily be as high as '10. Since the wind equation is given in polynomial form, it is easy to experiment with the manifold of all real solutions. In general, one finds four real solution branches, but only one is physical. The total energy E in the flow is not a free parameter, but is given by the occurrence of the fast magnetosonic point, which occurs in this case slightly beyond the light cylinder. The slow magnetosonic point would give an additional constraint e.g. for the total mass flow. The acceleration is clearly visible in the wind branch, it mainly occurs outside the light cylinder. The corresponding solution for the Lorentz factor is shown in Fig. 3. This demonstrates that not all of the energy is used up to accelerate the plasma, about 20% of the total energy stays in the Poynting flux. This is however completely different from plasma acceleration in spherically symmetric flux tubes which were originally discussed. In that case it is only possible to transform a small amount of Poynting energy into kinetic energy. In Fig. 4 we show the successive transformation of angular momentum in the magnetic field, given by the current funct $I(R)$, into specific angular momentum of the plasma. It is interesting that the specific angular momentum j of the plasma can be written in the form

$$j = R^2 \Omega = R^2 \frac{u^\phi}{\gamma} = R_{LC} c \frac{(1 - \epsilon)x^2 - M^2 \epsilon}{1 - M^2 - \epsilon}. \tag{154}$$

For $\sigma_* \to \infty$, or $\epsilon \to 1$ the specific angular momentum reaches the limiting value $R_L c$. Even for moderate magnetization, $\sigma_* = 10$, this theoretical limit is practically obtained. This has the important consequence that collimated matter outside the light cylinder rotates with a specific angular momentum roughly given by R_{LC} (Fig. 4).

In Fig. 5 we demonstrate the successive transformation of current I into spin of the plasma. The current I carried by MHD winds is essentially determined by the total angular momentum lost through the wind, $I \propto \eta L$. The expression for RB_ϕ can be written, by using $M^2 = 4\pi\mu\eta^2/n$ and $U_p = (\eta/\alpha n)B_p$ in the form, $\beta_p = u_p/\gamma$,

$$I = \alpha R B_\phi = \Omega_*(B_p R^2) \frac{M^2}{x^2 \beta_p(x)} \frac{x^2 - \epsilon}{1 - M^2 - \epsilon} = -\Omega_*(B_p R^2) \frac{1}{\beta_p(x)} i(x), \tag{155}$$

or as

$$I = -\Omega_* (B_p R^2)_* \frac{1}{\beta_p} \frac{i(x)}{\Phi_\Psi(x)} \tag{156}$$

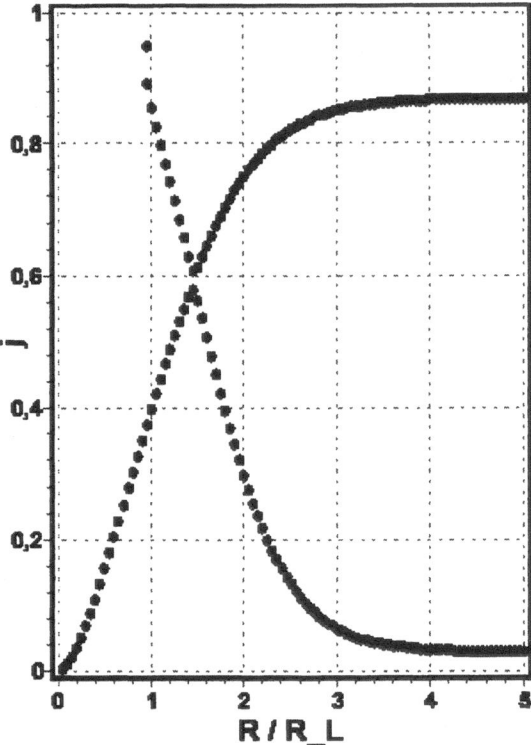

Figure 4 The specific angular momentum j in units of $R_L c$ for the cold wind solution as a function of the radius along the collimated flux surface. Even for a moderate magnetization, the specific angular momentum reaches practically its theoretical limit.

with the flux tube function Φ_Ψ as introduced previously. This relation demonstrates that the current in the jet is produced by the rotation of the central object and the magnetic flux contained in the flux tube, $I_* = -\Omega_*(B_p R^2)_*$. It is the essence of force–free models that this current injected into the flux surface at the footpoint is conserved along the flux surface, $I = I(\Psi)$. This is the essential difference between force–free magnetospheres and plasma flows including inertia, and poloidal current flows out of the surfaces around the light cylinder.

The Poynting flux $\vec{P}$ generated by accretion disks in quasars is the main energy source for the outflows. This can be written in the form

$$\vec{P}_p = \frac{\Omega_* I}{2\pi c}\, \vec{B}_p \,. \tag{157}$$

When integrating this Poynting flux over the entire surface of the central disk, we end up with a magnetic luminosity

$$L_{\mathrm{mag}} = \frac{1}{c}\, \Omega_* I_* \Psi_d \,, \tag{158}$$

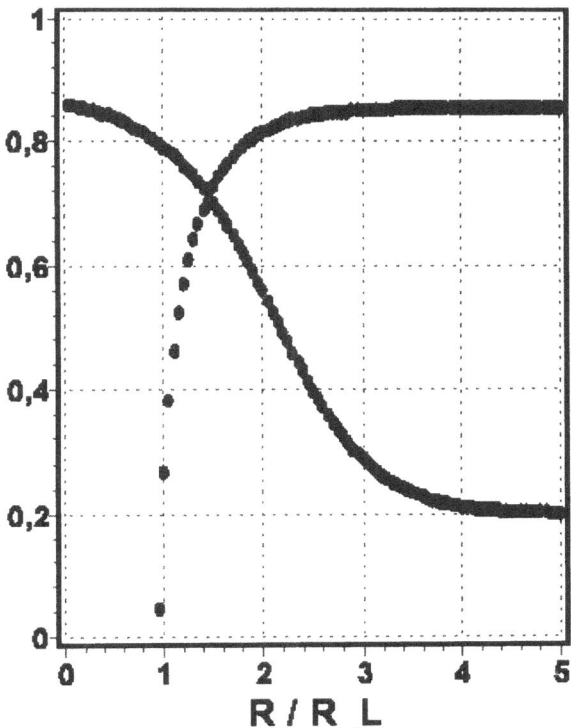

R / R_L

Figure 5 The characteristic poloidal current carried by a cold wind solution as a function of the radius along the collimated flux surface. Shown is the dimensionless function $i(x)/\beta_p(x)\Phi_\Psi(x)$. Current is successively used to spin up the plasma in the flux tube. The upper branch of the solution is a non–physical branch which crosses the wind solution at the fast magnetosonic point.

depending on the rotational state of the object, the total magnetic flux Ψ_d carried by the magnetosphere and the total current I_* driven by the rotation of the disk. This magnetic luminosity is a new source of energy which is usually neglected in the energy budgets. It is however an important source of energy for strongly magnetized and rapidly rotating objects.

The field rotation is related to the angular velocity Ω_H of the horizon which is a function of the angular momentum parameter $a_H < 1$ and the radius r_H of the horizon

$$\Omega_* \leq \Omega_H = \frac{1}{2} \frac{a_H c}{r_H} \simeq 10^{-4} \, \text{rad s}^{-1} \frac{10^9 \, M_\odot}{M_H} . \tag{159}$$

This rotation rate, together with the typical values for the magnetic flux Ψ_H covering a black hole and the typical current driven through the magnetosphere, provide the following estimate for the magnetic luminosity of quasars

$$L_{\text{mag}} \simeq 3 \times 10^{46} \, \text{erg s}^{-1} \frac{10^9 \, M_\odot}{M_H} \frac{\Psi_H}{10^{33} \, \text{Gauss cm}^2} \frac{I_H}{10^{18} \, \text{Amps}} . \tag{160}$$

This has to be compared with the kinetic luminosity of bright jets

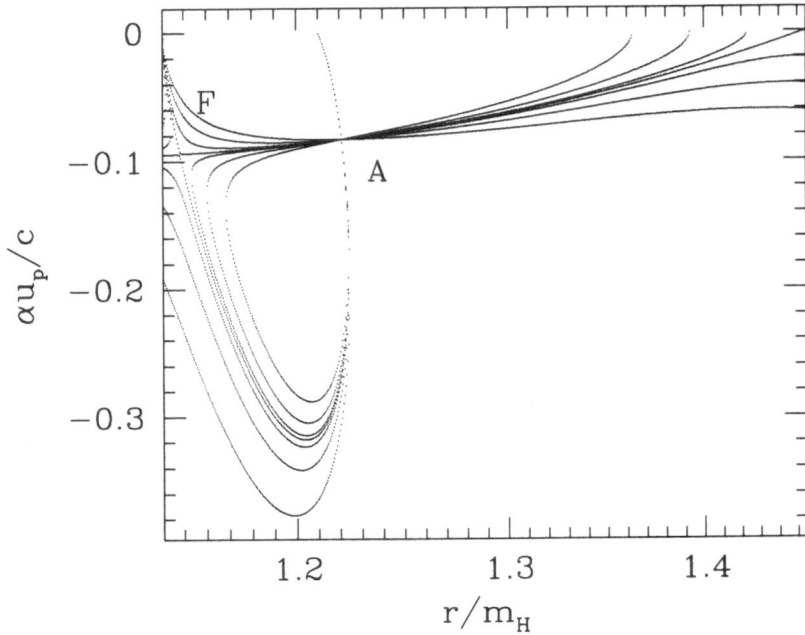

Figure 6 The accretion branches of the cold wind approximation on a rotating Kerr background (Camenzind and Englmaier 1997). For extremely rapidly rotating black holes, the Alfvén point (A) moves into the ergosphere ($r_A < 2M_H$) and the fast magnetosonic point (F) occurs immediately before the plasma falls through the horizon. In this case, the accretion carries a negative total angular momentum.

$$L_j = (\gamma_j - 1)\,\dot{M}_j c^2 \simeq 5 \times 10^{46}\,\mathrm{erg\,s}^{-1}\,\frac{\dot{M}_j}{0.1\,\mathrm{M}_\odot\,\mathrm{yr}^{-1}}\,\frac{\gamma_j}{10}\,. \qquad (161)$$

The mass outflow is typically a few percents of the mass inflow, or the accretion rate that determines the overall bolometric luminosity of quasars. *Therefore, we see that the magnetic luminosity of accretion disks around supermassive black holes can easily explain the kinetic luminosity in jets.*

5.6   Magnetic Disk Accretion and the Blandford–Znajek Process

As an example we consider here the accretion of a cold plasma towards a rotating black hole along some magnetic flux surface $\Psi = $ const (Camenzind and Englmaier 1997). As expected from the general discussion, this problem has three critical points. For plasma accretion from regions near the marginally stable orbit, the magnetization is probably of order unity, since the magnetic field is anchored in the inner part of the disk. The plasma flows through the slow magnetosonic point (not shown here), the Alfvén point, and crosses the fast magnetosonic point just before it enters into the horizon (Fig. 6). For causal reasons, all these critical points must occur, but relativistic MHD adjusts itself to this behavior.

The position of the Alfvén point is a question of the magnetization σ_* and the spin of the black hole. For rapidly rotating black holes, the Alfvén point can move into the

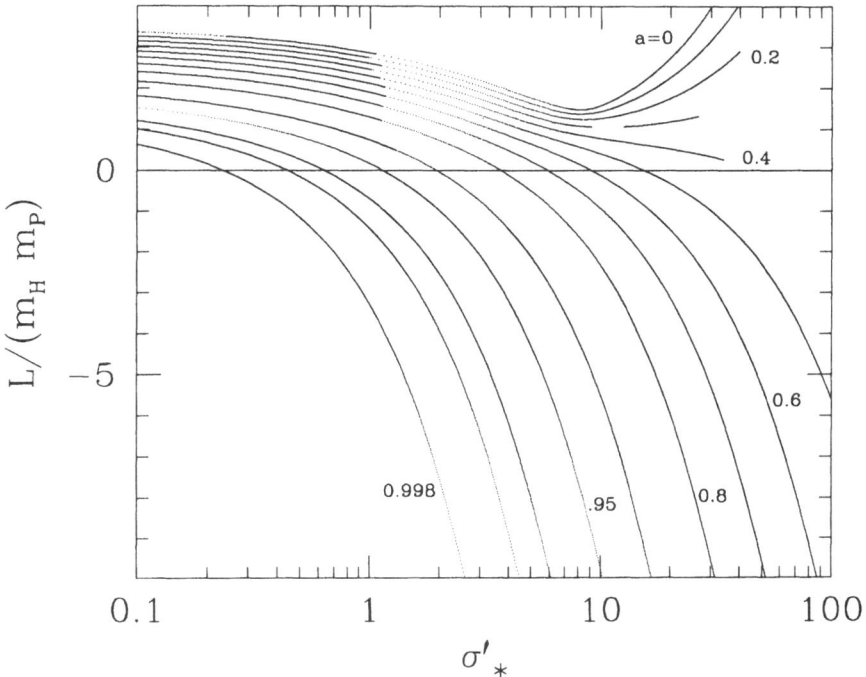

Figure 7 The total angular momentum in adiabatic magnetic accretion as a function of the magnetization in the disk (Camenzind and Englmaier 1997). The Kerr parameter is indicated for the individual curves. For slowly rotating black holes, the total angular momentum is always positive, and only for extremely rapidly rotating holes accretion with negative angular momentum can occur for a realistic magnetization of order unity.

ergosphere, and the total angular momentum L of the accretion flow can become negative (Camenzind and Englmaier 1997), indicating the possibility of extracting angular momentum from the hole (Fig. 5). This is a general relativistic effect which has no analog in stellar accretion. In particular, if the fields are rotating slower than the horizon, $0 < \Omega^F < \Omega_H$, solutions can occur with either negative angular momentum, or negative total energy, or both (for $a_H > 0.95 M_H$). Since the black hole rotates faster than the magnetic field lines, it can transfer angular momentum to the plasma via the gravitomagnetic coupling, and from there to the fields. This process would deposit angular momentum in the inner disk which must be carried off by some disk wind. This is probably the injection mechanism for the strong jet sources.

5.7 Disk Magnetospheres

In the case of stars or neutron stars, the magnetic flux is anchored in the stellar source itself. This is different in the case of black holes. The source of the magnetic flux is external to the black hole, most probably in the innermost part of the accretion disk itself (see Sect. 5.3). This means that we have a flux distribution $\Psi_d(R)$ along the disk surface. External to the disk, the flux comes into force–equilibrium with the boundary condition $\Psi = 0$ along the rotational axis. This force equilibrium can be derived from Ampère's equation

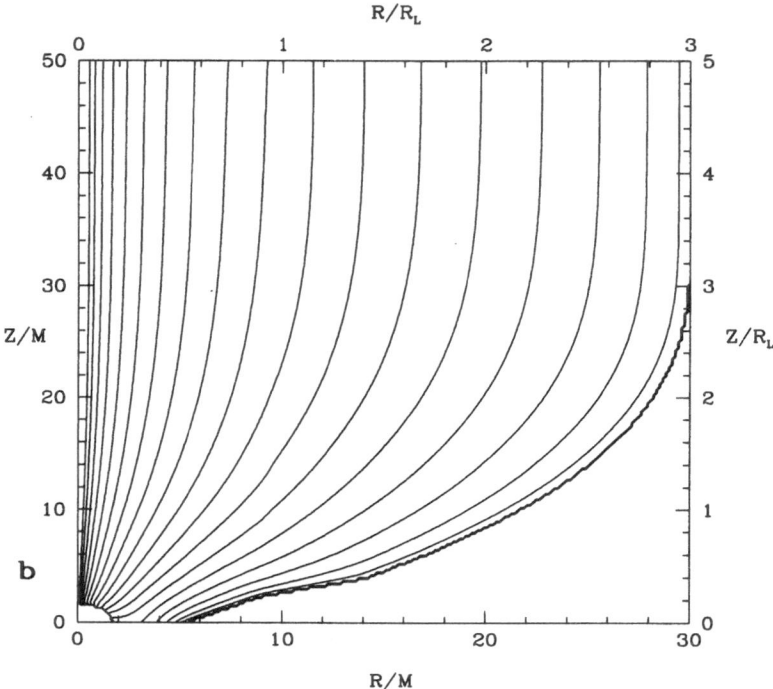

Figure 8 Jets are formed by collimated magnetic surfaces that are anchored in the inner part of the accretion disk around a rotating black hole. The contour lines are the poloidal magnetic fields of a force–free solution (Fendt 1997).

[for the derivation, see Okamoto (1992) and Fendt (1997)]. The resulting equation is known as pulsar equation for the poloidal flux function

$$\alpha \varpi^2 \nabla \cdot \left(\frac{D}{\alpha R^2} \nabla \Psi \right) + \varpi (\Omega^F - \omega) |\nabla \Psi|^2 \frac{d\Omega^F}{d\Psi} + 4I \frac{dI}{d\Psi} = 0. \qquad (162)$$

In this treatment, the current function $I(\Psi)$ is arbitrary. Since $I = I(\Psi)$, this represents a highly non–linear equation which can be solved with a suitable FE–solver (Camenzind 1987, Fendt et al. 1995, Fendt 1997). In general, only flux outside the inner light cylinder surface can escape towards the outer light cylinder surface, and therefore carry angular momentum away by means of some disk winds.

5.8 Collimated Disk Outflows and Jets in AGN

In accretion disks, the magnetization depends on the typical poloidal flux $\Psi_D = B_p R^2$, produced by the disk, as well as on the mass loss $\dot{M}_w$ from the disk surface

$$\sigma_* \simeq \frac{\Psi_D^2}{2\dot{M}_w c R_L^2} . \qquad (163)$$

The magnetic flux in the disk is limited by the total pressure in the disk [for the pressure in standard accretion disks, see, e.g., Straumann (1992), Sect. 9.2.4], $p \leq 1$

$$\Psi_D^2 \simeq \frac{2p}{I_{N+1}} \frac{\dot{M}_{acc}R_D^2}{\alpha_T} \sqrt{\frac{GM_H}{R_D}} \left(\frac{H}{R}\right)^{-1}. \tag{164}$$

The mass–loss rate in the disk wind is limited by the angular momentum loss in the wind

$$2\dot{M}_w R_A^2 \Omega^F \leq \dot{M}_{acc}R_D^2 \Omega_K(R_D) \tag{165}$$

which yields the ratio

$$\frac{\dot{M}_{acc}}{2\dot{M}_w} \geq \frac{R_A^2}{R_D^2}, \quad \Omega^F \simeq \Omega_K(R_D). \tag{166}$$

With this we get a lower limit on the magnetization

$$\boxed{\sigma_* \geq \frac{2p}{\alpha_T I_{N+1}} x_A^2 \sqrt{\frac{GM_H}{R_D}} \left(\frac{H}{R}\right)^{-1}.} \tag{167}$$

For accretion disks we have $H/R \leq 1$, $\alpha_T < 1$ *and, since* $x_A^2 \simeq 1$, *we get in general* $\sigma_* \geq 1$ *for accretion onto black holes. For equipartition fields,* $p \simeq 1$, $\sigma_* \simeq (1/\alpha_T) \simeq 10$. Under high accretion rates, outflows from the inner disk around black holes are always strongly magnetized. In contrast to this, we find for protostellar outflows $x_A^2 \simeq 10^{-6}$, $H/R \simeq 0.01$ and $GM_*/R_* \simeq 10^{-6}$, so that $\sigma_* \simeq 10^{-8}$, and the outflow is non–relativistic (Fendt et al. 1995).

The collimation of magnetized winds into a conical outflow is very similar to the plasma confinement in a z–pinch. The equilibrium condition for a nonrotating axisymmetric current carrying pinch configuration has to be generalized to include various relativistic effects (Appl and Camenzind 1993a,b)

$$\kappa \frac{B_p^2}{4\pi}(1 - M^2 - x^2) = (1 - x^2)\nabla_\perp \frac{B_p^2}{8\pi} + \nabla_\perp \frac{B_\phi^2}{8\pi} + \nabla_\perp P$$
$$- \frac{B_p^2 \Omega^F}{4\pi c^2}\nabla_\perp(\Omega^F R^2) + \left(\frac{\mu n j^2}{R^3} - \frac{I^2}{4\pi R^3}\right)(-\nabla_\perp R). \tag{168}$$

Pressure gradients acting perpendicular to the magnetic surfaces are in equilibrium with electric forces, centrifugal forces, pinch forces produced by poloidal currents, as well as curvature forces expressed on the left hand side. This latter force is modified by the Mach number and the light cylinder.

We discuss in the following the solutions for a cylindrical pinch, where the curvature forces vanish, $\kappa = 0$. In this case, the equilibrium condition reduces to its one–dimensional form

$$(1 - x^2)\frac{d}{dR}\frac{B_p^2}{8\pi} + \frac{d}{dR}\frac{B_\phi^2}{8\pi} + \frac{dP}{dR}$$
$$- \frac{B_p^2 R}{2\pi R_L^2} - \left(\frac{\mu n j^2}{R^3} - \frac{I^2}{4\pi R^3}\right) = 0. \tag{169}$$

The pressure of the toroidal field and the pinch force can be combined into one single expression. In addition, we normalize radii in units of the light cylinder R_L, $x = R/R_L$, and the poloidal magnetic field in units of its value on the central axis, B_0, $y = B_p^2/B_0^2$,

$$(1 - x^2)\frac{dy}{dx} - 4xy + \frac{1}{x^2}\frac{dI^2}{dx} + \frac{8\pi}{B_0^2}\frac{dP}{dx} - \frac{8\pi\mu n}{R_L^2 B_0^2}\frac{j^2}{x^3} = 0. \tag{170}$$

Figure 9 Jets are formed by collimated magnetic surfaces that are anchored in the inner part of the accretion disk.

Electric fields due to the rotation of the magnetic surfaces modify the classical pinch equilibrium in two ways. First, the equation has a singular point at the light cylinder, and secondly, the additional term $4xy$ is crucial for the form of the equilibrium. In contrast to force–free models (Appl and Camenzind 1993a,b), the specific angular momentum j of the plasma and the poloidal current I are not arbitrary, but follow from disk winds, as discussed in the previous section.

We discuss solutions of the relativistic pinch equilibrium neglecting the influence of the centrifugal force. In this case the equilibrium is determined by the condition

$$(1 - x^2)\frac{dy}{dx} - 4xy + \frac{1}{x^2}\frac{dI^2}{dx} + \frac{8\pi}{B_0^2}\frac{dP}{dx} = 0 \,. \tag{171}$$

The toroidal current follows to a good approximation, $i = 1$,

$$I = -\Omega^F(R^2 B_p)\frac{1}{\beta_p} = -B_0 R_L\left(x^2\sqrt{y}\right)\frac{1}{\beta_p} \tag{172}$$

and we prescribe the plasma pressure in terms of $P = \beta B_p^2/8\pi$ with a constant plasma beta. This leads finally to the equation, $\gamma_p = 1/\sqrt{1 - \beta_p^2}$,

$$\left(1 + \beta + \frac{x^2}{\gamma_p^2\beta_p^2}\right)\frac{dy}{dx} + \frac{4}{\gamma_p^2\beta_p^2}\,xy = 0 \,. \tag{173}$$

This homogeneous equation has the remarkably simple solution

$$y(x) = \frac{C}{(1 + \beta + x^2/x_c^2)^2} \tag{174}$$

with

$$x_c = \gamma_p \beta_p \quad , \quad R_c = \gamma_p \beta_p R_L \tag{175}$$

as the *core radius* of the jet. Inside the core, given by the light cylinder radius and the poloidal velocity U_p, the poloidal magnetic field is essentially homogeneous and decays as a monopole field outside the core, $B_p \propto 1/R^2$ for $R > R_c$

$$B_p = \frac{B_0}{1 + \beta + R^2/R_c^2} \, . \tag{176}$$

This is not unexpected since the jet is formed by a supermagnetosonic wind. The current function increases quadratically inside the core and saturates towards a constant value outside the core, $I_\infty \simeq B_0 R_L$. This is exactly the structure of the force–free solutions found previously (Appl and Camenzind 1993a,b). The constant C can be absorbed into the central value B_0.

The structure function $i(x)$ will modify this behavior somewhat. A stronger modification follows from the inclusion of the centrifugal force in the force–balance equation

$$\left(1 + \beta + \frac{x^2}{\gamma_p^2 \beta_p^2}\right) \frac{dy}{dx} + \frac{4}{\gamma_p^2 \beta_p^2} xy - \frac{2}{R_L^2 V_A^2} \frac{j^2}{x^3} = 0 \, . \tag{177}$$

The Alfvén velocity V_A is defined as $V_A^2(x) = B_0^2/4\pi\mu n$. The detailed analysis of such pinch solutions is still missing, but one can easily see that the centrifugal force will widen the magnetic field distribution. Jet solutions including the centrifugal force are wider than the force–free solutions discussed in Sect. 5.7.

6   Conclusions

Relativistic MHD is a very young field of research, compared, e.g., with the activities in solar physics. The relevance of this theory for rapidly rotating neutron stars and black holes has still to be worked out in the future. It is, however, my impression that we already made some progress in the last ten years. I hope that I have convinced everybody to use in the future relativistic MHD instead of the Newtonian one which must fail when outflows are driven by rapidly rotating objects. In this case, even general relativistic aspects are important as shown by the accretion solutions for rapidly rotating black holes.

Relativistic MHD is in particular important for the physics of accretion disks near black holes and neutron stars. Disk outflows, which always must occur in an astrophysical environment, can be collimated into more cylindrical shapes by the magnetic pinch force. Jets driven by accretion onto rotating black holes have radii which just scale with the Schwarzschild radius. In this sense, jets are found having radii of a few astronomical units in the Galactic center up to a few light years in the brightest quasars. These structures are now resolvable using VLBI in the millimeter regime.

On the theoretical side, many problems remain to be solved. One would be happy to have some analytical solutions of the non–linear Grad–Schlüter–Shafranov equation. In the near future, it will be possible to show the existence of self–collimated outflows for some broader class of outflows including differential rotation, centrifugal forces and plasma pressure. I think it is also a question of time to include differential rotation into the 2D solution procedures.

We are, however, far away from having a time–dependent solver for the propagation of relativistic MHD jets. Only with such a scheme will we be able to simulate the helical motion found in the jet of 3C 273, as well as the self–collimated cocoon structure of this most extreme jet outflow in the Universe. The plasma processes that accelerate the electrons up to energies of 10^{12} eV must be related to these instabilities. The Universe provides us with an extraordinary jet lab in form of the radio jets of quasars and radio galaxies.

Acknowledgments: Many of the results discussed in this review grew out of numerous discussions with my collaborators in Heidelberg. Special thanks go to Stefan Appl, Ramon Khanna, Jonathan Ferreira, Peter Englmaier, Christian Fendt, Oliver Dreissigacker, Gernot Paatz, Jochen Peitz and Stefan Spindeldreher. Projects on jets and AGN were funded by our Sonderforschungsbereich 328 in Heidelberg.

References

Appl, S. and Camenzind, M. (1993a): Self-Collimated Jets beyond the Light Cylinder. *Astron. Astrophys.*, **270**, 71

Appl, S. and Camenzind, M. (1993b): The Structure of Relativistic MHD Jets: A Solution to the Nonlinear Grad-Shafranov-Equation. *Astron. Astrophys.*, **274**, 699

Blandford, R.D. (1992): In: *Theory of Accretion Disks – 1* (eds. F. Meyer et al.). Kluwer, Dordrecht, p. 35

Blandford, R.D. and Rees, M. (1974): The "Twin-Exhaust" Model for Double Radio Sources. *Mon. Not. Roy. Astron. Soc.*, **169**, 395

Blandford, R.D. and Znajek, R.L (1977): Electromagnetic Extraction of Energy from Kerr Black Holes. *Mon. Not. Roy. Astron. Soc.*, **179**, 433

Blandford, R.D., Netzer, H., and Woltjer, L. (1993): In: *Physics of Active Galactic Nuclei.* Springer-Verlag, Berlin

Camenzind, M. (1986): Hydromagnetic Flows from Rapidly Rotating Compact Objects. I. Cold Relativistic Flows from Rapid Rotators. *Astron. Astrophys.*, **162**, 32

Camenzind, M. (1987): Hydromagnetic Flows from Rapidly Rotating Compact Objects. II. The Relativistic Axisymmetric Jet Equilibrium. *Astron. Astrophys.*, **184**, 341

Camenzind, M. (1995): Magnetic Fields and the Physics of Active Galactic Nuclei. *Rev. Mod. Astron.*, **8**, 201–233

Camenzind, M. (1996): Relativistic MHD Flows. In: *Solar and Astrophysical Magnetohydrodynamic Flows* (ed. K. Tsinganos). Kluwer, Dordrecht, 699

Camenzind, M. (1997): Les noyaux actifs de galaxies. *Lecture Notes in Physics*, **m46**. Springer-Verlag, Berlin

Camenzind, M. and Englmaier, P. (1997): Magnetic accretion onto rotating black holes. *Astron. Astrophys.*, submitted

Carter, B. (1979): The General theory of Mechanical, Electrodynamic and Thermodynamic Properties of Black Holes. In: *General Relativity, An Einstein Centenary Survey* (eds. S.W. Hawking and W. Israel). Cambridge Univ. Press, Cambridge

Chokshi, A. and Turner, E.L. (1992): Remnants of the Quasars. *Mon. Not. Roy. Astron. Soc.*, **259**, 421

Durrer, R. and Straumann, N. (1988): Some Applications of the 3+1 Formalism of General Relativity. *Helvetica Phys. Acta*, **61**, 1027

Faber, S.M., Tremaine, S., Ajhar, E.A. et al. (1997): The Centers of Early–Type Galaxies with HST. IV. Central Parameter Relations. *Astron. J.*, in press

Fendt, C. (1997): Collimated jet magnetospheres around rotating black holes. General relativistic force-free 2D equilibrium. *Astron. Astrophys.*, **319**, 1025

Fendt, C., Camenzind, M., and Appl, S. (1995): On the Collimation of Stellar Magnetospheres to Jets. I. Relativistic Force-Free 2D Equilibrium. *Astron. Astrophys.*, **300**, 791

Heyvaerts, J. and Norman, C. (1996): In: *Solar and Astrophysical Magnetohydrodynamic Flows* (ed. K. Tsinganos). Kluwer, Dordrecht

Khanna, R. and Camenzind, M. (1994): The Gravitomagnetic Dynamo Effect in Accretion Disks of Rotating Black Holes. *Astrophys. J. Lett.*, **435**, L129

Khanna, R. and Camenzind, M. (1996): The Gravitomagnetic Dynamo in Accretion Disks of Rotating Black Holes. *Astron. Astrophys.*, **307**, 665

Kormendy, J. and Richstone, D. (1995): Inward bound – the search for supermassive black-holes in galactic nuclei. *Ann. Rev. Astron. Astrophys.*, **33**, 581

Kormendy, J., Bender, R., Richstone, D., Ajhar, E.A., Dressler, A., Faber, S.M., Gebhardt, K., Grillmair, C., Lauer, T.R., and Tremaine, S. (1996): Hubble Space Telescope Spectroscopic Evidence for a 2x10^9 M$\odot$ Black Hole in NGC 3115. *Astrophys. J. Lett.*, **459**, L57

Kudoh, T. and Kaburaki, O. (1996): Resistive Magnetohydrodynamic Accretion Disks around Black Holes. *Astrophys. J.*, **460**, 199

Lynden-Bell, D. (1969): Galactic Nuclei as Collapsed Old Quasars. *Nature*, **223**, 690

Narayan, R. and Yi, I. (1995): Advection-Dominated Accretion: Underfed Black Holes and Neutron Stars. *Astrophys. J.*, **452**, 710

Okamoto, I. (1992): The Evolution of a Black Hole's Force-Free Magnetosphere. *Mon. Not. Roy. Astron. Soc.*, **254**, 192

Peitz, J. and Appl, S. (1997): Viscous accretion disks around rotating black holes. *Mon. Not. Roy. Astron. Soc.*, **286**, 681

Rawlings, S. and Saunders, R. (1991): Evidence for a Common Central-Engine Mechanism in all Extragalactic Radio Sources. *Nature*, **349**, 138

Rees, M.J. (1984): Black Hole Models for Active Galactic Nuclei. *Ann. Rev. Astron. Astrophys.*, **22**, 471

Rees, M.J., Begelman, M.C., Blandford, R.D., and Phinney, E.S. (1982): Ion-Supported Tori and The Origin of Radio Jets. *Nature*, **295**, 17

Salpeter, E.E. (1964): Accretion of Interstellar Matter by Massive Objects. *Astrophys. J.*, **140**, 796

Straumann, N. (1992): General Relativity and Relativistic Astrophysics. Springer-Verlag, Berlin

Thorne, K.S. and Macdonald, D.A. (1982): Electrodynamics in Curved Spacetime: 3+1 Formulation. *Mon. Not. Roy. Astron. Soc.*, **198**, 339

Thorne, K.S., Price, R.H., and Macdonald, D.A. (1986): Black Holes – The Membrane Paradigm. Yale University Press, New Haven

Wald, R.M. (1974): Black Hole in a Uniform Magnetic Field. *Phys. Rev.*, **D10**, 1680

Thin Accretion Disks around Black Holes

Harald Riffert

Institut für Theoretische Astrophysik
Universität Tübingen, Auf der Morgenstelle 10, D-72076 Tübingen
`riffert@tat.physik.uni-tuebingen.de`

Abstract

Active galactic nuclei (AGN) are powered by accretion and recent observations indicate the existence of accretion disks around a central black hole of 10^8 to 10^{10} M$_\odot$. A derivation and discussion of the basic equations for the structure of a thin accretion disk around a rotating black hole is presented. Starting from the relativistic hydrodynamics equations the disk structure is obtained if we assume a geometrically thin flow on (almost) geodesic orbits. The energy transport in the disk is due to radiation, and the equation of radiation transfer in the local rest frame of the gas is discussed in some detail. Finally we present a numerical calculation to obtain a disk spectrum in the UV and soft X-ray energy range.

1 Introduction

Accretion of matter is a very powerful source of energy for many astrophysical objects. If, for example, a mass flux $\dot{M}$ is accreted by a star of mass M, i.e., with Schwarzschild radius $R_s = 2GM/c^2$, and radius R the resulting accretion luminosity

$$L_A = \frac{GM\dot{M}}{R}\xi = \dot{M}c^2\frac{R_s}{2R}\xi \tag{1}$$

can be a significant fraction of the rest-mass flux $\dot{M}c^2$ provided the object is sufficiently compact ($R/R_s \sim 1$) and the gravitational energy gained can be converted into radiation with high efficiency ($\xi \sim 1$). Accretion is the common energy source for a wide range of astrophysical objects including cataclysmic variables (CVs), low-mass X-ray binaries, X-ray pulsars, X-ray burst sources, the various stellar black-hole candidates, and even the nuclei of active galaxies (AGN). The global structure of the accretion flow is strongly determined by the amount of (specific) angular momentum of the gas. A more or less spherical accretion is expected for low angular momentum flows, whereas a large specific angular momentum results in an accretion disk around the central star. Such disks have been studied in much detail both observationally and theoretically for close binary systems such as CVs where a white dwarf accretes hot gas from a normal companion star (Papaloizou et al. 1983, Meyer and Meyer-Hofmeister 1984). These investigations

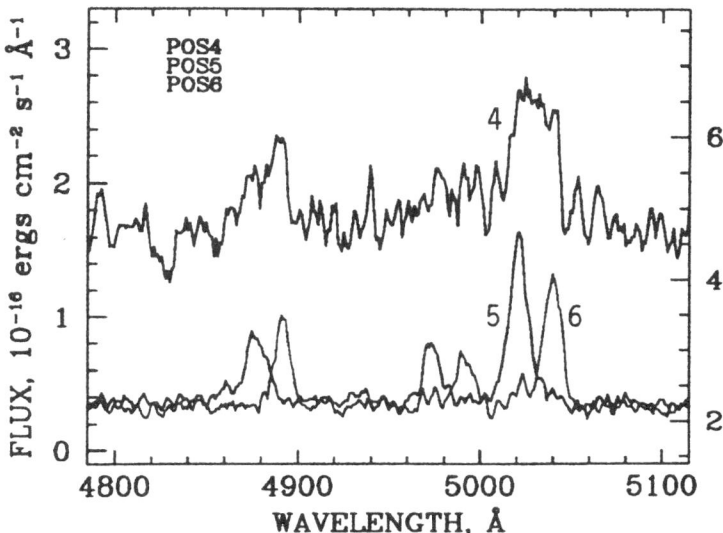

Figure 1 Emission-line spectrum from the center of the galaxy M87 observed with the Hubble-Space-Telescope. The Spectra 4, 5, and 6 were taken from three adjacent emission regions of 18 pc. The central specrum 4 shows Doppler-broadened emissions, whereas the spectra 5 and 6 are Doppler shifted with respect to each other. The interpretation of an emission from a Keplerian disk leads to a central mass of 2.4×10^9 M$_\odot$ (Harms et al. 1994).

have established a standard model of an accretion disk (see Frank et al. 1992) which is geometrically thin with a flow velocity that is essentially given by the Kepler velocity $v_K = \sqrt{GM/r}$ at each radius r from the central object. Viscous processes in the disk lead to a small radial inflow of matter and an outflow of angular momentum. It is believed that a fully developed turbulence, probably in connection with internal magnetic fields, is the source of viscosity. The viscous heat production is compensated by radiation and there exists a local equilibrium between viscous heating and radiative cooling (see also Shakura and Sunyaev 1973). The standard thin accretion disk is a very successful model to explain many observational details from cataclysmic variables, and the question is whether a disk structure also exists in the center of active galaxies and whether emission from a disk around a supermassive black hole is responsible for parts of the continuum spectra in AGN.

The large luminosity of quasars (typically 10^{46} erg/s) can only be explained by some accretion process and the observed X-ray variability of a few hours indicates a small emission region of about 10^{14} cm (Turner and Pounds 1989). This corresponds to the Schwarzschild radius of a black hole of 10^8 M$_\odot$. If the accretion efficiency ξ is large ($\xi \sim 1$), a mass flux of about 1 M$_\odot$ per year is sufficient to account for the observed luminosities. Spherical accretion, however, is very inefficient (Rees 1984), and thus, for theoretical reasons alone an accretion disk would be more satisfactory. The overall astrophysical scenario in an AGN, in particular the existence of dust- and molecular tori as well as of jets, also points towards an axisymmetric configuration in favor of a spherical one. The existence of a gas disk on a parsec scale in the inner part of active galaxies has been recently confirmed observationally by Doppler shifts of spectral lines. Harms et al. (1994)

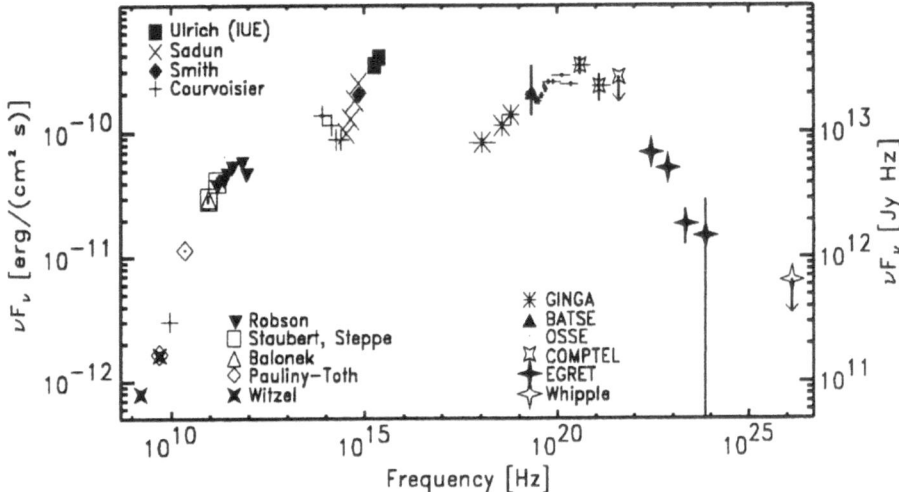

Figure 2   Broadband spectrum of the quasar 3C273 ranging from 10^9 Hz to 10^{26} Hz. The "Big Blue Bump" is the emission maximum between 3×10^{14} Hz and 10^{17} Hz (Lichti et al. 1994).

observed the giant elliptical galaxy M87 in the Virgo cluster with the Hubble-Space-Telescope and found red and blue shifted lines in an 18 parsec central region which indicates the existence of a supermassive black hole of 2.4×10^9 $M_\odot$ (see Fig. 1). The best evidence for parsec-scale accretion disks comes from observations of discrete spots of water-maser emission in AGN such as NGC 4945, NGC 1068, or NGC 4258 (Greenhill et al. 1995, 1996, 1997, Miyoshi et al. 1995).

Since the gas in this region carries a large amount of angular momentum it is very likely that the accretion disk extends down to the scale of the central black hole, i.e., the Schwarzschild radius R_s. The typical AGN spectrum observed over many orders of magnitude in energy (Fig. 2) contains a strong emission maximum in the UV and soft X-ray range (from 3×10^{14} Hz to 10^{17} Hz) which is known as the "Big Blue Bump" (Neugebauer et al. 1987, Sanders et al. 1989); the soft X-ray part of this structure is sometimes called the "Soft Excess" (Turner and Pounds 1989). This spectral feature is thought to be due to thermal emission from a thin accretion disk around the central black hole (Shields 1978, Arnauld et al. 1985, Wilkes and Elvis 1987). A number of analytical and numerical calculations have been performed in order to determine the spectral properties of accretion disk models that are essentially extensions of the above mentioned standard thin CV disk for the case of AGN. Considering a 10^8 $M_\odot$ black hole and a mass flux of 1.75 $M_\odot$/a typical parameters at 10 R_s are (Collin-Souffrin 1994)

$$\rho = 3 \times 10^{-14} \text{ gcm}^{-3}, \qquad T = 10^5 \text{ K}, \qquad \frac{P_r}{P_g} = 10^6 \qquad (2)$$

(ρ = mass density, T = temperature, P_r = radiation pressure, P_g = gas pressure). In contrast to the CV disks, here radiation pressure is dominant, and Compton scattering contributes strongly to the total opacity. In addition, the disk structure is modified by general relativistic effects in the vicinity of a black hole.

The relativistic generalization of the standard disk goes back to the fundamental papers of Novikov and Thorne (1973) and Page and Thorne (1974) where the structure of a thin disk is calculated in the equatorial plane of the Kerr metric describing a rotating black hole with mass M and specific angular momentum a. A simple way to obtain an emitted spectrum is to assume that the disk radiates locally like a blackbody, however, realistic spectra differ strongly from the blackbody approximation. Thus, in the last years a number of quite detailed calculations were performed to determine the vertical and radial disk structure together with the spectral radiative transfer (Ross et al. 1992, Shimura and Takahara 1993, Yamada et al. 1994, Dörrer et al. 1996).

In this article we give an overview of the basic equations that describe the structure of a relativistic thin accretion disk around a Kerr black hole. The following section treats the hydrodynamics of the gas flow in the disk. In section 3 radiative transfer is considered and an example of a numerical calculation of the spectral emission properties of a thin AGN disk is presented.

2 The Disk Structure Equations

2.1 Basic Assumptions

We consider a gas flow around a rotating black hole of mass M and specific angular momentum a. Usually the total mass of gas in the flow is much smaller than M, thus one can safely assume that the metric of the entire system is a Kerr metric given by M and a, which can be written in terms of the usual Boyer-Lindquist coordinates (t, ϕ, r, θ) (Misner et al. 1973) (we use $G = c = h = 1$ throughout this paper)

$$ds^2 = -\frac{\Sigma\Delta}{A}dt^2 + \frac{A\sin^2\theta}{\Sigma}[d\phi - \omega dt]^2 + \frac{\Sigma}{\Delta}dr^2 + \Sigma d\theta^2 \tag{3}$$

where

$$\Delta = r^2 + a^2 - 2Mr \;, \;\; \Sigma = r^2 + a^2\cos^2\theta \;, \;\; A = (r^2 + a^2)^2 - a^2\Delta\sin^2\theta \;, \;\; \omega = \frac{2aMr}{A} \;. \tag{4}$$

This metric shows a horizon at $r_+ = M + \sqrt{M^2 - a^2}$ and a so-called static limit $r_0 = M + \sqrt{M^2 - a^2\cos^2\theta}$. Observers with r and θ fixed and angular velocity $\Omega = d\phi/dt$ (stationary observers) must corotate with the hole in the range $r_+ < r < r_0$; Ω is limited by

$$\Omega_- < \Omega < \Omega_+ \quad \text{with} \quad \Omega_\pm = \omega \pm \frac{\sqrt{\Delta\Sigma}}{A\sin\theta} \;,$$

and $\Omega = \omega$ corresponds to the zero-angular momentum orbit.

The dynamics of the gas is determined by the energy momentum balance

$$T^{\mu\nu}_{\;;\nu} = 0$$

where the semicolon indicates a covariant derivative. Since both viscous processes and energy transport by radiation or convection are important, here the energy-momentum tensor $T^{\mu\nu}$ must include corresponding terms

$$T^{\mu\nu} = [\rho + \varepsilon + P]U^\mu U^\nu + Pg^{\mu\nu} + S^{\mu\nu} + U^\mu q^\nu + U^\nu q^\mu. \tag{5}$$

Here ρ is the rest-mass density, ε stands for the thermal energy density, P is the pressure and U^μ is the 4-velocity of the flow with the normalization $U_\mu U^\mu = -1$. Since radiation pressure is important for AGN disks, P and ϵ contain radiation contributions

$$P = P_g + P_r \,, \quad \varepsilon = \varepsilon_g + J \,.$$

The viscous part of $T^{\mu\nu}$ is given by $S^{\mu\nu}$, and q^ν is an energy-flux 4-vector; they fulfill the additional constraints

$$U_\mu S^{\mu\nu} = 0 \qquad U_\mu q^\mu = 0 \,. \tag{6}$$

As in the case of non-relativistic hydrodynamics $S^{\mu\nu}$ can be expressed in terms of velocity gradients

$$S^{\mu\nu} = \eta \left[h^{\mu\lambda} U^\nu{}_{;\lambda} + h^{\nu\lambda} U^\mu{}_{;\lambda} \right] + \left(\zeta - \frac{2}{3}\eta \right) U^\lambda{}_{;\lambda} h^{\mu\nu} \tag{7}$$

where $h^{\mu\nu} = U^\mu U^\nu + g^{\mu\nu}$ is a projection tensor projecting onto the 3-space orthogonal to U^μ; the scalars η and ζ indicate the two viscosity coefficients (Landau and Lifschitz 1991). The above energy momentum balance has to be augmented by a number of additional equations. First, baryon conservation is expressed by the continuity equation

$$(\rho U^\mu)_{;\mu} = 0 \,, \tag{8}$$

furthermore an equation of state is required, the source of viscosity (i.e., η and ζ) must be specified, and finally one needs a description of the overall energy transport mechanism. So far the above equations are very general, valid for any gas flow around the black hole. In the following we make a number of simplifying assumptions that will lead us to the structure equations of a thin accretion disk. First, let the flow cover only a small region close to the equatorial plane of the above coordinate system. We therefore introduce cylindrical polar coordinates (t, ϕ, r, z) such that $z = 0$ corresponds to the equatorial plane. The metric tensor $g_{\mu\nu}$ can be expanded in powers of z/r (Riffert and Herold 1995), and we obtain to second order

$$
\begin{aligned}
g_{00} &= -1 + \frac{2M}{r} - \frac{Mz^2}{r^3}\left(1 + \frac{2a^2}{r^2}\right) \\[2mm]
g_{01} &= -\frac{2aM}{r} + \frac{aMz^2}{r^3}\left(3 + \frac{2a^2}{r^2}\right) = g_{10} \\[2mm]
g_{11} &= r^2 + a^2 + \frac{2Ma^2}{r} - \frac{a^2 z^2}{r^2}\left(1 + \frac{5M}{r} + \frac{2Ma^2}{r^3}\right) \\[2mm]
g_{22} &= \frac{1}{\mathcal{A}} - \frac{z^2}{r^2 \mathcal{A}^2}\left[\frac{M}{r}\left(3 - \frac{4M}{r}\right) - \frac{a^2}{r^2}\left(3 - \frac{6M}{r} + \frac{2a^2}{r^2}\right)\right] \\[2mm]
g_{23} &= \frac{z}{r\mathcal{A}}\left(\frac{2M}{r} - \frac{a^2}{r^2}\right) = g_{32} \\[2mm]
g_{33} &= 1 + \frac{z^2}{r^2 \mathcal{A}}\left(\frac{2M}{r} - \frac{2Ma^2}{r^3} + \frac{a^4}{r^4}\right)
\end{aligned} \tag{9}
$$

where

$$\mathcal{A} = 1 - \frac{2M}{r} + \frac{a^2}{r^2} \tag{10}$$

(the remaining components of $g_{\mu\nu}$ vanish). The assumption of a thin disk means that its proper height $H(r)$ is everywhere much smaller than r ($H(r) \ll r$). Consequently, the radial gradients of scalar functions such as the mass density, the pressure, or the energy density are much smaller than the vertical gradients

$$\frac{P_{,r}}{P_{,z}} \sim \frac{z}{r} \ll 1 \quad \text{etc. ,}$$

and the vertical energy flux is much larger than the radial flux $q_z \gg q_r$. Next, only stationary and axisymmetric flows are considered, i.e., for any function f we have

$$\frac{\partial f}{\partial t} = \frac{\partial f}{\partial \phi} = 0 , \quad \text{or} \quad f = f(r,z) .$$

Note that $\partial/\partial t$ and $\partial/\partial \phi$ are the Killing vectors of the Kerr metric.
A further essential simplification is achieved by assuming that the gas motion in the azimuthal direction is the dominant part of the velocity field. For a non-relativistic disk this leads to

$$V_\phi \gg V_r \gg V_z$$

where $\boldsymbol{V} = (V_\phi, V_r, V_z)$ is the velocity 3-vector, and the last inequality is a consequence of the continuity equation

$$\frac{\partial}{\partial r}(\rho V_r) + \frac{\partial}{\partial z}(\rho V_z) = 0 \quad \Rightarrow \quad \frac{\rho V_r}{r} \sim \frac{\rho V_z}{z} \quad \Rightarrow \quad \frac{V_z}{V_r} \sim \frac{z}{r} \ll 1 .$$

Thus the disk gas moves essentially on Keplerian orbits, i.e., $V_\phi \approx V_K = \sqrt{GM/r}$ with a small radial drift velocity V_r and an even smaller vertical velocity V_z, and it follows

$$\frac{V_\phi}{V_r} \sim \frac{r^2}{H^2} \tag{11}$$

(Frank et al. 1992). For a relativistic flow we accordingly consider the components U^0 and U^1 of the 4-velocity as "large" components.

2.2 Momentum Balance

Projecting the energy-momentum balance $T^{\mu\nu}_{;\nu} = 0$ with $h^i_\mu = U^i U_\mu + \delta^i_\mu$ and considering non-relativistic matter, i.e., $\rho \gg P$, $\rho \gg \epsilon$, we get

$$\left[U^i U^\mu + g^{i\mu} \right] P_{,\mu} + \rho U^\mu U^i_{;\mu} + S^{i\mu}_{;\mu} + U^i U_\nu S^{i\mu}_{;\mu}$$
$$+ q^i U^\mu_{;\mu} + q^\mu U^i_{;\mu} + U^\mu q^i_{;\mu} - U^i U^\mu q^\nu U_{\nu;\mu} = 0 . \tag{12}$$

Taking the dominant part of the radial component of this equation ($i = 2$) we are left with

$$U^0 U^0 \, \Gamma^2_{00} + 2 \, U^0 U^1 \, \Gamma^2_{01} + U^1 U^1 \, \Gamma^2_{11} = 0 \tag{13}$$

with the Christoffel symbols $\Gamma^\alpha_{\beta\gamma}$ calculated in the equatorial plane $z = 0$. Together with the normalization $U_\mu U^\mu = -1$ the large components of U^μ can be calculated

$$U^0 = \frac{1}{\sqrt{B}} \left(1 + \frac{a}{r} \sqrt{\frac{M}{r}} \right) \qquad U^1 = \frac{1}{\sqrt{B}} \sqrt{\frac{M}{r^3}} \tag{14}$$

with

$$\mathcal{B} = 1 - \frac{3M}{r} + \frac{2a}{r}\sqrt{\frac{M}{r}} \quad . \tag{15}$$

Since eq. (13) is equivalent to the dominant part of the geodesic equation

$$U^i = \frac{dx^i}{ds} \quad , \qquad \frac{dU^i}{ds} + \Gamma^i_{\mu\nu}U^\mu U^\nu = 0 \qquad \text{(for } i = 2\text{)} \ , \tag{16}$$

it follows that the gas in the disk moves essentially on geodesic orbits around the black hole, and for large radii ($M/r \to 0$) rU^1 approaches the non-relativistic Kepler velocity

$$rU^1 = \sqrt{\frac{M}{r}} + O\left(r^{-3/2}\right) \quad .$$

The velocity field $\tilde{U} := (U^0, U^1, 0, 0)$ with U^0 and U^1 given by (14) allows for the definition of a local corotating reference frame which is a good approximation to the local rest-frame of matter (LRFM) even for small but finite values of z. It is characterized by an orthonormal tetrad $\boldsymbol{b}_\alpha$ with $\boldsymbol{b}_0 \equiv \tilde{U}$ and $\boldsymbol{b}_\mu \cdot \boldsymbol{b}_\nu = \eta_{\mu\nu}$ where $\eta_{\mu\nu}$ is the Minkowski metric. Let L^α_β mediate the basis transformation from the coordinate basis $\boldsymbol{e}_\alpha \equiv \partial/\partial x^\alpha$ to the tetrad $\boldsymbol{b}_\beta$

$$\boldsymbol{e}_\alpha = L^\beta_{\ \alpha}\boldsymbol{b}_\beta \quad , \qquad \boldsymbol{b}_\beta = \hat{L}^\alpha_{\ \beta}\boldsymbol{e}_\alpha \tag{17}$$

($\hat{L}$ is the inverse of L). It will be useful to write the disk structure equations in terms of the tetrad components of the viscous tensor $S^{\mu\nu}$ and the flux vector q^μ

$$-t^{\mu\nu} = L^\mu_{\ \alpha} \, L^\nu_{\ \beta} \, S^{\alpha\beta} \quad , \qquad F^\mu = L^\mu_{\ \alpha} \, q^\alpha \tag{18}$$

(note that $F^0 = 0$, $S^{0\mu} = 0$ because of (6)). The transformations L and $\hat{L}$ can be calculated explicitly (see also Novikov and Thorne 1973)

$$L = \begin{pmatrix} -U_0 & -U_1 & 0 & 0 \\ L^1_{\ 0} & L^1_{\ 1} & 0 & 0 \\ 0 & 0 & \frac{1}{\sqrt{\mathcal{A}}} & 0 \\ 0 & 0 & 0 & 1 \end{pmatrix} \qquad \hat{L} = \begin{pmatrix} U^0 & \hat{L}^0_{\ 1} & 0 & 0 \\ U^1 & \hat{L}^1_{\ 1} & 0 & 0 \\ 0 & 0 & \sqrt{\mathcal{A}} & 0 \\ 0 & 0 & 0 & 1 \end{pmatrix} \tag{19}$$

where $L^1_{\ 0}$, $L^1_{\ 1}$, $\hat{L}^0_{\ 1}$, and $\hat{L}^1_{\ 1}$ are functions of r, M, a

$$L^1_{\ 0} = -\sqrt{\frac{\mathcal{A}}{\mathcal{B}}}\sqrt{\frac{M}{r}} \qquad\qquad \hat{L}^0_{\ 1} = \frac{1}{\sqrt{\mathcal{A}\mathcal{B}}}\sqrt{\frac{M}{r}}\left(1 + \frac{a^2}{r^2} - \frac{2a}{r}\sqrt{\frac{M}{r}}\right)$$

$$L^1_{\ 1} = r\sqrt{\frac{\mathcal{A}}{\mathcal{B}}}\left(1 + \frac{a}{r}\sqrt{\frac{M}{r}}\right) \qquad \hat{L}^1_{\ 1} = \frac{1}{r\sqrt{\mathcal{A}\mathcal{B}}}\left(1 - \frac{2M}{r} + \frac{a}{r}\sqrt{\frac{M}{r}}\right) \ . \tag{20}$$

Next we calculate the leading terms of the vertical momentum balance (eq. (12) for $i = 3$), and we get

$$\frac{\partial P}{\partial z} = -\rho z \frac{M}{r^3}\frac{\mathcal{C}}{\mathcal{B}} \tag{21}$$

where

$$C = 1 - \frac{4a}{r}\sqrt{\frac{M}{r}} + \frac{3a^2}{r^2} \ . \tag{22}$$

Vertically there is a hydrostatic equilibrium, i.e., gravity from the central black hole is balanced by the pressure gradient. Note that this relation is first order in z/r and in order to calculate the right-hand side second order terms of the metric are required. The last part of the momentum equation (12) that remains to be investigated is the azimuthal component ($i = 1$) which corresponds to the angular momentum balance. Again, taking the leading terms only we obtain for the tetrad components of $S^{\mu\nu}$

$$\frac{\partial t_{r\phi}}{\partial r} + C_0(r)\frac{\partial t_{\phi z}}{\partial z} = C_1(r)t_{r\phi} + C_2(r)\, r\rho U^2 \tag{23}$$

where

$$C_1(r) = -\frac{2}{r\mathcal{A}}\left(1 - \frac{M}{r}\right) \ , \quad C_2(r) = -\frac{1}{2r^2\mathcal{AB}}\sqrt{\frac{M}{r}}\left(1 - \frac{6M}{r} + \frac{8a}{r}\sqrt{\frac{M}{r}} - \frac{3a^2}{r^2}\right) \tag{24}$$

($C_0(r)$ is not needed for the following calculations). In order to eliminate the $t_{\phi z}$-component we integrate over the height of the disk; with

$$< t_{r\phi} > = \int_{-H}^{+H} t_{r\phi}\, dz \quad \text{and} \quad < \rho U^2 > = \int_{-H}^{+H} \rho U^2\, dz \tag{25}$$

we obtain an ordinary differential equation for $< t_{r\phi} >$

$$\frac{\partial}{\partial r} < t_{r\phi} > = C_1(r) < t_{r\phi} > + C_2(r)\, r < \rho U^2 > \ , \tag{26}$$

and the last term can be expressed by the total mass flux $\dot{M}$ through the disk using the equation of continuity (8)

$$\frac{1}{r}\frac{\partial}{\partial r}(r\rho U^2) + \frac{\partial}{\partial z}(\rho U^3) = 0 \quad \Rightarrow \quad r < \rho U^2 > = -\frac{\dot{M}}{2\pi} = \text{constant} \ . \tag{27}$$

Thus, the solution of the above differential equation for $< t_{r\phi} >$ reads

$$< t_{r\phi} > = \frac{\dot{M}}{4\pi r^2\mathcal{A}}\int_{r_i}^{r}\sqrt{\frac{M}{x}}\,\frac{x^2 - 6Mx + 8a\sqrt{Mx} - 3a^2}{x^2 - 3Mx + 2a\sqrt{Mx}}\, dx \ , \tag{28}$$

where r_i is an integration constant denoting a radius $r = r_i$ such that $< t_{r\phi} >|_{r_i} = 0$. For any value of a the denominator in the integral has a real zero at $x = x_0$ where

$$M \leq x_0 \leq 3M \ . \tag{29}$$

Thus, the integral exists only if $r_i > x_0$, i.e., the boundary condition cannot be imposed at arbitrary values of r_i, in particular not at the horizon. This is a consequence of the standard thin disk approximation with a Keplerian orbital velocity given by eq. (14). Other rotation laws lead to different accretion flow topologies where even the boundary condition $< t_{r\phi} > = 0$ (no torque condition) can be chosen at the horizon (Abramowicz and Prasanna 1990, Peitz and Appl 1997). Within the framework of the standard thin disk model, however, the inner radius r_i is taken at the radius of the last stable circular orbit given by the only real solution of

$$r_i^2 - 6Mr_i + 8a\sqrt{Mr_i} - 3a^2 = 0 \qquad (30)$$

with $r_i \geq M$; note that the integral (28) now always exists and we obtain

$$<t_{r\phi}> = \frac{\dot{M}}{2\pi}\sqrt{\frac{M}{r^3}}\frac{\mathcal{D}}{\mathcal{A}} \quad, \qquad (31)$$

where

$$\mathcal{D} = \frac{1}{2\sqrt{r}}\int_{r_i}^r \frac{x^2 - 6Mx + 8a\sqrt{Mx} - 3a^2}{x^2 - 3Mx + 2a\sqrt{Mx}}\frac{dx}{\sqrt{x}} \quad ; \qquad (32)$$

it is possible to express $\mathcal{D}$ explicitly in terms of elementary functions (see Page and Thorne 1974). Equation (31) shows that viscous processes are responsible for the mass transport through the disk.

2.3 Energy Balance

An energy balance equation is obtained by projecting the divergence of $T^{\mu\nu}$ onto the 4-velocity U_μ

$$U_\mu T^{\mu\nu}{}_{;\nu} = 0 \quad \Rightarrow \quad U^\mu \varepsilon_{,\mu} + (P+\varepsilon)U^\mu{}_{;\mu} + S^{\mu\nu}U_{\mu;\nu} + q^\mu{}_{;\mu} + q^\mu U^\nu U_{\mu;\nu} = 0 \qquad (33)$$

For a thin Keplerian disk the leading terms of this relation describe a balance between viscous heating and the vertical energy loss

$$S^{\mu\nu}U_{\mu;\nu} + q^\mu{}_{;\mu} \cong 0 \quad , \qquad (34)$$

however, if z/r is not very small the advective part $U^\mu\varepsilon_{,\mu}$ becomes the dominant cooling contribution. To make this more explicit we compare the advective and the radiative terms for a radiation dominated non-relativistic flow

$$\frac{\boldsymbol{V}\cdot\nabla\varepsilon}{\nabla\cdot\boldsymbol{F}} \sim \frac{H}{r}\frac{V_r}{V_\phi}\frac{V_\phi}{c}\frac{cJ}{F_z} \qquad (35)$$

and the last ratio follows from a vertical diffusion equation

$$F_z = -\frac{c}{3\bar{\kappa}\rho}\frac{\partial J}{\partial z} \quad \Rightarrow \quad J \sim \frac{2}{c}F(1+\tau)$$

where $\bar{\kappa}$ is an average opacity and the optical depth over the disk height is defined as

$$\tau = \int_0^H \bar{\kappa}\rho\,dz \quad .$$

Thus, from eq. (35)

$$\frac{\boldsymbol{V}\cdot\nabla\varepsilon}{\nabla\cdot\boldsymbol{F}} \sim 2\left(\frac{H}{r}\right)^3 \frac{V_\phi}{c}(1+\tau)$$

which is small only if $H \ll r$, otherwise the flow structure changes into an advection dominated slim disk (see Abramowicz et al. 1995).

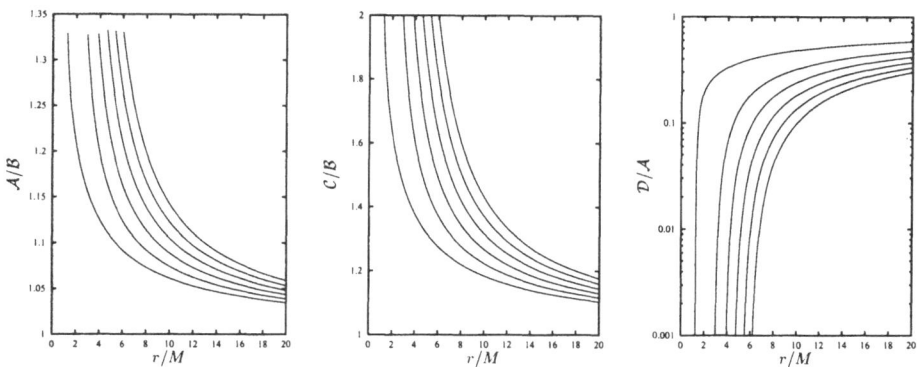

Figure 3   Relativistic correction factors; from left to right: $\mathcal{A}/\mathcal{B}$ (see eqs. (38) and (39)), $\mathcal{C}/\mathcal{B}$ (from the hydrostatic equilibrium (21), and $\mathcal{D}/\mathcal{A}$ (from eq. (31)). The various curves in each graph correspond to different angular momenta a/M of the black hole (from right to left: $a/M = 0$, 0.2, 0.4, 0.6, 0.8, 1.0).

For a thin disk eq. (34) reads in terms of the various tetrad components

$$\frac{\partial F_z}{\partial z} = \frac{3}{2}\sqrt{\frac{M}{r^3}}\frac{\mathcal{A}}{\mathcal{B}}\, t_{r\phi} \ . \tag{36}$$

Integration over the disk height leads to

$$F_z(H) = \frac{3}{4}\sqrt{\frac{M}{r^3}}\frac{\mathcal{A}}{\mathcal{B}} < t_{r\phi} > = \frac{3M\dot{M}}{8\pi r^3}\frac{\mathcal{D}}{\mathcal{B}} \ . \tag{37}$$

Thus, the energy loss at each radius r is given by the mass flux $\dot{M}$. The last relation that needs to be calculated for the disk structure is eq. (7) which expresses the tensor component $t_{r\phi}$ in terms of the dynamical viscosity η (assuming $\zeta = 0$)

$$t_{r\phi} = \frac{3\eta}{2}\sqrt{\frac{M}{r^3}}\frac{\mathcal{A}}{\mathcal{B}} \ , \tag{38}$$

and from that

$$\frac{\partial F_z}{\partial z} = \frac{9\eta}{4}\frac{M}{r^3}\frac{\mathcal{A}^2}{\mathcal{B}^2} \ . \tag{39}$$

2.4   Summary

The basic disk structure is so far given by a differential equation for the hydrostatic equilibrium (21), and another one for the heating-cooling relation (39) together with the boundary condition (37) that includes the angular momentum balance (31). Both differential equations contain only derivatives with respect to z, i.e., the radius r enters as a parameter in this system. Thus the two-dimensional problem of the stationary axisymmetric disk structure has been reduced to a (1+1)-dimensional set of equations. This is very similar to the non-relativistic case, the only difference appears through the relativistic functions $\mathcal{A}$, $\mathcal{B}$, $\mathcal{C}$, and $\mathcal{D}$ (eqs. (10), (15), (22), (32)) which all depend on r, M, and a, and which approach unity for large values of r. Figure 3 shows three different relativistic factors that appear in the disk structure equations.

Of course, the above two differential equations are not sufficient to determine the disk structure completely. First an energy transport equation is required. In the standard CV disk model energy can be transported either by convection or by radiative diffusion depending on the temperature and the density. For the (radiation dominated) AGN disks convection is usually negligible, especially in the inner parts, i.e., near the black hole. We therefore consider only radiation transport which will be explored in some detail in the next section. The real unknown quantity in any accretion disk model, however, is the source of viscosity, and the situation here is completely equivalent to the non-relativistic standard model. Since the molecular viscosity is too small by orders of magnitude to produce any significant mass flux, it is generally assumed that some kind of turbulence is present in the disk, probably in connection with magnetic fields. Let l_{turb} and V_{turb} be the typical length scale and velocity of a turbulent eddie then a turbulent viscosity can be estimated

$$\eta_{\mathrm{turb}} = \rho\, l_{\mathrm{turb}} V_{\mathrm{turb}} \quad,$$

and assuming that V_{turb} is limited by the local sound speed and taking $l_{\mathrm{turb}} \approx H$ leads to the famous Shakura and Sunyaev α-disk model where

$$t_{r\phi} = \alpha P \qquad \text{with} \qquad \alpha < 1 \tag{40}$$

(Shakura and Sunyaev 1973). The dimensionless parameter α takes up all the quantitative uncertainties in the above estimate of the turbulent viscosity. It can in principle depend on other dimensionless parameters of the model such as H/r or the optical depth τ, however, in most model calculations α is assumed to be constant. Now the disk model contains 4 freely adjustable parameters: M, a, $\dot{M}$, and α.

3 Radiative Transfer

The thin relativistic disk is characterized by a local balance of viscous heating and radiative cooling. The transport of radiation is described by a Boltzmann equation for a photon distribution function $f(x^\alpha, p^\beta)$ depending on the coordinates x^α and the photon 4-momentum p^β; f is related to the usual specific intensity I_ν through $f = I_\nu c^2/h^4\nu^3$. Lindquist (1966) has derived a relativistic formulation of the transfer equation by assuming that between pointlike events such as emission, absorption, or scattering photons move on null-geodesics

$$p^\alpha \frac{\partial f}{\partial x^\alpha} + \Gamma^i_{\alpha\beta}\, p^\alpha p^\beta \frac{\partial f}{\partial p^i} = Q \quad. \tag{41}$$

The index "i" on the momentum derivatives takes only 3 independent values since the momentum components are not all independent because of $p^\mu p_\mu = 0$. Here Q contains all the above interactions of photons and matter which are obtained from transition-probability calculations in atomic or molecular physics or QED. These calculations are most easily performed in the local rest frame of matter (LRFM), and consequently the corresponding cross-sections have their simplest form in the LRFM. We thus rewrite the transfer equation (41) in terms of the local tetrad components $\hat{p}^\alpha = L^\alpha_\beta p^\beta$

$$\hat{L}^\lambda_\alpha\, \hat{p}^\alpha \frac{\partial \hat{f}}{\partial x^\lambda} + \hat{\Gamma}^i_{\alpha\beta}\, \hat{p}^\alpha \hat{p}^\beta \frac{\partial \hat{f}}{\partial \hat{p}^i} = \hat{Q} \quad. \tag{42}$$

Here $\hat{f}$ is a function of x^α and $\hat{p}^\beta$, and $\hat{Q}$ contains the photon-matter interactions in the usual form (see Pomraning 1973, Mihalas 1978); an example for $\hat{Q}$ is given in Sect. 3.3. The connection coefficients $\hat{\Gamma}^i_{\mu\nu}$ can be calculated from the Ricci rotation coefficients (Misner et al. 1973) or directly from the Christoffel symbols of the Kerr metric (see Riffert 1986)

$$\hat{\Gamma}^i_{\alpha\beta} = L^i_\lambda \hat{L}^\gamma_\beta \left(\hat{L}^\sigma_\alpha \, \Gamma^\lambda_{\sigma\gamma} + \frac{\partial}{\partial x^\gamma} \hat{L}^\lambda_\alpha \right) \quad .$$

We proceed by expressing the 4-momentum $\hat{p}^\alpha$ through the frequency ν and unit 3-vector $\boldsymbol{n}$ for the photon propagation direction

$$\hat{p}^\alpha = \nu(1,\boldsymbol{n}) \quad , \qquad \boldsymbol{n} = \left(\sqrt{1-\mu^2} \sin\chi, \, \sqrt{1-\mu^2} \cos\chi, \, \mu \right)$$

where μ is the cosine of angle between the photon direction and the z-axis, and χ is a corresponding azimuthal propagation angle (note that ν, μ, and χ are variables measured in the LRFM). From that

$$
\begin{aligned}
m^i \frac{\partial \hat{f}}{\partial \hat{p}^i} &= m^0 \frac{\partial \hat{f}}{\partial \nu} + \left[m^0 \sqrt{1-\mu^2} - m^1 \sin\chi - m^2 \cos\chi \right] \frac{\sqrt{1-\mu^2}}{\mu\nu} \frac{\partial \hat{f}}{\partial \mu} \\
&\quad + \frac{-m^1 \cos\chi + m^2 \sin\chi}{\nu\sqrt{1-\mu^2}} \frac{\partial \hat{f}}{\partial \chi}
\end{aligned}
\tag{43}
$$

with

$$m^i = \hat{\Gamma}^i_{\alpha\beta} \, \hat{p}^\alpha \hat{p}^\beta = \nu^2 \left[\hat{\Gamma}^i_{00} + (\hat{\Gamma}^i_{0k} + \hat{\Gamma}^i_{k0}) n^k + \hat{\Gamma}^i_{jk} n^j n^k \right] \quad .$$

An explicit calculation of this expression using the expanded metric (9) to lowest order in z/r leads to the stationary transfer equation (i.e., for $\partial \hat{f}/\partial t = 0$):

$$
\begin{aligned}
\nu \sqrt{\mathcal{A}} \sqrt{1-\mu^2} \cos\chi \frac{\partial \hat{f}}{\partial r} &+ \nu\mu \frac{\partial \hat{f}}{\partial z} + \frac{\nu}{r\sqrt{\mathcal{B}}} \left[\sqrt{\frac{M}{r}} + \frac{\mathcal{E}}{\sqrt{\mathcal{A}}} \sqrt{1-\mu^2} \sin\chi \right] \frac{\partial \hat{f}}{\partial \phi} \\
&+ \frac{3\mathcal{A}}{2\mathcal{B}} \sqrt{\frac{M}{r^3}} (1-\mu^2) \sin\chi \cos\chi \, \nu^2 \frac{\partial \hat{f}}{\partial \nu} \\
&- \left[\frac{3\mathcal{A}}{2\mathcal{B}} \sqrt{\frac{M}{r^3}} \sqrt{1-\mu^2} \sin\chi + \frac{\mathcal{A}-1}{r\sqrt{\mathcal{A}}} \right] \nu\mu \sqrt{1-\mu^2} \cos\chi \frac{\partial \hat{f}}{\partial \mu} \\
+ \left[\frac{3\mathcal{A}}{2\mathcal{B}} \sqrt{\frac{M}{r^3}} \cos^2\chi \right. &- \frac{\sin\chi}{r\sqrt{\mathcal{A}}\sqrt{1-\mu^2}} \left. \left(1 - \frac{M}{r} - \mu^2 \mathcal{F} \right) - 2\sqrt{\frac{M}{r^3}} \right] \nu \frac{\partial \hat{f}}{\partial \chi} = \hat{Q} \tag{44}
\end{aligned}
$$

where

$$\mathcal{E} = 1 - \frac{2M}{r} + \frac{a}{r}\sqrt{\frac{M}{r}} \qquad\qquad \mathcal{F} = 1 + \frac{M}{r} - \frac{a^2}{r^2} \quad .$$

For an axisymmetric radiation field we have

$$\frac{\partial \hat{f}}{\partial \phi} = 0 \quad , \qquad \int_0^{2\pi} \sin\chi \, \hat{f} \, d\chi = 0 \quad , \tag{45}$$

and the last condition indicates that there is no radiation flux in the ϕ-direction. As pointed out in the last section in a thin disk model the z-component of the radiation flux is much larger than the radial component. Thus one can assume that $\hat{f}$ is approximately independent of χ. This assumption breaks down only in the optically thin regimes close to the inner and outer edge of the disk where the photons are able to escape rendering the photon distribution anisotropic. It turns out that this is an essential simplification because after integrating the transfer equation over χ all that is left is

$$\mu\nu^3\frac{\partial\hat{f}}{\partial z}\equiv\mu\frac{\partial\hat{I}_\nu}{\partial z}=\nu^2\hat{Q}\quad\text{with}\quad\hat{I}_\nu=\hat{I}_\nu(r,z,\nu,\mu)\tag{46}$$

where we have used the fact that $\hat{Q}$ is also independent of χ. The connection between this transfer equation and the disk structure equations from the previous section is established by noting that the radiative energy density J, the pressure P_r, and the energy flux F_z are just moments of $\hat{I}_\nu$

$$J=\int_0^\infty\int_{4\pi}\hat{I}_\nu\,d\mu d\chi d\nu\ ,\quad P_r=\int_0^\infty\int_{4\pi}\mu^2\hat{I}_\nu\,d\mu d\chi d\nu\ ,\quad F_z=\int_0^\infty\int_{4\pi}\mu\hat{I}_\nu\,d\mu d\chi d\nu\ .\tag{47}$$

Thus, integrating the transfer equation and using eq. (39) we obtain

$$\frac{\partial}{\partial z}\int_0^\infty\int_{4\pi}\mu\hat{I}_\nu\,d\mu d\chi d\nu=\frac{\partial F_z}{\partial z}=\int_0^\infty\int_{4\pi}\nu^2\hat{Q}\,d\mu d\chi d\nu=\frac{9\eta}{4}\frac{M}{r^3}\frac{\mathcal{A}^2}{\mathcal{B}^2}\tag{48}$$

with the boundary condition (see eq. (37))

$$\int_0^\infty\int_{4\pi}\mu\hat{I}_\nu\,d\mu d\chi d\nu\bigg|_{z=H}=\frac{3M\dot{M}}{8\pi r^3}\frac{\mathcal{D}}{\mathcal{B}}\ .\tag{49}$$

The above transfer equation has to be solved for every radius r; this requires boundary conditions in the midplane of the disk (at $z=0$) and at $z=H$. For symmetry reasons $\hat{I}_\nu$ is isotropic at $z=0$, and at $z=H$ a given flux is assumed to enter the disk

$$\hat{I}_\nu=I_0(\nu,\mu)\qquad\text{at }z=H\quad\text{for }\mu<0\ ;\tag{50}$$

if no flux enters the disk then $I_0=0$. As a result the coupling of the transfer equation to the disk structure relations leads to a system of differential and integral equations. The integral relation (48) is a modification of the radiation equilibrium condition encountered in stellar atmosphere models (Mihalas 1978), which determines the gas temperature, however, the flux F_z in a stellar atmosphere is constant whereas here viscosity leads to a volume heating.

3.1   Diffusion Approximation

So far the specific intensity is still a function of four variables, and this number can be reduced to three by introducing the diffusion approximation of radiative transfer (see Pomraning 1973). Assuming that $\hat{I}_\nu$ is essentially isotropic it can be approximated by

$$\hat{I}_\nu\cong\frac{1}{4\pi}\left(J_\nu+3\mu F_\nu\right)\tag{51}$$

where the spectral energy density and the spectral flux

$$J_\nu = \int_{4\pi} \hat{I}_\nu \, d\mu \quad , \qquad F_\nu = \int_{4\pi} \mu \hat{I}_\nu \, d\mu \qquad (52)$$

depend only on (r, z, ν). A consequence of this assumption is a relation between the energy density and the pressure

$$J = \int_0^\infty J_\nu \, d\nu = 3P_r \quad . \qquad (53)$$

The transfer equation is replaced by two moment equations for J_ν and F_ν

$$\begin{aligned}
\frac{\partial F_\nu}{\partial z} &= \nu^2 \int_{4\pi} \hat{Q} \, d\mu d\chi \\
\frac{1}{3} \frac{\partial J_\nu}{\partial z} &= \nu^2 \int_{4\pi} \mu \hat{Q} \, d\mu d\chi \quad .
\end{aligned} \qquad (54)$$

The boundary condition at $z = 0$ reads $F_\nu(z = 0) = 0$ whereas at $z = H$ a Marshak condition is used

$$\frac{1}{2} J_\nu - \frac{1}{4} F_\nu = F_0 \quad , \qquad F_0 = 2\pi \int_{-1}^0 \mu I_0 \, d\mu \quad . \qquad (55)$$

Of course, the integral relations (48) and (49) remain unchanged.

3.2 Transfer Function

The solution of the radiative transfer together with the disk structure equations delivers the radiated spectral flux in the LRFM at each radius r. Observationally accessible, however, is only the total flux at large distances integrated over the entire disk. The transition from the LRFM intensity to the observed intensity has to include two effects, namely (1) the Doppler boost due to the disk rotation and (2) the relativistic light deflection and redshift in the Kerr metric of the black hole. Both effects can be cast into a transfer function f_T which acts as an integration kernel to calculate the observed spectral flux (Cunningham 1975, Speith et al. 1995)

$$F_{\text{obs}}(\nu_o) = \frac{\pi}{D^2} \int_{r_i}^\infty \int_0^1 \frac{g^2}{\sqrt{g^\star(1 - g^\star)}} f_T(g^\star, r, \Theta) \, \hat{I}_\nu(r, H, \nu, \mu) \, dg^\star \, r dr \qquad (56)$$

Here D is the distance from the disk to the observer, $\nu_o = g\nu$ is the observed frequency, Θ denotes the inclination angle of the disk, and g is a redshift function depending on r, $g^\star$, and Θ. A visualization of the disk's "shape" and the redshift seen by a distant observer is shown in Fig. 4.

3.3 Numerical Calculation of Disk Spectra

In order to obtain a disk spectrum the structure equations have to be solved numerically together with the radiative transfer. First a physical model for the gas in the disk is required, and the easiest assumption is to consider a fully ionized proton-electron plasma. Consequently, the relevant interactions between the matter and the radiation are bremsstrahlung and Compton scattering. The right-hand side of eq. (42) then reads

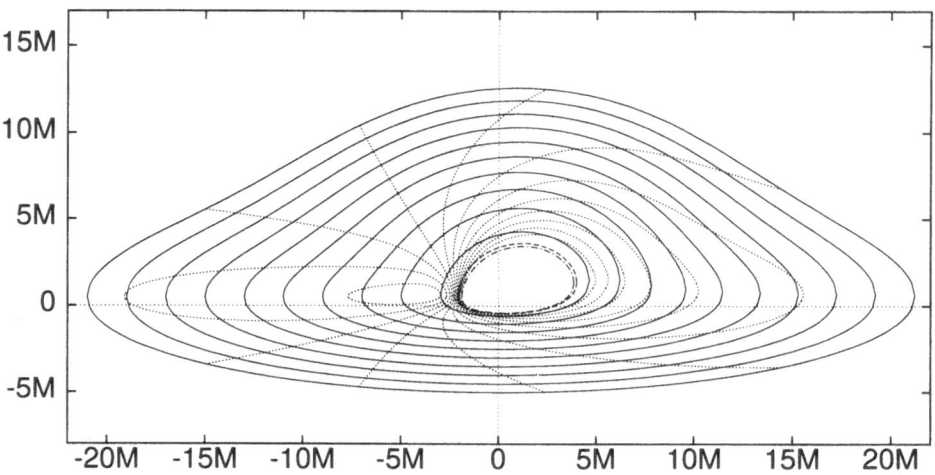

Figure 4 Visualisation of relativistic effects in the vicinity of a rotating black hole with angular momentum $a/M = 0.998$. The closed curves show the distortion of circles in the equatorial plane (i.e., lines of r = constant) as seen by a distant observer. The dotted lines crossing these curves are lines of constant redshift. The inclination angle of the disk is $\Theta = 75.5°$. The disk is seen face on for $\Theta = 0°$ (Speith et al. 1995).

$$\nu^2 \hat{Q} = \kappa_A \rho \left[B_\nu - \hat{I}_\nu \right] + \kappa_T \rho \frac{\nu}{m_e} \frac{\partial}{\partial \nu} \left[T \nu B_\nu \frac{\partial}{\partial \nu} \frac{\hat{I}_\nu}{B_\nu} + \frac{\hat{I}_\nu}{2\nu^2} \left(\hat{I}_\nu - B_\nu \right) \right] \quad . \tag{57}$$

Here κ_A is the bremsstrahlung opacity, $\kappa_T = 0.4 \text{ cm}^2/\text{g}$ stands for the Thomson opacity, and m_e is the electron mass. The differential operator describes the Compton scattering in the Fokker-Planck approximation (see Rybicki and Lightman 1979, Pomraning 1973) which is valid for non-relativistic energies ($h\nu \ll m_e c^2$) and temperatures ($kT \ll m_e c^2$). Equation (57) contains the assumption of a local thermodynamic equilibrium (LTE) where the emission of photons is proportional to $\kappa_A B_\nu$, and

$$B_\nu = \frac{2\nu^3}{e^{\nu/T} - 1}$$

is the Planck function with the local temperature $T = T(r,z)$. The relation (57) can be inserted into the transfer equation using the diffusion approximation. Thus from (54) and (48)

$$\frac{\partial F_\nu}{\partial z} = \kappa_A \rho \left[4\pi B_\nu - J_\nu \right] + \kappa_T \rho \frac{\nu}{m_e} \frac{\partial}{\partial \nu} \left[T \nu B_\nu \frac{\partial}{\partial \nu} \frac{J_\nu}{B_\nu} + \frac{J_\nu}{8\pi \nu^2} \left(J_\nu - 4\pi B_\nu \right) \right]$$

$$\frac{\partial J_\nu}{\partial z} = -3 \left(\kappa_A + \kappa_T \right) \rho F_\nu$$

$$\frac{9\eta}{4} \frac{M}{r^3} \frac{\mathcal{A}^2}{\mathcal{B}^2} = \rho \int_0^\infty \kappa_A \left[4\pi B_\nu - J_\nu \right] d\nu + \frac{\kappa_T \rho}{m_e} \int_0^\infty \left[(4T - \nu)J_\nu - \frac{J_\nu^2}{8\pi \nu^2} \right] d\nu \quad .$$

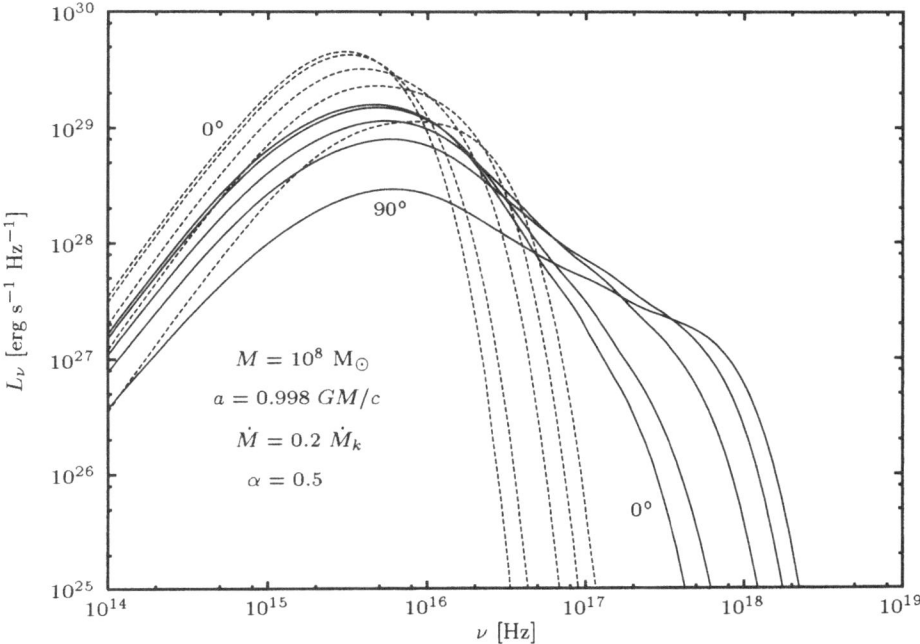

Figure 5 Calculated luminosity integrated over the entire disk for various observation angles: $\Theta = \{0°, 30°, 60°, 75°, 90°\}$. The solid lines indicate the actual emitted flux, the dotted lines correspond to a superposition of local blackbody spectra with the same flux.

Solving this set of equations together with the disk structure for a particular radius r leads to the vertical profiles of all functions such as the density, the temperature, the pressure etc. Of course, the spectral properties are also obtained, among other things the locally emitted flux, i.e., $F_\nu(r, z = H)$. The transfer function (56) is then used to integrate over the disk producing an observable spectrum for any chosen inclination angle Θ. A result of such a calculation is shown in Fig. 5 for a black hole of 10^8 M$_\odot$. Most of the flux is emitted in the UV energy range (a few times 10^{15} Hz) which is a consequence of having typical gas temperatures of 10^5 K close to the inner boundary of the disk where most of the flux is emitted. However, the spectra strongly deviate from a superposition of local blackbodies. In fact, there is a high-energy tail up to the X-ray range (2.4×10^{17} Hz corresponds to 1 keV) which is due to two effects: first, close to the inner disk boundary the local spectra contain an excess of high-energy photons, and second, the Doppler boost due to the disk rotation shifts up the photon frequencies. As a result the spectra obtained from a model of a thin accretion disk around a rotating black hole contains an energy maximum in the frequency range of the big blue bump. In addition they show a significant soft X-ray emission which can be fitted to observed X-ray data in order to obtain best-fit model parameters for M, a, $\dot{M}$, α, and Θ (Brunner et al. 1997). An example of such a fit is shown in Fig.6.

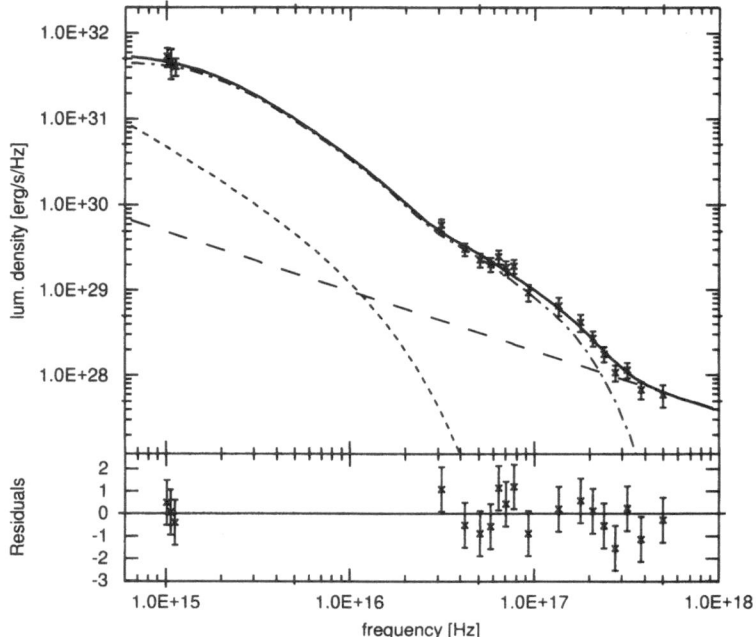

Figure 6 UV to X-ray spectrum of the radio-quiet quasar B 201. The crosses indicate ROSAT and IUE data points including 1σ error bars. The solid line is the model prediction. The contributions from the accretion disk (dot-dashed) and the hard X-ray (long dashes) and IR (short dashes) power laws are shown separately. The best-fit parameters are: $M = 4.9 \times 10^8$ $M_\odot$, $a = 0$, $\dot{M} = 0.46$ $M_\odot/a$, $\alpha = 0.95$, and $\cos\Theta = 0.76$ (from Brunner et al. 1997).

References

Abramowicz, M.A., Chen, X., Kato, S., Lasota, J.P., and Regev, O. (1995): Thermal Equilibria of Accretion Disks. *Astrophys. J. Lett.*, **438**, L37

Abramowicz, M.A. and Prasanna, A.R. (1990): Centrifugal-force reversal near a Schwarzschild Black hole. *Mon. Not. R. Astron. Soc.*, **245**, 720

Arnaud, K.A., Branduardi-Raymont, G., Culhane, J.L., Fabian, A.C., Hazard, C., Mc Glynn, T.A., Shafer, R.A., Tennant, A.F., and Ward, M.J. (1985): EXOSAT observations of a strong soft X-Ray excess in MKN 841. *Mon. Not. R. Astron. Soc.*, **217**, 105

Brunner, H., Müller, C., Friedrich, P., Dörrer, T., Staubert, R., and Riffert, H. (1997): UV to X-ray spectra of radio-quiet quasars. Comparison with accretion disk models. *Astron Astrophys.*, in press

Collin-Souffrin, S. (1994): An empirical approach to accretion disks in AGN. In: *Theory of Accretion Disks – 2* (eds. W.J. Duschl, J. Frank, F. Meyer, E. Meyer-Hofmeister, and W.M. Tscharnuter). *Nato ASI Series*, **417**, Kluwer, Dordrecht

Cunningham, C.T. (1975): The Effects of Redshifts and focusing on the Spectrum of an Accretion Disk around a Kerr Black Hole. *Astrophys. J.*, **202**, 788

Dörrer, T., Riffert, H., Staubert, R., and Ruder, H. (1996): Vertical structure and spectrum of accretion disks in active galactic nuclei. *Astron Astrophys.*, **311**, 69

Frank, J., King, A., and Raine, D. (1992): *Accretion Power*. Cambridge University Press, Cambridge

Greenhill, L.J., Jiang, D.R., Moran, J.M., Reid, M.J., Lo, K.Y., and Claussen, M.J. (1995): Detection of a subparsec Diameter Disk in the Nucleus of NGC 4258. *Astrophys. J.*, **440**, 619

Greenhill, L.J., Gwinn, C.R., Antonucci, R., and Barvainis, R. (1996): VLBI Imaging of Water Maser Emission from the Nuclear Torus of NGC 1068. *Astrophys. J. Lett.*, **472**, L21

Greenhill, L.J., Moran, J.M., and Herrnstein, J.R. (1997): The Distribution of H_2O Maser Emission in the Nucleus of NGC 4945. *Astrophys. J. Lett.*, in press

Harms, R.J., Ford, H.C., Tsvetanov, Z.I., Harting, G.F., Dressler, L.L., Kriss, G.A., Bohlin, R., Davidsen, A.F., Margon, B., and Kochhar, A.K. (1994): HST FOS Spectroscopy of M87: Evidence for a Disk of Ionized Gas around a Massive Black Hole. *Astrophys. J. Lett.*, **435**, L35

Landau, L.D. and Lifschitz, E.M. (1991): *Hydrodynamik*. Akademie Verlag, Berlin

Lichti, G.G, Balonek, T., Courvoisier, T.J.-L., Johnson, N., McConnell, M., von Montigny, C., Paciesas, W., Robson, E.I., Sadun, A., Schalinski, C., Smith, A.G., Staubert, R., Steppe, H., Swanenburg, B.N., Turner, M.J.L., Ulrich, M.-H., and Williams, O.R. (1994): Simultaneous Observations of the Continuum Emission of the Quasar 3C273 from Radio to γ-Ray Energies. In: *Multi-Wavelength Continuum Emission of AGN* (eds. T.J.-L. Courvoisier and A. Blecha). IAU Symp., **159**. Kluwer, Dordrecht

Lindquist, R.W. (1966): Relativistic Transport Theory. *Ann. Phys.*, **37**, 487

Meyer, F. and Meyer-Hofmeister, E. (1984): Outbursts in dwarf novae accretion disks. *Astron. Astrophys.*, **132**, 143

Mihalas, D. (1978): *Stellar Atmospheres*. Freeman, New York

Misner, C.W., Thorne, K.S., and Wheeler, J.A. (1973): *Gravitation*. Freeman, San Francisco

Miyoshi, M., Moran, J.M., Herrnstein, J.R., Greenhill, L.J., Nakai, N., Diamond, P., and Inoue, M. (1995): Evidence for a black hole from high rotation velocities in a sub-parsec region of NGC 4258. *Nature*, **373**, 127

Neugebauer, G., Green, R.F., Matthews, K., Schmidt, M., Soifer, B.T., and Bennet, J. (1987): Conitinuum Energy Distribution of Quasars in the Palomar-Green Survey. *Astrophys. J. Supp.*, **63**, 515

Novikov, I.D. and Thorne, K.S. (1973): Astrophysics of Black Holes. In: *Black Holes* (eds. C. DeWitt and B.S. DeWitt). Gordon & Breach, New York, p. 344

Page, D.N. and Thorne, K.S. (1974): Disk-accretion onto a Black Hole. I. Time-averaged Structure of Accretion Disk. *Mon. Not. R. Astron. Soc.*, **191**, 499

Papaloizou, J.C.B, Faulkner, J., and Lin, D.N.C. (1983): On the evolution of accretion disc flow in cataclysmic variables – II. The existence and nature of collective relaxation oscillations in dwarf nova systems. *Mon. Not. R. Astron. Soc.*, **205**, 487

Peitz J. and Appl, S. (1997): Viscous accretion discs around rotating black holes. *Mon. Not. R. Astron. Soc.*, in press

Pomraning, G.C. (1973): *The Equations of Radiation Hydrodynamics*. Pergamon Press, Oxford

Rees, M.J. (1984): Black Hole Models for Active Galactic Nuclei. *Annual Rev. Astron. Astrophys.*, **22**, 471

Riffert, H. (1986): A General Eulerian Formulation of the Comoving-frame Equation of Radiative Transfer. *Astrophys. J.*, **310**, 729

Riffert, H. and Herold H. (1995): Relativistic Accretion Disk Structure Revisited. *Astrophys. J.*, **450**, 508

Ross, R.R., Fabian, A.C., and Mineshige, S. (1992): The spectra of accretion discs in active galactic nuclei. *Mon. Not. R. Astron. Soc.*, **258**, 189

Rybicki, G.B and Lightman, A.P. (1979): *Radiative Processes in Astrophysics.* John Wiley & Sons, New York

Sanders, D.B., Phinney, E.S., Neugebauer, G., Soifer, B.T., and Matthews, K. (1989): Continuum Energy Distributions of Quasars: Shapes and Origins. *Astrophys. J.*, **347**, 29

Shakura, N.I. and Sunyaev, R.A. (1973): Black Holes in Binary Systems. Observational Appearance. *Astron. Astrophys.*, **24**, 337

Shields, G.A. (1978): Thermal continuum from accretion disks in quasars. *Nature*, **272**, 706

Shimura, T. and Takahara, F. (1993): Vertical structure and emission spectrum of an accretion disk around a massive black hole. *Astrophys. J.*, **419**, 78

Speith, R., Riffert, H., and Ruder, H. (1995): The photon transfer function for accretion disks around a Kerr black hole. *Comp. Phys. Comm.*, **88**, 109

Turner, T.J. and Pounds, K.A. (1989): The EXOSAT spectral survey of AGN. *Mon. Not. R. Astron. Soc.*, **240**, 833

Wilkes, B.J. and Elvis, M. (1987): Quasar Energy Distributions. I. Soft X-Ray Spectra of Quasars. *Astrophys. J.*, **323**, 243

Yamada, T.T., Mineshige, S., Ross, R.R., and Fukue, J. (1994): Spectrum of a Relativistic Accretion Disk in Active Galactic Nuclei. *Publ. Astron. Soc. Japan*, **46**, 553

Low-Frequency Oscillations of Relativistic Accretion Disks

James R. Ipser

Physics Department
University of Florida, Gainsville, FL 32611, USA
ipser@possum.phys.ufl.edu

Abstract

A recently developed formalism for computing the pulsations of rapidly rotating fluid configurations in general relativity is reviewed. A summary is given of results for modes of a relativistic accretion disk around a Kerr black hole. The disk's self-gravity is neglected, and attention is focused on the existence of modes with very low frequencies and their possible connection with the quasi-periodic-oscillation phenomena exhibited by certain black-hole candidates. The calculations reviewed indicate that in many cases thin relativistic disks exhibit warped precession modes with very low frequencies in the observed range. Effects associated with the dragging of inertial frames play a crucial role in determining the values of the frequencies. These effects allow an explanation, in terms of a common mechanism, for the full range of frequencies of observed quasi-periodic oscillations.

1 Introduction

The theory of general relativity has become a standard tool in a variety of research areas of modern physics and astrophysics. These would include, for example, quantum gravity, experimental gravity, cosmology, and relativistic astrophysics. The area of relativistic astrophysics, the main focus of our interest here, includes the study of black holes, compact stars, accretion disks around compact objects, and compact star clusters. Of particular interest is the structure and evolution of these relativistic systems, their interactions with their environments, their characteristic emissions, and their observable features.

In connection with the above topics, this paper will concentrate on aspects of the pulsations of relativistic systems that exhibit rapid rotation. In the past, much work has been done on the theory of the pulsations of black holes and of nonrotating, spherical stars. Rather little work has been carried out, however, on the pulsations of rapidly rotating relativistic systems other than black holes.

What is meant by the term "rapidly rotating relativistic system"? First, a system is relativistic if its Schwarzschild radius is an appreciable fraction, say at least a tenth or so, of its physical radius. This requires that

$$\frac{GM}{c^2 R} \gtrsim 0.1 \,, \tag{1}$$

where G is Newton's constant, M is the mass of the system, c is the speed of light, and R is a measure of the effective radius, or size, of the system. Second, a system is rapidly rotating if the effective centrifugal force associated with the system's rotation is an appreciable fraction of the typical gravitational force. This requires that

$$\frac{v^2}{R} \gtrsim 0.1 \frac{GM}{R^2} \,, \tag{2}$$

where v is the typical rotational speed of the system. This relation guarantees that centrifugal effects will strongly influence the structure and shape of the system. Combining eqs. (1) and (2) yields the condition

$$\frac{v^2}{c^2} \gtrsim 0.1 \frac{GM}{c^2 R} \gtrsim 0.01 \tag{3}$$

required of a rapidly rotating relativistic system.

Note that the large rotational speed of a rapidly rotating relativistic system needs not be induced by the self-gravity of the system itself. If it is, the system exhibits strong self-gravity. An example is a rapidly spinning neutron star. If it is not, the system exhibits weak self-gravity relative to the dominant source of gravity. An example is a relatively low-mass accretion disk around a black hole or a neutron star.

In this paper we shall discuss the adiabatic pulsations of rapidly rotating relativistic systems composed of perfect fluid. More specifically, we shall examine the theory of adiabatic pulsations of such systems within the context of a formalism developed recently by Ipser and Lindblom (1992). This formalism simplifies considerably the task of analyzing pulsations of rotating systems.

Section 2 reviews the general features of the formalism and its application to relativistic systems. Section 3 focuses on application of the formalism to relativistic accretion disks. It summarizes the construction of simple equilibrium models of relativistic disks, and describes how the Ipser-Lindblom formalism is used to study certain low-frequency pulsations of disks. These pulsation modes provide a possible explanation of the quasi-periodic oscillations exhibited by a variety of low-mass x-ray binaries thought to contain black holes. Section 4 describes the properties of the low-frequency modes, the way in which they are influenced by the dragging of inertial frames in general relativity, and their possible connection with observed low-frequency phenomena in systems containing accretion disks. Section 5 closes the discussion with a few concluding remarks.

2   Formalism for Analyzing Relativistic Pulsations

We restrict attention to relativistic systems composed of perfect fluid, with stress-energy tensors of the form ($G = c = 1$ throughout the remainder of the paper)

$$T^{ab} = \varrho u^a u^b + p q^{ab} \tag{4}$$

where ϱ is the fluid energy density, p is the pressure, and u^a is the fluid four-velocity. Also, the tensor $q^{ab} = g^{ab} + u^a u^b$, and it projects orthogonally to the metric g_{ab}. At equilibrium, the fluid four-velocity is taken to have the form

$$u^a = \gamma(t^a + \Omega\varphi^a), \tag{5}$$

where t^a and φ^a are the Killing vectors associated with the stationary and rotational symmetries of the unperturbed spacetime. The scalar γ is the redshift factor and Ω is the fluid angular velocity. In terms of time and angular coordinates, t and φ, and Bardeen-Wagoner cylindrical coordinates ϖ and z, the equilibrium spacetime metric has the form [see, e.g., Butterworth and Ipser (1976)]

$$ds^2 = -e^{2\nu}dt^2 + B^2 e^{-2\nu}\varpi^2(d\varphi - \alpha dt)^2 + e^{2(\zeta-\nu)}(d\varpi^2 + dz^2). \tag{6}$$

The metric functions ν, B, α, and ζ are independent of t and φ.

Our focus is on the pulsations of axisymmetric rotating equilibrium systems. Hence we look at the linear perturbations of rotating systems away from their equilibrium states. This involves writing down the perturbations of the energy-momentum conservation laws and of the Einstein field equation. The perturbation of the conservation laws (i.e. of the vanishing of the divergence of the stress energy tensor) yields equations of the form

$$(\varrho + p)(q^a{}_b u^c\nabla_c\delta\hat{u}^b + \delta\hat{u}^b\nabla_b u^a) + (\delta\varrho + \delta p)u^b\nabla_b u^a + q^{ab}\nabla_b\delta p = (\varrho + p)\delta F^a, \tag{7}$$

$$u^a\nabla_a\delta\varrho + \delta\hat{u}^a\nabla_a\varrho + (\varrho + p)\nabla_a\delta\hat{u}^a = -\frac{1}{2}(\varrho + p)u^a\nabla_a(q^{bc}\delta g_{bc}). \tag{8}$$

Here $\delta\hat{u}^a \equiv q^a{}_b\delta u^b$, the symbol δ denotes the Eulerian perturbation of a quantity, and ∇_a is the equilibrium covariant derivative. It is sufficient to note that the tensor δF^a depends linearly on the metric perturbations. Its exact form is not needed here. It should be noted, however, that it is through it that eqs. (7) and (8) are coupled to the perturbed Einstein equation

$$\delta G^{ab} = 8\pi\delta T^{ab}. \tag{9}$$

As they stand, eqs. (7) and (8), in particular, look rather complicated. It turns out, however, that one can actually use eq. (7) to simplify the analyses considerably for normal modes, which by definition have temporal and azimuthal dependence of the form $e^{i\omega t + im\varphi}$, where ω is the frequency and m is the azimuthal angular index. One achieves this simplification by noticing, for the normal modes of an axisymmetric equilibrium configuration, the validity of the key identity

$$q^b{}_d u^c\nabla_c\delta\hat{u}^d - \delta\hat{u}^c\nabla_c\delta u^b = i(\omega + m\Omega)\gamma\delta\hat{u}^b - \gamma q^b{}_d \varphi^d\delta\hat{u}^c\nabla_c\Omega. \tag{10}$$

This identity, when used in eq. (7), transforms that equation into one that is linear in $\delta\hat{u}^a$! This fortunate circumstance enables one to completely eliminate the dependent variable $\delta\hat{u}^a$ from the problem, by simply solving for it algebraically in terms of the remaining perturbation variables. The details of the procedure are spelled out in detail in Ipser and Lindblom (1992). We note here that the resulting expression for $\delta\hat{u}^a$ is of the form

$$\delta\hat{u}^a = iQ^{ab}[\nabla_b\delta U + (\varrho + p)A_b\delta U - \delta F_b], \tag{11}$$

where the tensors Q^{ab} and A^a depend only on the equilibrium configuration, and where

$$\delta U \equiv \delta p/(\varrho + p). \tag{12}$$

The exact forms of the tensors Q^{ab} and A^a are given in Ipser and Lindblom (1992). It follows that the set of dependent variables has been reduced to δU and δg_{ab}. All of the hydrodynamical information involving the pulsations is carried by the variable δU.

Our present interest is in the pulsations of an accretion disk with negligible mass in comparison with that of the central object (e.g., black hole or neutron star) about which it rotates. Hence we are behooved to adopt the Cowling approximation, which neglects the self-gravity of the perturbed configuration. In other words, we view the perturbed configuration as one that pulsates in a fixed passive gravitational field (that of the massive central object, in the case of an accretion disk). This vastly simplifies the analyses, because the metric perturbations δg_{ab} vanish in this approximation. It follows that the whole problem then reduces to that of solving a single pulsation equation of the form

$$\nabla_a[(\varrho+p)Q^{ab}\nabla_b\delta U] - Q^{ab}\nabla_a p\nabla_b\delta U + \Psi\delta U = 0\,, \tag{13}$$

where Ψ is a function given in eq. (31) of Ipser and Lindblom (1992).

3 Equilibrium Relativistic Disks

The relativistic disks whose pulsations we shall study are geometrically thin and axisymmetric, consisting of perfect fluid that rotates in the axisymmetric gravitational field of a central compact object.

Our assumption of a geometrically thin disk permits one to expand equations in powers of z/ϖ and to keep only the lowest order terms. It turns out to be convenient to work with the Kepler angular velocity Ω_K of a geodesic circular orbit in the equatorial plane of the central object. This quantity is determined by the condition

$$g_{tt,\varpi} + 2g_{t\varphi,\varpi}\Omega_K + g_{\varphi\varphi,\varpi}\Omega_K^2 = 0 \tag{14}$$

at $z = 0$. A comma, as used in this equation, denotes a partial derivative. In terms of the Kepler angular velocity, the relativistic Euler equation takes a form reminiscent of its Newtonian analogue. That is, the radial and axial components are

$$\frac{\varrho_{,\varpi}}{\varrho+p} = \frac{p_{,\varpi}}{c_s^2(\varrho+p)} = \frac{1}{2}\frac{\gamma^2}{c_s^2}[(\Omega_K^2-\Omega^2)g_{\varphi\varphi,\varpi} + 2(\Omega_K-\Omega)g_{t\varphi,\varpi}] \tag{15}$$

and

$$\frac{\varrho_{,z}}{\varrho+p} = \frac{p_{,z}}{c_s^2(\varrho+p)} = \frac{\gamma^2}{2c_s^2}(g_{tt,z} + 2\Omega g_{t\varphi,z} + \Omega^2 g_{\varphi\varphi,z}) \equiv -\gamma^2 e^{2(\zeta-\nu)}\frac{\Omega_p^2}{c_s^2}z\,, \tag{16}$$

respectively. The quantity Ω_p is the angular velocity of vertical oscillations perpendicular to the disk plane, as exhibited by a particle that has been perturbed vertically from a geodesic circular orbit. When the gravitational field is spherically symmetric, $\Omega_p = \Omega_K$. In the rotating case, frame dragging breaks this symmetry and $\Omega_p < \Omega_K$. It turns out that this symmetry breaking plays a crucial role in determining the frequencies of the pulsation modes of interest to us here. In the thin-disk approximation, the coefficients in the above equilibrium equations are evaluated at $z = 0$. These equations determine the equilibrium structure of the disk once the pressure-density relation and the angular velocity Ω have been specified.

Having constructed an equilibrium disk, we can turn to using eq. (13) to study the modes of pulsation of the disk. In a first application of eq. (13), we shall look for modes with very low frequencies, which are of interest in connection with observations of certain black-hole candidates, as will be discussed below. In the spherical limit, a circular orbit, when perturbed in the vertical direction, yields another circular orbit, which is also time independent. Hence such a perturbation corresponds to a zero-frequency perturbation. In the presence of pressure, if a whole group of circular orbits (the disk itself) is perturbed vertically, one might expect a slow time variation to be induced. Such tilt modes exhibit dipole symmetry. This motivates us to search for low-frequency modes with azimuthal index $m = 1$. We next assume that the pressure is a unique function of the density (barotropic condition), and the eigenfunction δu has the form

$$\delta u \approx \sigma \gamma z F(\varpi), \tag{17}$$

where $\sigma \equiv \omega + m\Omega$, the frequency of pulsation as measured in the frame rotating with the fluid. Modes of this form describe a disk that tilts, warps, and precesses about the rotation axis. For such modes, a straightforward, but somewhat tedious, procedure (Ipser 1996) demonstrates that the pulsation eq. (13) takes on the rather simple form

$$\frac{d}{d\varpi}\left(\varrho^* \frac{dF}{d\varpi}\right) + \frac{(\sigma^2 - \kappa^2)}{\sigma^2}\left\{e^{2(\zeta-\nu)}(\sigma^2 - \Omega_p^2)\frac{\gamma^2 \varrho^*}{c_s^2} + \left[\frac{\sigma\gamma J}{\varpi\beta} + \frac{\Lambda(\sigma\gamma)_{,\varpi}}{\sigma\gamma}\right]\varrho^*_{,\varpi}\right\}F \approx 0,$$
$$\tag{18}$$

where

$$\varrho = \varrho^*(\varpi)\exp(\frac{-z^{*2}}{2}), \quad z^* \equiv \frac{z}{H_z}, \quad H_z \equiv e^{-\zeta+\nu}\frac{c_s}{\Omega_p\gamma}. \tag{19}$$

In these equations, $c_s^2 = dp/d\rho$, the squared sound speed; κ is the radial epicyclic frequency (Ipser 1996), eq. (16); and J and Λ are certain functions of the equilibrium disk [see Ipser (1996), Section 3.2].

We are interested in modes that are trapped, or confined, within the inner regions of a disk, where most of the binding energy is liberated. This leads to the outer boundary condition that physical quantities like the pulsation energy die out in the outer regions of the disk. The inner boundary condition takes the form $F = 0$ at the inner edge of the disk, which we take to lie at the last stable circular orbit. This inner condition on F is a consequence of the requirement that the Lagrangian perturbation of the pressure vanish at the inner edge of the disk.

The central sources of present interest are black holes, rotating in general so that frame-dragging effects can be explored. Transforming the standard Boyer-Lindquist description of the Kerr geometry (Chandrasekhar 1983) yields the following expressions for the metric functions:

$$e^{2\nu} = R^2 \frac{\Delta}{\Sigma^2}, \qquad B^2 = \frac{\Delta}{\varpi^2 + z^2}, \qquad \alpha = \frac{2aMr_{BL}}{\Sigma^2},$$

$$e^{2(\zeta-\nu)} = \frac{R^2}{\varpi^2 + z^2}, \quad \Delta = r_{BL}^2 - 2Mr_{BL} + a^2, \quad R^2 = r_{BL}^2 + a^2 \cos^2\Theta,$$

$$\Sigma^2 = (r_{BL}^2 + a^2)^2 - a^2\Delta\sin^2\Theta, \quad r_{BL} = \frac{1}{4r}[(2r + M)^2 - a^2],$$

$$r = \sqrt{\varpi^2 + z^2}, \qquad \sin\Theta = \frac{\varpi}{r}. \tag{20}$$

Here M is the mass of the black hole, and a is the angular momentum per unit mass.

4   Results for Disks around Kerr Black Holes

The rotation law that we use to determine the equilibrium disk's structure is of the form

$$\Omega^2 = \Omega_K^2 \left[1 - \frac{\beta}{2M^2}(r_{BL} - r_{BL,min})(r_{BL} - r_2)(r_{BL}/r_2)^{-4} \right],$$

$$r_{BL,min}^2 - 6r_{BL,min} + 8ar_{BL,min}^{1/2} - 3a^2 = 0, \tag{21}$$

where β and r_2 are specifiable constants. This form generalizes to relativity the rotation law used in earlier pseudo-Newtonian calculations. [Pseudo-Newtonian disks involve a modified Newtonian-like gravitational potential that mimics certain relativistic effects connected with the stability of orbits; see Ipser (1994).] It permits exploration of the extent to which properties of pseudo-Newtonian disks reproduce those of relativistic disks.

A comparison of pseudo-Newtonian disks and relativistic disks is provided by Tab. 1, which exhibits frequencies of trapped modes for the case of spherical (nonrotating) black holes and for disks with $c_s = 10^{-3}$ at the inner edge of the disk. The first two columns define the rotation law (21). The column label N denotes the number of nodes in the radial eigenfunction F. The label $r_E/2M$ denotes the characteristic radius at which a mode begins to die out exponentially. The last two columns exhibit the frequencies $f = \omega/2\pi$ of relativistic modes and of corresponding pseudo-Newtonian modes. The frequencies are all positive, signifying that the modes are all retrograde in this limit. That is, in the

Corresponding Frequencies of Pseudo-Newtonian
and Relativistic Modes $(a = 0)$

$r_2/2M$	$\beta/c_{s,in}^2$	N	$r_E/2M$	$f \times (M/M_\odot)(Hz)$ relativistic$(a=0)$	pseudo$-$Newtonian
3.5	100	1	4	0.46	0.92
3.5	100	2	4.5	0.94	1.9
3.5	10			no trapped mode	
4.5	100	1	5	0.96	1.8
4.5	100	2	5	1.8	3.6
4.5	10	1	5.5	0.086	0.18
4.5	10	2	6	0.18	0.34
5.0	100	1	5.5	1.3	2.5

Table 1   The column labels are defined in the text. The speed of sound has the value 10^{-3} at the inner edge of the disk. The central black hole is nonrotating.

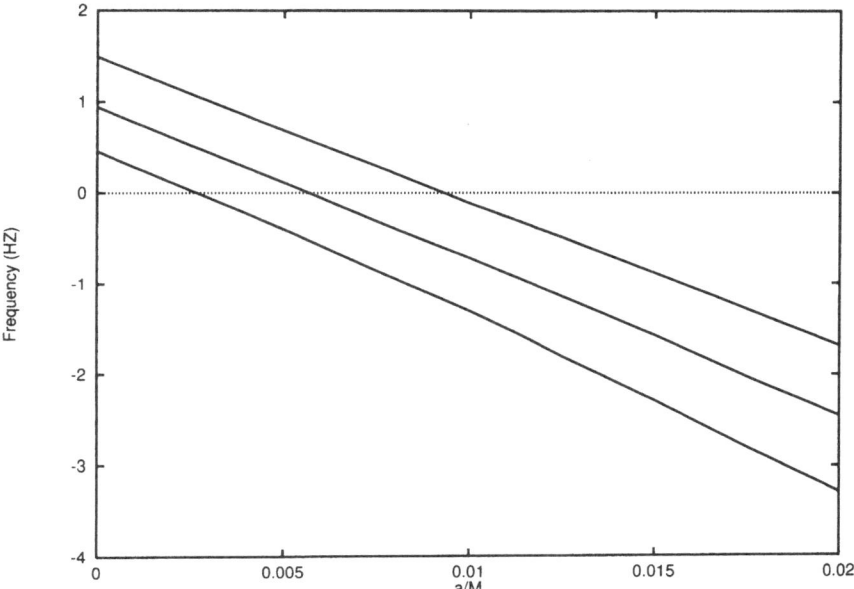

Figure 1 The lowest three modes of a disk with $c_{s,\text{in}} = 10^{-3}$ (value of speed of sound at inner edge of the disk) and with rotation-law parameter values $r_2/2M = 2.5$ and $\beta/c_{s,\text{in}}^2 = 10^2$. The vertical scale is the frequency in Hz multiplied by the black hole mass in solar units.

spherical black hole limit, the modes rotate in the direction opposite to the direction of disk rotation. It is also evident that the relativistic frequencies of warped precession modes in this limit are consistently smaller than their pseudo-Newtonian counterparts by about a factor of two. This can be traced to relativistic redshift effects. In most other ways, the pseudo-Newtonian modes mimic fairly accurately the relativistic modes, in the spherical black hole limit.

This similarity in the spherical limit is misleading, however. Significant differences, associated with the dragging of inertial frames in general relativity, emerge when the central black hole is allowed to rotate (in the same direction as the disk rotates). The perpendicular oscillation frequency Ω_p plays a key role here. Due to frame-dragging effects, the rate of rotation as sensed locally becomes progressively smaller than the rate as seen at infinity. And since Ω_p is roughly given by the former quantity, it follows that Ω_p also becomes progressively smaller than Ω or Ω_K. Study of the form of eq. (18) then leads to the expectation that frame-dragging decreases the relativistic frequency ω below its nonrotating limit value by an amount $\bar{\delta}\omega$ satisfying

$$\frac{\bar{\delta}\omega}{2\pi}\frac{M}{M_\odot} \sim -\frac{2aM}{r^3}\frac{M/M_\odot}{2\pi} \sim -10^2\,\frac{a}{M}\ \text{Hz}\,. \tag{22}$$

According to eq. (22), frame-dragging effects, even for small amounts of rotation of the central black hole, can decrease the frequencies of modes by large percentages. In addition, these effects should also be able to drag modes through zero frequency so that they become prograde. That this is precisely what happens is evident in Fig. 1, which exhibits modes of a disk with $c_{s,\text{in}} = 10^{-3}$, where $c_{s,\text{in}}$ is the speed of sound at the inner

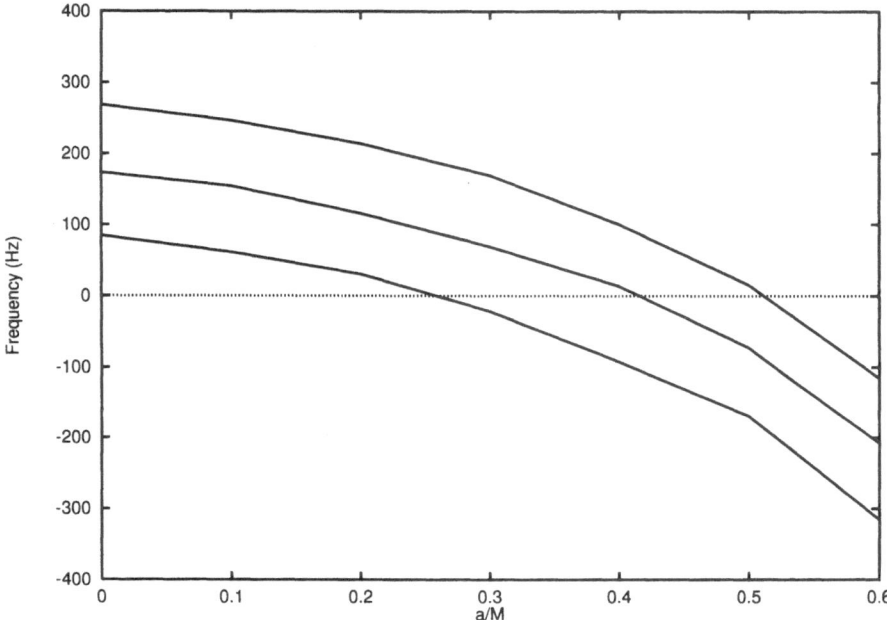

Figure 2 The lowest three modes of a disk with $c_{s,\mathrm{in}}^2 = 10^{-2}, r_2/2M = 3.5, \beta/c_{s,\mathrm{in}}^2 = 3 \times 10^2$. The vertical scale is the frequency in Hz multiplied by the black hole mass in solar units.

edge of the disk. Figure 1 exhibits the frequencies of the lowest three trapped modes of the disk. These modes have one, two, and three nodes, respectively. It is evident that the lowest mode is dragged through zero frequency, and becomes prograde, at $a/m \approx 0.005$. This agrees well with the order-of-magnitude estimate (22). It is also evident that another mode is dragged through zero frequency for each increase of a/M by an amount ~ 0.005. It turns out that all existing trapped modes become prograde for $a/M \gtrsim 0.5$. Hence for almost all smaller values of a/M, there is a mode with frequency quite close to zero.

The values of a/M at which the lowest modes pass through zero frequency can be made larger by increasing the disk temperature. This is evident in Fig. 2, which exhibits the frequencies of the lowest three modes in a hot disk with $c_{s,\mathrm{in}} = 10^{-2}$. In this disk, the lowest three modes pass through zero frequency at $a/M \approx 0.3$, 0.4, and 0.5, respectively. These results provide strong evidence that relativistic accretion disks around black holes can support warped precession modes with very low frequencies. For black holes of the order of a solar mass with disk temperatures of the order of 10^7K, the frequencies range from the order of 10 Hz down to 0. The frequencies scale inversely with the mass of the black hole and are approximately proportional to the squared sound speed and to the temperature. The dragging of inertial frames permits frequencies arbitrarily close to zero to be obtained. Hence it matters not at all whether the observed frequencies are relatively high, as in the case of the 6 Hz quasi-periodic oscillations observed in the power spectra of $GX\ 339 - 4$ (Miyamoto et al. 1991), or low, as in the case of the 0.04 Hz oscillations observed in Cyg X-1 (Vikhlinin et al. 1993) and in the case of $GRO\ J0422 + 32$ (Pietsch et al. 1993). In all such cases, frame-dragging effects allow an explanation in terms of a common mechanism.

One can argue that the problem of constructing a model with a mode frequency as small as desired is not a fine-tuning problem. Slight adjustments of the rotation law, and/or of the black-hole rotation rate, will yield a zero-frequency mode for $0.005 \leq a/M \leq 0.5$.

5 Concluding Remarks

The results discussed above provide an example of the applicability of the Ipser-Lindblom formalism to the study of relativistic pulsations. In the context of accretion disks, the calculations lend support to the interesting possibility that warped precession modes of relativistic accretion disks are the source of the phenomenon of quasi-periodic oscillations of certain x-ray binaries. This possibility should be especially exciting for the relativist, since it involves a phenomenon in which the dragging of inertial frames plays a vital role. It is hoped that applications of the formalism to other problems, such as that of the pulsations of rotating relativistic stars, will uncover additional interesting effects associated with relativistic gravity.

It is a pleasure to thank Lee Lindblom for important and illuminating comments and suggestions. This research was supported in part by grants from NASA (NAGW-4864) and from NSF (PHY-9408910).

References

Butterworth, E.M. and Ipser, J.R. (1976): On the Structure and Stability of Rapidly Rotating Fluid Bodies in General Relativity. I. The Numerical Method for Computing Structure and Its Application to Uniformly Rotating Homogeneous Bodies. *Astrophys. J.*, **204**, 200–223

Chandrasekhar, S. (1983): *The Mathematical Theory of Black Holes*. Oxford Clarendon Press

Ipser, J.R. (1994): Low-Frequency Modes and Nonbarotropic Effects in Pseudo-Newtonian Accretion Disks. *Astrophys. J.*, **435**, 767–774

Ipser, J.R. (1996): Relativistic Accretion Disks: Low-Frequency Modes and Frame Dragging. *Astrophys. J.*, **458**, 508–513

Ipser, J.R. and Lindblom, L. (1992): On the Pulsations of Relativistic Accretion Disks and Rotating Stars: the Cowling Approximation. *Astrophys. J.*, **389**, 392–399

Miyamoto, S., Kimura, K., and Kitamoto, S. (1991): X-Ray Variability of *GX* 339 – 4 in Its Very High State. *Astrophys. J.*, **383**, 784–807

Pietsch, W., Harberl, F., Gehrels, N., and Petre, R. (1993): A ROSAT Observation of the Black Hole Candidate *GRO J*0422 + 32. *Astron. Astrophys.*, **273**, L11-L14

Vikhlinin, A., Churazov, E., Gilfanov, M., Sunyaev, R., Dychkov, A., Khavenson, N., Kremnev, R., Sukhanov, K., Ballet, J., Laurent, P., Salotti, L., Claret, A., Olive, J.F., Denis, M., Mandrou, P., and Roques, J.P. (1993): Discovery of a Low-Frequency Broad Quasi-Periodic Oscillation Peak in the Power Density Spectrum of Cygnus X-1 with GRANAT/SIGMA. *Astrophys. J.*, **424**, 395–400

Relativistic Radiation Hydrodynamics and Shocks in Gamma-Ray Bursts

Peter Mészáros

Department of Astronomy
525 Davey Lab, Pennsylvania State University, University Park, PA 16802, USA
nnp@astro.psu.edu

Abstract

The arguments for a relativistic fireball model of gamma-ray bursts are reviewed, and an outline is given of the relativistic impulsive and wind flow formulations, in which dissipative shocks are thought to produce the nonthermal spectra observed. We discuss the main results obtained from analytic approximations, as well as the questions of radiative efficiency, spectrum formation, and the late evolution of impulsive or wind flows leading to gamma-ray burst remnants. We also discuss some recent numerical calculations of external deceleration shocks.

1   Motivation for Relativistic Dissipative Flows

Gamma-Ray Bursts (GRB), from observational limits on the isotropicity of their distribution [see, e.g. Fishman and Meegan (1995)] must be distributed either in a very extended galactic corona (typical distances $D \gtrsim 200$ Kpc) or else at cosmological distances. In either case, from causality considerations the dimensions must be of order $\lesssim 10^7$ cm, and the luminosities must be much higher than a solar Eddington limit. Since most of the spectral energy is observed above 0.5 MeV, the optical depth against $\gamma\gamma \to e^{\pm}$ is large, and an $e^{\pm}, \gamma$ fireball is expected. Due to the highly super-Eddington luminosity, this fireball must expand (e.g. Cavallo and Rees 1978). Since in many bursts one also observes photons representing a large fraction of the total energy at $\epsilon_\gamma \gtrsim 1 GeV$, somehow the flow must be able to avoid degrading these photons [$\gamma\gamma \to e^{\pm}$ would lead, in a stationary or slowly expanding flow, to photons just below 0.511 MeV, e.g. Harding and Baring (1994)]. In order to avoid this, it seems inescapable that the flow must be expanding with a very high Lorentz factor, since in this case relative angle at which the photons collide is less than Γ^{-1} and the threshold for the pair production is effectively diminished, so photons with energy

$$\epsilon_{\gamma,\mathrm{MeV}} \lesssim 10^4 \epsilon_{t,\mathrm{MeV}}^{-1} \Gamma_2^2 \qquad (1)$$

are able to escape, where Γ_2 is the bulk Lorentz factor in units of 10^2, and ϵ_t are the target photons, e.g. Mészáros (1995). Thus, simply from observations and general physical considerations, a relativistically expanding fireball is expected, independently of whether the GRB are at galactic corona or cosmological distances.

In addition, the γ-ray spectrum observed is generally a broken power law (see Band et al. 1993), i.e., highly nonthermal. The optically thick $e^{\pm}\gamma$ fireball cannot produce such a spectrum. In addition, the expansion would lead to a conversion of internal energy into kinetic energy of expansion, so even after the fireball becomes optically thin, it would be highly inefficient, most of the energy being in the kinetic energy of the associated protons, rather than in photons. The most likely way to achieve a nonthermal spectrum in an energetically efficient manner is if the kinetic energy of the flow is re-converted into random energy via shocks, after the flow has become optically thin. This is a plausible scenario, in which two cases can be distinguished. In the first case (a) the expanding fireball runs into an external medium [the ISM, or a pre-ejected stellar wind, Rees and Mészáros (1992), Mészáros and Rees (1993a)]. The second possibility (b) is that (even before such external shocks occur) internal shocks develop in the relativistic wind itself, faster portions of the flow catching up with the slower portions (Rees and Mészáros 1994, Paczyński and Xu 1994).

2   Energetic, Temporal and Spectral Aspects

The expanding optically thick fireball initially expands, and the flow becomes increasingly beamed into a forward cone of solid angle Γ^{-2}. In the comoving frame, however, the flow must still be isotropic, and one can see that this frame must therefore move with $\Gamma \propto r$, while the comoving temperature drops as $\propto r^{-1}$. However, if the initial impulsive event had an energy E_o and rest mass M_o in the form of entrained baryons, Γ cannot increase beyond $\Gamma_{max} \sim \eta \sim E_o/M_o c^2$, which is achieved at a radius $r/r_l \sim \eta$, beyond which the flow continues to coast with $\Gamma \sim \eta \sim$ constant (e.g. Mészáros et al. 1993):

$$\Gamma \sim \begin{cases} (r/r_l) , & \text{for } r/r_l \lesssim \eta \\ \eta , & \text{for } r/r_l \gtrsim \eta . \end{cases} \tag{2}$$

The comoving density varies as r^{-3} until $r \sim \eta r_l$, as r^{-2} until $r \sim \eta^2 r_l$ and again as r^{-3} beyond that. The matter in the expanding gas appears, to the external observer, as a geometrically thin shell of matter. For an initial isotropic distribution of expansion velocities of the matter within the initial radius r_l, the lab-frame width of the shell remains initially equal to r_l, until the velocity dispersion $\Delta v \sim (c-v)/c \sim \Gamma^{-2} c \sim \eta^{-2} c$ results in a spatial dispersion $\Delta r \sim r\Delta v/c \sim r\Gamma^{-2}$ which increases linearly with r and eventually, beyond the "broadening radius" $r_b \sim r_l \eta^2$, this dominates the initial spatial dispersion r_l. Thus the lab-frame width is

$$\Delta r \sim \max\left[r_l , r/\Gamma^2\right] \sim \begin{cases} r_l , & \text{for } r/r_l \lesssim \eta^2 \\ r/\Gamma^2 , & \text{for } r/r_l \gtrsim \eta^2 . \end{cases} \tag{3}$$

External shocks will occur in such an impulsive outflow of total energy E_o in an external medium of average particle density n_o, which begin to decelerate the fireball at the radius

$$r_{\text{dec}} \sim 10^{16} n_o^{1/3} E_{51}^{1/3} \eta_3^{-2/3} \text{ cm} , \tag{4}$$

where the lab-frame energy of the swept-up external matter ($\Gamma^2 m_p c^2$ per proton) equals the initial energy E_o of the fireball. The typical observer-frame dynamic time of the shock (assuming the cooling time is shorter than this) is $t_{\rm dec} \sim r_{\rm dec}/c\Gamma^2 \sim$ seconds, for typical parameters, and $t_b \sim t_{\rm dec}$ would be the burst duration (the impulsive assumption requires that the initial energy input occur in a time shorter than $t_{\rm dyn}$). Variability on timescales shorter than $t_{\rm dec}$ may occur on the cooling timescale or on the dynamic timescale for inhomogeneities in the external medium (Mészáros and Rees 1993a, Sari et al. 1996), but generally this is not ideal for reproducing highly variable profiles (Sari and Piran 1997). However, it can reproduce bursts with several peaks (Panaitescu and Mészáros 1997, in prep.) and may therefore be applicable to the class of long, smooth bursts (Lamb et al. 1993).

The same behavior $\Gamma \propto r$ with comoving temperature $\propto r^{-1}$, followed by saturation $\Gamma_{\rm max} \sim \eta$ at the same radius $r/r_l \sim \eta$ occurs in a wind scenario, if one assumes that a lab-frame luminosity L_o and mass outflow $\dot{M}_o$ are injected at $r \sim r_l$ and continuously maintained over a time t_w; here $\eta = L_o/\dot{M}_o c^2$ (see Paczyński 1990). In such a wind model, internal shocks will occur at a radius

$$r_{\rm dis} \sim ct_{\rm var}\eta^2 \sim 3 \times 10^{14} t_{\rm var} \eta_2^2 \text{ cm}, \qquad (5)$$

where shells of different energies $\Delta\eta \sim \eta$ initially separated by ct_v (where $t_v \leq t_w$ is the timescale of typical variations in the energy at r_l) catch up with each other. These types of models (Rees and Mészáros 1994) have the advantage that they allow an arbitrarily complicated light curve, the shortest variation timescale $t_{\rm var} \gtrsim 10^{-3}$ s being limited only by the dynamic timescale at r_l, where the energy input may be expected to vary chaotically. In order for internal shocks to occur above the wind photosphere $r_{ph} \sim \dot{M}\kappa/(4\pi c\Gamma^2) = 1.2 \times 10^{12} L_{51}\eta_2^{-3}$ cm, but also at radii greater than the saturation radius (so that most of the energy does not come out in the photospheric quasi-thermal radiation component) one needs to have $3 \times 10^1 L_{51}^{1/5} t_{\rm var}^{-1/5} \lesssim \eta 10^2 L_{51}^{1/4} t_{\rm var}^{-1/4}$ (see also Sari et al. 1996).

3   Shock Physics, Radiative Efficiency and Remnants

Optically thin shocks in the relativistic outflow are ideal sites for accelerating electrons to a power law relativistic energy distribution, which is random in the gas comoving frame, e.g. via the Fermi mechanism. Such nonthermal electron distributions can lead via the synchrotron and inverse Compton (IC) mechanisms to nonthermal photon spectra similar to those observed [e.g. Mészáros et al. (1994) for impulsive shock spectra, Papathanassiou and Mészáros (1996) for wind spectra; see also Mészáros and Rees (1994)]. Basically, two types of models are possible: those where the "break" in the 50 KeV - 2 MeV range is due to the synchrotron characteristic energy, or those where it is due to the IC upscattering of a lower energy break (typically at optical energies) which itself is due to synchrotron. The latter obviously requires smaller magnetic fields and smaller electron minimum energies γ_m. Above γ_m shock acceleration is assumed to provide the electron power law responsible for the flattish νF_ν spectrum characteristic of bursts: an electron index $p \sim -3$ reproduces this well. The burst spectra can satisfy the X-ray paucity condition (i.e. the observation that generally $F_x \lesssim 10^{-2} F_\gamma$), since below the break one expects a spectrum $\nu F_\nu \propto \nu^{4/3}$ (and spectra flatter than this can be easily obtained in an inhomogeneous magnetic field, or for a spatially varying bulk Lorentz factor). However, one does expect significant emission at GeV energies, and modest but detectable fluxes in the X-ray and optical ranges are predicted (Mészáros and Rees 1993b, Mészáros et al. 1994, Papathanassiou and Mészáros 1996).

The overall efficiency of radiation from such shocks is controlled by the mechanical efficiency of transfer of energy between baryons and electrons in the shocks (since the latter are the radiators, while the former carry most of the energy and momentum). If all electrons entering the shock are accelerated to a power law distribution, the non-thermal electron minimum energy γ_m must satisfy $1 \leq (\gamma_m/\Gamma) \leq (m_p/m_e) \sim 10^3$. The RHS limit provides the maximal radiative efficiency, and implies that electrons achieve essentially equipartition of energy with post-shock protons (as opposed to the LHS limit, which is the usual "adiabatic" shock limit where the electrons only achieve the same random velocity as the post-shock protons). In AGN, where shocks and nonthermal spectra are also observed, it is known that very high ($\gtrsim 30\%$) mechanical efficiencies are achieved. For GRB, and also for AGN, a specific mechanism has been proposed for achieving such high efficiencies (Bykov and Mészáros 1996). This is based on the expected presence in such energetic nonthermal sources of both relativistic and transrelativistic shocks which are accompanied by violent motions of moderately relativistic plasma. Very general considerations indicate that repeated scattering from MHD turbulent cells, or between multiple shocks, leads to a modified electron power law spectrum. The electron energy index is very hard, $\propto \gamma^{-1}$ or flatter. A simplified way to see this is that the acceleration time $t_{acc} \sim l/c$ (where here $l \ll L$ is the typical inter-shock or MHD scale, much smaller than the acceleration region scale L) is much shorter than the escape time $t_{esc} \sim N^{1/2}l/c$, so the accelerated particle energy spectrum implied by the usual Fermi mechanism $\phi(E) \propto E^{-(1+t_{acc}/t_{esc})} \propto E^{-1}$ is flatter than in the single shock acceleration picture (where $t_{esc} \sim t_{acc}$ and the index is near -2). The actual physics is more complicated than this, but the above crude argument gives the qualitative result. The flat lepton spectrum extends, from energy conservation, up to a comoving frame break energy which for an index -1 is $\gamma_* \sim \xi(m_p/m_e)\Gamma/\zeta$, and becomes steeper above that (where ξ is the wave or magnetic field energy as a fraction of the post-shock proton thermal energy, and ζ is the lepton injection fraction). In the example of gamma-ray bursts the Lorentz factor reaches $\gamma_* \sim 10^3$ for $e^\pm$ accelerated by the internal shock ensemble on subhydrodynamical time scales. For pairs accelerated on hydrodynamical timescales in the external shocks similarly hard spectra are obtained, and the break Lorentz factor can be as high as $\gamma_* \lesssim 10^5$. This scenario provides a plausible way to solve the crucial efficiency problem of GRB and other nonthermal high energy sources, namely the efficient transfer of energy from the proton flow to an appropriate nonthermal lepton component.

An interesting prediction of GRB models is that they should leave behind a cooling remnant. By analogy with the SNR nomenclature of supernova remnants, these may be called GRBR. The energy, typical cooling timescale and spectrum of the remnant will be different for galactic corona and for cosmological models, and within the latter, will depend on whether the burst was caused by external or internal shocks, and whether the initial burst spectral break was due to synchrotron or IC (i.e. high or low magnetic field). In general, after the GRB has occurred (on a timescale t_{dec} for external shocks, or $t_w \geq t_v$ for internal shocks), the ejecta has lost most of its initial total energy. For external models, the remnant ejecta continues to expand at a decelerating speed characterized by $\Gamma^2(r) \cdot \rho_{ext} \cdot r^3 \sim$ constant, or $\Gamma(r) \propto r^{-3/2}$. For internal models, it continues to expand at $\Gamma \sim \alpha\eta \sim$ constant, where $\alpha < 1$, up to r_{dec} where an external shock occurs, and as

$\Gamma \propto r^{-3/2}$ beyond that. The cooling GRBR spectrum is simply the GRB spectrum, but with a spectral break gradually shifting to lower energies and lower flux levels, as a power law in time (Mészáros and Rees 1997). Models based on a GRB break at MeV energies from synchrotron require a higher field, and in internal shock models (where the shock occurs closer in and the field is even higher) the cooling is so fast that no detectable soft wavelength afterglow is expected after the GRB. However for GRB models where the internal shock MeV break is due to IC, or for external shock models where the break is due to either synchrotron or IC, the cooling is slower, and fading but detectable optical fluxes (optical magnitudes in the range 15 to 20) are expected for times up to an hour after the GRB event.

4 Numerical Models of Relativistic GRB Flows and Shocks

There has been a reasonable amount of detailed numerical work on relativistic flows with modest bulk Lorentz factors of $1 \lesssim \Gamma \lesssim 10 - 20$. The free expansion of a hot gas with very high Lorentz factors ($\Gamma \sim 10^2 - 10^3$) has been numerically studied for GRB situations (Mészáros et al. 1993, see also Piran et al. 1993). A finite difference Eulerian scheme with adaptive mesh gives good results for values of Γ up to $\sim 10^5$, provided no steep gradients are present. Relativistic shocks have been numerically resolved for AGN examples in flows up to $\Gamma \sim 22$ (e.g. Marti et al. 1995), while over limited regions of space shocks in flows up to $\Gamma \sim 2000$ have been followed (Romero et al. 1996). Numerical calculations which are able to follow the full development of shocks at $\Gamma \sim 100 - 500$ over times several orders of magnitude larger than the system's crossing time were first published by Panaitescu et al. (1997) who used an Eulerian hybrid code, in the context of GRB external shock models. This code uses a finite difference scheme over most of the flow where gradients are gradual, and a Glimm algorithm based on an exact Riemann solver near the shocks. An example of the development of a relativistic external GRB shock is shown in Fig. 1. Details of the numerical procedure and tests are given in Wen et al. (1997). Calculations were carried out both in the adiabatic and in the energy loss ($t_{\rm cool} \leq t_{\rm dyn}$ limits, which essentially confirm earlier analytic estimates (e.g. Mészáros and Rees 1993a) about the typical timescales, energies and spectra. These calculations provide new information on the detailed light curves expected from external shocks. The light curves are dominated by the forward shock, if one assumes that the shocked external matter and shocked ejecta are equally efficient at generating turbulent magnetic fields at the expense of proton thermal energy. Generally, if the magnetic field and the radiated energy loss in the comoving gas frame are on average isotropic and homogeneous throughout the shocked volume, the light curves are smooth and present a FRED (fast rise-[quasi] exponential decay) profile. It is known that a fraction of the total GRB population (mainly of longer durations) is characterized by smooth profiles, and simple external shocks can be representative of many of these bursts. Bursts with a few peaks may be accommodated by external shocks at the expense of (modest) additional assumptions. However, complicated or highly variable light curves appear to be compatible only with internal shock models (Rees and Mészáros 1994).

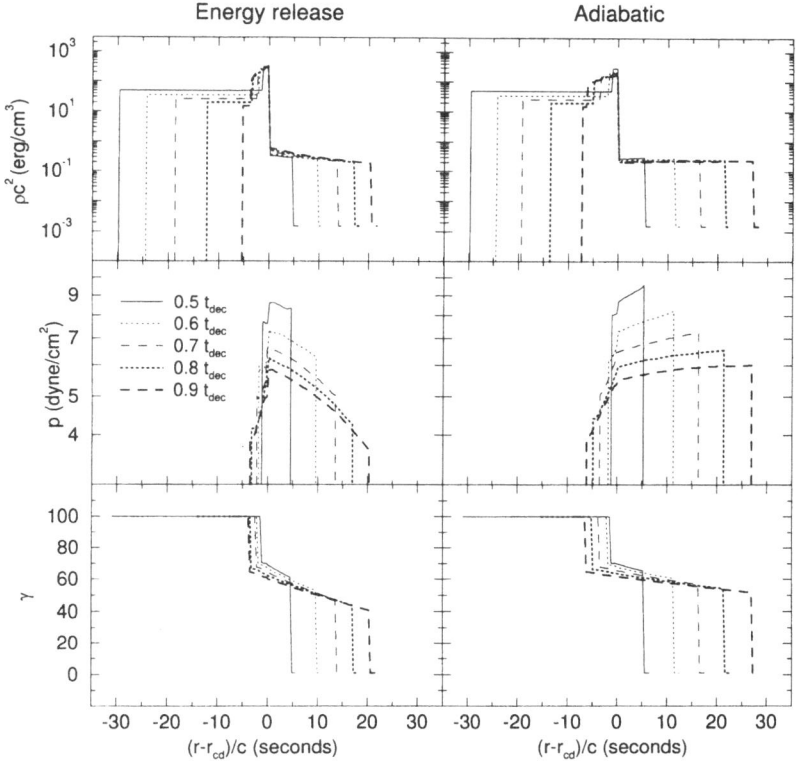

Figure 1 Density, pressure and flow Lorentz factor for $\Gamma = 100$, $E = 10^{51}$ ergs, $r_0 = 10^8$ cm and $n = 1\ \mathrm{cm}^{-3}$, at times indicated in the legend. The left column shows these profiles for the $0.25\ t_{\mathrm{dec}}$ energy release time-scale, while the right column is for the adiabatic interaction. The structure is much thinner than its curvature radius and the position inside it is indicated relative to the contact discontinuity. Negative values correspond to the inner shell, positive values to the outer shell. Note that in the adiabatic interaction the outer shell is less dense and more extended, and that the gradients in density, pressure and Lorentz factor are smaller. After $t = 0.9\ t_{\mathrm{dec}}$ the reverse shock crosses the inner shell, in both cases.

References

Band, D., Palmer, D., Teegarden, B., Cline, T., Briggs, M., Paciesas, W., Pendleton, G., Fishman, G., Kouveliotou, C., Meegan, C., Wilson, R., and Lestrade, P. (1993): Batse Observations of Gamma-Ray Bursts in Spectra. *Astrophys. J.*, **413**, 28

Bykov, A. and Mészáros, P. (1996): Electron acceleration and efficiency in nonthermal Gamma-ray sources. *Astrophys. J.*, **461**, L37

Cavallo, G. and Rees, M.J. (1978): A qualitative study of cosmic fireballs and γ-ray bursts. *Mon. Not. Roy. Astron. Soc.*, **183**, 359

Fishman, G. and Meegan, C. (1995): Gamma Ray Bursts. *Ann. Rev. Astron. Astrophys.*, **33**, 415

Harding, A.K. and Baring, M.G. (1994). In: *Gamma-ray Bursts* (eds. G. Fishman, J. Breinerd, and K. Hurley), p. 520 (AIP 307, NY)

Lamb, D.Q., Graziani, C., and Smith, I. (1993): Evidence for two distinct morphological classes of gamma-ray bursts from their short scale variability. *Astrophys. J.*, **413**, L11

Marti, J.M., Müller, E., Font, J.A., and Ibáñez, J.M. (1995): Morphology and Dynamics of Highly Supersonic Relativistic Jets. *Astrophys. J.*, **448**, L105

Mészáros, P. (1995): In: *Proc. of the 17th Texas Symp. on Relativistic Astrophysics and Cosmology* (eds. H. Böhringer, G.E. Morfill, and J.E. Trümper). p. 440, N.Y. Acad Sci.

Mészáros, P. and Rees, M.J. (1993a): Relativistic fireballs and their impact on external matter: models for cosmological gamma-ray bursts. *Astrophys. J.*, **405**, 278

Mészáros, P. and Rees, M.J. (1993b): Gamma-ray Bursts: Multiwaveband Spectral Predictions from Blast Wave Models. *Astrophys. J. Lett.*, **418**, L59

Mészáros, P. and Rees, M.J. (1994): Delayed GeV Emission from Cosmological Gamma-ray Bursts. *Mon. Not. Roy. Astron. Soc.*, **269**, L41

Mészáros, P. and Rees, M.J. (1997): Optical and Long Wavelength Afterglow from cosmological Gamma-Ray Bursts. *Astrophys. J.*, **476**, 232

Mészáros, P., Laguna, P., and Rees, M.J. (1993): Gasdynamics of relativistically expanding gamma-ray sources: kinematics, energetics, magnetic fields, and efficiency. *Astrophys. J.*, **415**, 181

Mészáros, P., Rees, M.J., and Papathanassiou, H. (1994): Spectral Properties of Blast Wave Models of Gamma-ray Burst Sources. *Astrophys. J.*, **432**, 578

Paczyński, B. (1990): Super-Eddington winds from neutron stars. *Astrophys. J.*, **363**, 218

Paczyński, B. and Xu, G. (1994): Neutrino bursts from Gamma-ray bursts. *Astrophys. J.*, **427**, 708

Panaitescu, A., Wen, L., Laguna, P., and Mészáros, P. (1997): Impact of relativistic fireballs on external matter: numerical models for cosmological γ-ray bursts. *Astrophys. J.* (in press)

Papathanassiou, H. and Mészáros, P. (1996): Spectra of Unsteady Wind Models of Gamma-Ray Bursts. *Astrophys. J. Lett.*, **471**, L91

Piran, T., Shemi, A., and Narayan, R. (1993): Hydrodynamics of relativistic fireballs. *Mon. Not. Roy. Astron. Soc.*, **263**, 861

Rees, M.J. and Mészáros, P. (1992): Relativistic fireballs: energy conversion and time-scales. *Mon. Not. Roy. Astron. Soc.*, **258**, 41

Rees, M.J. and Mészáros, P. (1994): Unsteady outflow models for cosmological Gamma-ray bursts. *Astrophys. J. Lett.*, **430**, L93

Romero, J.V., Ibáñez, J.M., Marti, J.M., and Miralles, J.A. (1996): A new spherically symmetric general relativistic hydrodynamic code. *Astrophys. J.*, **462**, 839

Sari, R., Narayan, R., and Piran, T. (1996): Cooling time scales and temporal structure of gamma-ray bursts. *Astrophys. J.* (in press)

Sari, R. and Piran, T. (1997): Cosmological GRBs: internal vs. external shocks. *Astrophys. J.* (submitted)

Wen, L., Panaitescu, A., and Laguna, P. (1997): A shock-patching code for ultra-relativistic fluid flows. *Astrophys. J.* (submitted)

Instabilities of Rotating Neutron Stars

Lee Lindblom

Theoretical Astrophysics 130-33
California Institute of Technology, Pasadena, CA 91125, USA
lindblom@tapir.caltech.edu

Abstract

This paper reviews the analysis of instabilities of rapidly rotating stellar models. Particular emphasis is given here to those instabilities driven by dissipative processes (e.g. viscosity and gravitational radiation emission) that are expected to play a significant role in influencing the observable properties of rotating neutron stars.

1 Introduction

Observations are beginning to provide a wealth of information about rapidly rotating stars in which relativistic effects play an important role. Measurements of the periods of pulsars show that neutron stars can rotate with periods as short as 1.56ms (Backer et al. 1989). Measurements of the orbital elements of binary systems containing pulsars now give accurate determinations of the masses of about ten neutron stars (Thorsett et al. 1993). To understand the meaning of these (and other related) observations, the appropriate theoretical tools must be developed for analyzing the structures and stability of rapidly rotating relativistic stellar models. The techniques for constructing and analyzing equilibrium models from a given equation of state are now well understood (Friedman et al. 1986, Cook et al. 1994). Thus it is relatively easy to compare the observable macroscopic properties (e.g. masses, angular velocities, etc.) of these models with the observations. The inverse problem of determining the poorly known equation of state from the observable properties of relativistic stars is only beginning to be understood however (Lindblom 1992).

In addition to a thorough understanding of the structures of equilibrium stellar models, the stability of these models must also be understood in order to interpret the observations. Stability theory is required, for example, to determine the ranges of masses and angular velocities present in stable (and thus physically possible) stars. This paper reviews the theory of the stability of rapidly rotating stellar models in both the Newtonian theory and general relativity. The emphasis here is on the effects of dissipation on the stability of these stars. Instabilities driven by dissipative processes may well determine the maximum rotation rates of neutron stars. The discussion here also attempts to point out issues and questions on which further analysis is needed.

Figure 1 Non-radial modes of rotating stars emit gravitational radiation. These modes are driven unstable in stars rotating sufficiently rapidly that the pattern speed changes from counter-rotating to co-rotating.

Our understanding of the stability of rotating stars was (until quite recently) based entirely on the analysis of the uniform density rigidly rotating stellar models: the Maclaurin spheroids. It has been known for over a century that rapidly rotating Maclaurin spheroids are subject to an instability driven by viscosity (Thompson and Tait 1883). This instability causes a rapidly rotating Maclaurin spheroid to evolve into a rigidly rotating but non-axisymmetric configuration such as a Jacobi ellipsoid (Roberts and Stewartson 1963, Press and Teukolsky 1973). This type of instability is referred to as *secular* since it is driven by dissipative forces in the star. The Maclaurin spheroids are also subject to a second type of secular instability that is driven by gravitational radiation (Chandrasekhar 1970a,b). This instability causes the Maclaurin spheroid to evolve into a stationary but non-axisymmetric configuration such as a Dedekind ellipsoid (Detweiler and Lindblom 1977). Maclaurin spheroids with very large angular momenta (about 1.7 times that required to trigger the viscous secular instability) are also subject to a *dynamical* instability that is driven by purely hydrodynamical forces (Chandrasekhar 1969). Secular instabilities grow on time scales proportional to the strength of the dissipative process. These secular time scales are generally much longer than the characteristic hydrodynamic time scale of the system. Dynamical instabilities grow on time scales that are comparable to the hydrodynamic time scale.

The gravitational radiation driven secular instability is of particular interest in the study of neutron stars. Neutron stars have comparatively strong gravitational fields and many of their modes couple strongly to gravitational radiation. Thus, the time scale on which the gravitational radiation instability can act in these stars is relatively short. Further, the gravitational radiation instability was shown by Friedman (1978) and Friedman and Schutz (1978) to be generic: every rotating perfect fluid star has some mode that is driven unstable by this mechanism. The physical nature of this instability mechanism can be visualized as follows. Consider the perturbation of a slowly rotating neutron star depicted in Fig. 1. The surface of the unperturbed star (viewed from above the rotation axis) is depicted as a dashed line and the perturbed surface by a solid line in this figure. The star rotates in a clockwise direction while the perturbation—rather like a wave that propagates along the surface of the star—moves in the counter-clockwise direction. This perturbation carries negative (relative to the direction of the unperturbed star's rotation) angular momentum, and emits negative angular momentum gravitational radiation to infinity. As gravitational radiation removes energy and angular momentum from the perturbed fluid, the fluid motion is damped away.

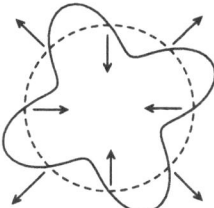

Figure 2 Non-radial modes are strongly sheared and thus strongly damped by internal fluid dissipation (e.g. viscosity).

Consider now the analogous mode in a very rapidly rotating neutron star. If the star rotates rapidly enough the pattern speed of the mode as seen from infinity would change from negative (counter-clockwise) to positive (clockwise). Under this circumstance the mode would emit positive angular momentum gravitational radiation, since it would be seen from infinity to propagate in the same direction as the star's rotation. The hydrodynamic waves themselves, however, carry negative angular momentum since they still propagate (relative to the fluid in the star) in the direction opposite the star's rotation. Angular momentum can only be conserved for this perturbation by increasing its amplitude in order to decrease its angular momentum. Thus any counter-rotating mode becomes unstable to this mechanism when the star's angular velocity reverses its propagation direction as seen from infinity. Friedman and Schutz (1978) and Friedman (1978) have shown that this happens to some mode in every rotating perfect fluid star. Contrary to the argument above, however, all rotating neutron stars are not unstable. Figure 2 illustrates why. It depicts the same perturbation as shown in Fig. 1 with the velocities of selected fluid elements near the surface indicated. At any given instant of time, those fluid elements on the leading edge of the perturbation must move radially outward as the wave travels under them. Similarly those fluid elements on the trailing edge of the wave move radially inward as the wave passes them by. This motion is highly sheared and consequently internal fluid dissipation (e.g. viscosity) tends to damp out this type of motion. Those modes which are unstable to gravitational radiation emission in very slowly rotating stars are very high order multipole modes. (Only in these modes is the pattern speed slow enough to be overcome by the star's rotation.) These modes are very strongly sheared and couple strongly to viscosity. In contrast, gravitational radiation couples only weakly to the higher order multipole moments. Thus, the presence of any viscosity will completely suppress the gravitational radiation instability in sufficiently slowly rotating stars (Lindblom and Hiscock 1983).

The secular instabilities of primary interest here all involve non-axisymmetric perturbations of rotating stars. While there exist interesting axisymmetric instabilities (e.g. those that determine the maximum and minimum masses of neutron stars) these have been relatively well understood for some time now. The interested reader is referred to the literature for a discussion of the specialized analytical techniques developed to study this type of instability [e.g. for a recent review see Lindblom (1996)]. The primary focus here is on the study of the stability of non-axisymmetric perturbations of rapidly rotating stars—a subject that is far from completely understood even now.[1] During the past

---

[1] Most of the analysis described here also applies to the particular case of axisymmetric perturbations, but that special case is not emphasized.

two decades new mathematical techniques have been developed that make it possible to analyze the stability of such perturbations in rapidly rotating stellar models composed of fluid with any realistic equation of state. These recent developments are described and reviewed in this paper. Section 2 describes certain general criteria for evaluating the stability of rotating stars with respect to general non-axisymmetric perturbations. These criteria can be applied without any specific knowledge about the properties of the normal modes of these stars. Section 3 presents the analytical techniques needed to analyze the non-axisymmetric normal modes of rotating stars in the Newtonian theory and in general relativity. For certain types of instabilities the only tools presently available for analyzing instability are the normal modes. Section 4 completes the discussion of the normal modes by showing how dissipation effects their evolution and stability.

2 General Stability Criteria

The stability of stars with respect to non-axisymmetric perturbations is an interesting and difficult subject, and this has been the focus of most of the research effort in this area in recent years. For the case of non-rotating stars, the situation turned out to be remarkably simple. In this case the stability of the star is determined completely by the quantity $S(r)$ defined by

$$S(r) = \frac{dp}{dr} - \left(\frac{\partial p}{\partial \rho}\right)_s \frac{d\rho}{dr}, \tag{1}$$

where ρ, p, and s are the density, pressure, and specific entropy of the fluid in the star, and r is the radial spherical coordinate. When S is positive the adiabatic exchange of fluid masses at different "elevations" within the star requires the addition of energy to the system (Schwarzschild 1958). When S is negative in some region, however, the energy of the configuration can be lowered by re-arranging the fluid. In this region, consequently, the stellar fluid is unstable to convection. It has been shown that the condition $S > 0$ everywhere within the star is the necessary and sufficient condition for the stability of the non-radial modes of Newtonian stellar models (Lebovitz 1965). In general relativity theory it has also been shown that the non-radial outgoing modes are stable if $S > 0$ throughout the star (Detweiler and Ipser 1973). The proof that this is also a necessary condition for stability in general relativity theory has not been completed to date.

Simple local stability conditions analogous to eq. (1) have not been found and probably do not exist for rotating stars. A few global conditions have been found, however, and these have been extremely helpful in understanding a number of interesting instabilities in rotating stars. These global conditions determine the stability of rotating stars from the properties of certain non-local functionals of the perturbations. Perhaps the most important example of such a functional is the energy E. This functional can be expressed as a Hermitian quadratic form in the perturbation fields integrated over the volume of the star.[2] For perturbations that satisfy dissipation-free (e.g. no viscosity) evolution equations the energy E is conserved for all perturbations. Thus E is not a useful tool for diagnosing the presence of dynamical instabilities. When the effects of dissipation are considered, however, the energy functional E evolves with time; and in

---

[2] The explicit expression for this functional is long, complicated, and not particularly enlightening. The interested reader is referred to the literature: Friedman and Schutz (1978) and Friedman (1978).

some circumstances it decreases monotonically for all fluid perturbations. Under these conditions E can be used to diagnose secular instabilities. If E is positive for all possible perturbations then the star is stable. The evolution equations in this case may only change E by decreasing its value toward its lower bound, zero. This ensures that the perturbation remains bounded (at least in an $\mathcal{L}^2$ sense). If the energy E were negative for some perturbation, however, then E would have no lower bound. The evolution equations would cause a perturbation with negative E to decrease without bound and the star would be unstable. The emission of gravitational radiation by a stellar perturbation causes the energy functional E to decrease. Thus E can be used to test the secular stability of rotating stars with respect to the emission of gravitational radiation. When the functional E for rotating stars is examined in detail a remarkable fact emerges: every rotating star is unstable to the emission of gravitational radiation (Friedman and Schutz 1978, Friedman 1978). That is, there exists some perturbation in every rotating star for which E is negative. An example of such a perturbation is illustrated in Fig. 1 above. When the star rotates sufficiently rapidly that the pattern speed of the wave becomes co-rotating, then the energy functional E becomes negative. The analysis of Friedman and Schutz (1978) and Friedman (1978) shows that a perturbation with negative energy E can be found for any rotating star simply by choosing the wavelength of the perturbation to be sufficiently small. Thus, every rotating star is unstable to the emission of gravitational radiation.

A closely related functional $\tilde{E}$, which represents the energy of a perturbation as measured in the co-rotating reference frame of the star, has also been useful for diagnosing instabilities in rotating stars. For Newtonian stellar models this functional has an extremely simple form:

$$\tilde{E} = \tfrac{1}{2} \int \left(\rho \, \delta v_a^* \delta v^a + \frac{\delta \rho^* \delta p}{\rho} - \delta \rho^* \delta \Phi \right) d^3x, \tag{2}$$

where ρ is the mass density, and δv^a, $\delta \rho$, δp, and $\delta \Phi$ are the perturbations in the fluid velocity, density, pressure, and gravitational potential respectively. An analogous functional is also known in the general relativistic case (Lindblom and Hiscock 1983). $\tilde{E}$ is conserved for fluid perturbations that satisfy dissipation-free evolution equations, hence it is not a useful diagnostic of dynamical instabilities. Internal fluid dissipation causes $\tilde{E}$ to decrease with time. Thus, $\tilde{E}$ can be used to diagnose secular instabilities that are driven by viscous forces in rotating stars. The study of this functional has revealed that thermal conductivity and bulk viscosity can cause the same type of secular instability as shear viscosity in rotating stars (Lindblom 1979).

The use of these energy functionals to diagnose instabilities is based on the expectation that any negative energy perturbation will grow without bound and thus represent an instability. While this is believed to be the case for each of the energy functionals discussed above, the careful mathematical analysis needed to establish this has only been completed to date for the Newtonian $\tilde{E}$ in a star having viscosity and thermal conductivity but no interaction with gravitational radiation. In this case it has been shown that $\tilde{E}$ is strictly decreasing with time unless $\tilde{E}$ vanishes (Lindblom 1983). This shows that a necessary condition for stability is that $\tilde{E} \geq 0$ for all fluid perturbations.

The effects of gravitational radiation cause the functional E to decrease with time while viscous effects cause $\tilde{E}$ to decrease. Unfortunately, neither functional is decreasing for every perturbation when both viscous and gravitational radiation effects are considered simultaneously. Thus in general neither functional (nor any known combination of them) can be used to diagnose these secular instabilities except in special cases. For very slowly rotating stars the waves with negative E that are subject to the gravitational radiation driven secular instability have very short wavelengths. These waves couple only weakly to gravitational radiation but very strongly to viscosity. Under these conditions it has been shown that the functional $\tilde{E}$ is a decreasing function of time while E is not. Thus, $\tilde{E}$ may be used to evaluate the secular stability of these perturbations while E may not. This analysis reveals that any amount of viscosity suppresses the gravitational radiation driven secular instability in sufficiently slowly rotating stars (Lindblom and Hiscock 1983).

3   Normal Modes

The analysis of the energy functional stability criteria discussed in Sect. 2 has revealed that gravitational radiation tends to make all rotating stars unstable, while viscous forces tend to suppress this instability. Unfortunately there is no known functional that always decreases with time when all of the relevant dissipative forces are present together. Thus no generally applicable test for the stability of rotating stars is presently available at all. The study of the stability of rotating stars has been directed therefore toward the study of the normal modes of rotating stars: solutions of the perturbation equations having time dependence $e^{i\omega t}$. This analysis provides a sufficient test for instability: the instability of one mode proves that the star is unstable.[3] Even the analysis of the normal modes of rotating stars turns out to be a rather difficult and interesting subject however. Considerable progress has been made in transforming this problem into a more tractable form in recent years. The analysis that leads to this simplification is simple and elegant, and so it is presented here in some detail for the simplest case of Newtonian stellar models.

In real stars the effects of dissipation are rather weak in that dissipative effects occur on time scales that are much longer than the dynamical time scale. Under these conditions it is possible to ignore the effects of dissipation as a first approximation. In this section the discussion is confined therefore to the simpler problem of the dissipation-free modes of rotating stars. The techniques for evaluating the effects of dissipation are discussed in Sect. 4.

The equations that govern the perturbations of a dissipation-free self-gravitating Newtonian fluid are given by

$$\partial_t \delta\rho + v^a \nabla_a \delta\rho + \nabla_a(\rho \delta v^a) = 0, \tag{3}$$

$$\partial_t \delta v^a + v^b \nabla_b \delta v^a + \delta v^b \nabla_b v^a = -\nabla^a \left(\frac{\delta p}{\rho} - \delta\Phi \right), \tag{4}$$

and

$$\nabla^a \nabla_a \delta\Phi = -4\pi G \delta\rho, \tag{5}$$

---

[3] Lacking a proof of the completeness of the normal modes, however, stability of all normal modes does not prove that the star is stable.

where any quantity preceded by δ represents the (Eulerian) perturbation of that quantity, while those not preceded by δ represent equilibrium values. In these equations ρ, p, Φ, and v^a represent the mass density, pressure, gravitational potential, and the fluid velocity. This system of equations is completed by specifying the thermodynamic relationship between the perturbed pressure and density. For simplicity here the equation of state is taken to be barotropic so that

$$\delta p = \frac{dp}{d\rho}\delta\rho. \tag{6}$$

The unperturbed equilibrium stellar model is assumed here to be rigidly rotating, i.e. $v^a = \Omega\varphi^a$ where Ω is the (constant) angular velocity and φ^a is the vector field representing rotations about the z^a axis.

The equations (3)–(6) that describe the perturbations of rotating stars constitute a complicated sixth-order system for the five independent components of the perturbation fields $(\delta\rho, \delta v^a, \delta\Phi)$. The solutions to these equations are known analytically only for the perturbations of uniform density stars (Bryan 1889) and have only been directly solved numerically for more realistic models quite recently (Yoshida and Eriguchi 1995). Rather than attempt to solve these equations directly, two different approaches have been devised to reduce the complexity of the equations by analytical means. The first approach introduces a potential ξ^a, the Lagrangian displacement, for the velocity perturbation:

$$\delta v^a = \partial_t \xi^a + v^b \nabla_b \xi^a - \xi^b \nabla_b v^a. \tag{7}$$

Using this potential the perturbed continuity equation (3) can be solved analytically: $\delta\rho = -\nabla_a(\rho\xi^a)$. This substitution reduces the number of independent perturbation fields to four, $(\xi^a, \delta\Phi)$, and reduces the equations that must be solved to the system (4)–(6). One nice feature of this representation of the equations is the existence of a Lagrangian from which the equations in this form may be derived (Lynden-Bell and Ostriker 1967). Unfortunately this representation also increases the order of the system of differential equations from sixth to eighth. For the purposes of actually solving the equations, this transformation does not offer much simplification. The equations have only been solved in this form (to my knowledge) numerically for the special case of axisymmetric normal modes (Clement 1981).

A second analytical transformation has been found that does significantly simplify the perturbation equations (Ipser and Managan 1985). This transformation is limited to perturbations which are normal modes with angular dependence $e^{im\varphi}$, where φ is measured about the rotation axis of the star. For this case eq. (4) reduces to

$$\left[i(\omega + m\Omega)\delta_{ab} + 2\nabla_b v_a\right]\delta v^b = -\nabla_a\left(\frac{\delta p}{\rho} - \delta\Phi\right), \tag{8}$$

where δ_{ab} represents the three-dimensional Euclidean metric (i.e., the identity matrix in Cartesian coordinates). This equation is *algebraic* in the velocity perturbation δv^a and can be solved analytically:

$$\delta v^a = iQ^{ab}\nabla_b\delta U, \tag{9}$$

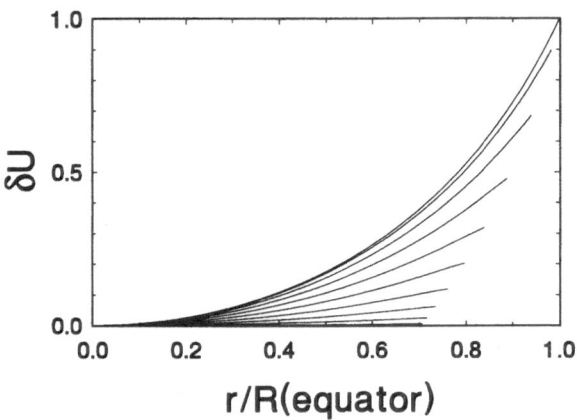

Figure 3 The eigenfunction δU for the $l = m = 3$ mode of a rapidly rotating Newtonian stellar model. Each curve represents the radial dependence of the eigenfunction along one angular spoke.

where δU is defined as

$$\delta U = \frac{\delta p}{\rho} - \delta \Phi \tag{10}$$

and Q^{ab} is the tensor

$$Q^{ab} = \frac{1}{(\omega + m\Omega)^2 - 4\Omega^2} \left[(\omega + m\Omega)\delta^{ab} - \frac{4\Omega^2}{\omega + m\Omega} z^a z^b - 2i\nabla^a v^b \right]. \tag{11}$$

Using equation (9) to replace δv^a in the remaining perturbation equations reduces the system to a pair of second-order equations for the scalar potentials δU and $\delta \Phi$:

$$\nabla_a \left(\rho Q^{ab} \nabla_b \delta U \right) = -(\omega + m\Omega)\rho \frac{d\rho}{dp}(\delta U + \delta \Phi), \tag{12}$$

$$\nabla^a \nabla_a \delta \Phi = -4\pi G \rho \frac{d\rho}{dp}(\delta U + \delta \Phi). \tag{13}$$

This transformation has reduced the equations for the modes of rotating stars to this relatively simple fourth-order system for the two scalar potentials (δU, $\delta \Phi$). These equations constitute a reasonably standard eigenvalue problem with eigenvalue ω. The tensor Q^{ab} in eq. (12) is positive definite if $(\omega + m\Omega)^2 > 4\Omega^2$, so the equation is elliptic for sufficiently slowly rotating stars. These equations can be solved for the eigenfunctions δU and $\delta \Phi$ and the eigenvalue ω using fairly standard numerical techniques (Ipser and Lindblom 1989, 1990). Figure 3 illustrates a typical eigenfunction δU for an $m = 3$ mode of a rapidly rotating Newtonian stellar model. Figure 4 illustrates the angular velocity dependence of the eigenvalue ω for two different sets of modes (Ipser and Lindblom 1990, Skinner and Lindblom 1996). The frequencies in Fig. 4 are displayed in terms of the dimensionless function $\alpha_m(\Omega)$,

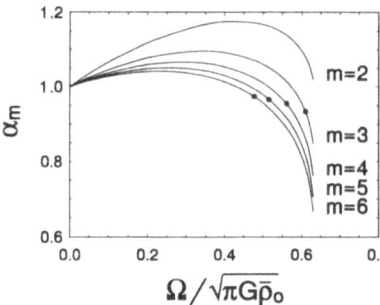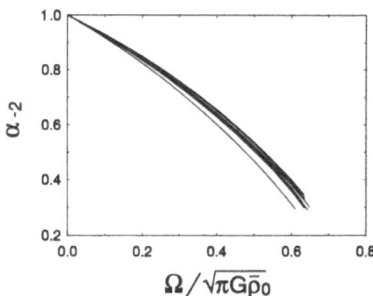

Figure 4 The angular velocity dependence of the frequencies of the modes of rapidly rotating Newtonian stellar models. The graph on the left gives the frequencies of the $l = m$ modes for $n = 1$ polytropic stellar models. The graph on the right gives the frequencies of the $l = -m = 2$ modes for stellar models constructed from thirteen realistic equations of state.

$$\alpha_m(\Omega) = \frac{\omega(\Omega) + m\Omega}{\omega(0)}, \tag{14}$$

which is normalized so that $\alpha_m(0) = 1$ for non-rotating stars.

Once the eigenfunctions δU and $\delta\Phi$ are determined, then every other physical property of the stellar oscillation may be determined from them. Equation (9) gives the velocity perturbation δv^a in terms of δU, while the density perturbation $\delta\rho$ is given by

$$\delta\rho = \rho \frac{d\rho}{dp}(\delta U + \delta\Phi), \tag{15}$$

and the Lagrangian displacement ξ^a by

$$\xi^a = \frac{Q^{ab}\nabla_b \delta U}{\omega + m\Omega}. \tag{16}$$

The particular version of the equations presented here, eqs. (12)–(13), is for the special case of barotropic perturbations of rigidly rotating stellar models. This approach can also be used to reduce the equations for the general adiabatic perturbations of differentially rotating stellar models without any restriction (e.g. barotropic) on the equation of state (Ipser and Lindblom 1991a). The equations in the general case are somewhat more complicated but remain, like eqs. (12)–(13), a fourth-order system for the two functions δU and $\delta\Phi$.

The problem of evaluating the modes of rapidly rotating stars has been rendered considerably simpler by the transformation that leads to eqs. (12)–(13). Nevertheless, there are some interesting questions that remain unresolved. The tensor Q^{ab} that appears in eq. (12) is positive definite whenever $(\omega + m\Omega)^2 > 4\Omega^2$. In this case eq. (12) is elliptic and can be solved using standard numerical techniques (Ipser and Lindblom 1990). This condition is always satisfied in non-rotating stars; however, in more rapidly rotating models it may be violated. When this condition is violated eq. (12) becomes hyperbolic yet the physical solutions must still satisfy Dirchlet boundary conditions. Little appears to be known about hyperbolic eigenvalue problems of this kind. Numerical techniques

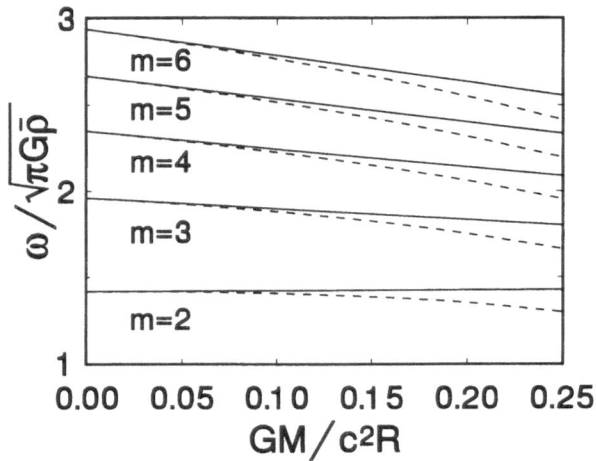

Figure 5 The post-Newtonian values for the frequencies of the modes of non-rotating stars (solid lines) are compared with the exact general relativistic values (dashed lines) for stars with different values of GM/c^2R.

based on a variational principle have been devised which give solutions to the equations even in this case however (Skinner and Lindblom 1996, Managan 1986). The change in signature of this equation does not appear to be connected to the onset of a physical instability. The physical significance of this change and the meaning of the characteristic surfaces that appear in eq. (12) are presently unknown.

In neutron stars the gravitational fields are rather strong and general relativistic effects significantly influence the structures and the dynamics. Thus it is of considerable interest to extend the analysis of the modes of rotating stars into the domain of general relativity theory. Unfortunately, this problem is extremely difficult. The chief obstacle is the coupling of these modes to gravitational radiation. In general relativity theory a star may oscillate at any frequency at all! If gravitational radiation of a given frequency were directed toward a star, then the star would oscillate at that frequency. The definition of normal modes for general relativistic stars must be refined therefore to include as an additional boundary condition that there be no incoming gravitational radiation. These solutions are referred to as the outgoing modes. This boundary condition is difficult to enforce because it must be done far away from the star in the wave zone of the gravitational radiation. This is reasonably easy to deal with in the case of non-rotating stars where the spacetime outside the unperturbed star is simply the Schwarzschild geometry (Thorne 1969, Lindblom and Detweiler 1983, Detweiler and Lindblom 1985). In rotating stars, however, the spacetimes outside the stars are only known numerically and only on rather small numerical grids. A practical method for imposing the outgoing radiation boundary condition on such spacetimes has not yet been devised. Fortunately there is a middle ground. The post-Newtonian approximation to general relativity provides a reasonably accurate description of the spacetimes associated with neutron stars. At the lowest orders the dynamics in the post-Newtonian approximation does not couple to gravitational radiation. Thus the problems associated with the outgoing radiation boundary condition does not arise in a post-Newtonian description of the modes of rotating stars.

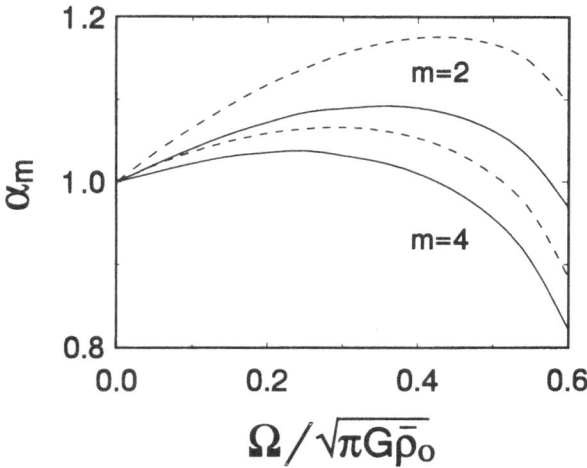

Figure 6 The angular velocity dependence of the frequencies of the modes of rotating stars. The post-Newtonian values for the frequencies (solid lines) are compared with Newtonian values (dashed lines) for stars with different angular velocities.

It is reasonably straightforward to extend the Newtonian analysis of the modes of rotating stars to the post-Newtonian theory (Cutler 1991, Cutler and Lindblom 1992). The oscillations of post-Newtonian stars are determined completely by the post-Newtonian corrections to the mode functions δU and $\delta \Phi$. These post-Newtonian eigenfunctions are determined by solving a pair of second-order equations having the same differential structures as eqs. (12)–(13) plus inhomogeneous terms that depend on δU and $\delta \Phi$ (and on the Newtonian and post-Newtonian structures of the equilibrium star). The post-Newtonian corrections to the frequency of a mode can be determined from the integrability condition for these pulsation equations, without solving the post-Newtonian pulsation equations at all! There exists an explicit formula for the post-Newtonian frequency that depends on δU and $\delta \Phi$ as well as the Newtonian and post-Newtonian structures of the star (Cutler and Lindblom 1992). As is typical of post-Newtonian analyses, this formula is extremely complicated (and unenlightening). However, it is straightforward to evaluate the needed integrals numerically and so determine the frequencies of the modes in this approximation. Figure 5 compares the frequencies of several modes of non-rotating stars computed in this post-Newtonian approximation with the exact general relativistic values. The post-Newtonian approximation for the frequencies of $1.4 M_{\odot}$ neutron stars agree with the exact general relativistic frequencies to within about 4%. In comparison, the Newtonian frequencies agree with the exact values only to within about 12% for these same neutron stars. Figure 6 illustrates the angular velocity dependencies of the frequencies of the modes of rotating stars in both the Newtonian and post-Newtonian approximations for stars with $GM/c^2R = 0.2$. The post-Newtonian frequencies for these modes differ from the Newtonian values by about 10%.

The analysis of the modes of rotating stars in full general relativity theory is far less complete. But, the general equations for these modes have been derived and a certain amount of analysis has been done with them. The general relativistic version of the Lagrangian displacement has been used to transform the equations into a simpler and more canonical form (Friedman and Schutz 1975). These equations have been very useful for analyzing the effects of general relativity on the secular instabilities of rotating stars (Friedman 1978, Lindblom and Hiscock 1983). These equations have never been solved (even numerically), however, except in the case of non-rotating stars (Thorne 1969, Lindblom and Detweiler 1983). The general relativistic version of the transformation that leads to eq. (9) has also been found. For modes with angular dependence $e^{im\varphi}$ the perturbed conservation laws, $\delta(\nabla_a T^{ab}) = 0$, can be solved analytically for the perturbed four velocity δu^a in terms of a scalar potential δU, defined by

$$\delta U = \frac{\delta p}{\rho + p}, \tag{17}$$

and the perturbed metric tensor δg_{ab} (Ipser and Lindblom 1992). The resulting equation for δu^a,

$$\delta u^a = iQ^{ab}\nabla_b \delta U + \delta F^a(\delta g_{cd}), \tag{18}$$

is the relativistic analog of eq. (9). The vector δF^a that appears in eq. (18) depends on the metric perturbation δg_{ab} and the functions that describe the unperturbed star. The tensor Q^{ab} depends on the geometry of the unperturbed star and the frequency of the mode ω. This Q^{ab} is simply the relativistic generalization of eq. (11). There is also a general relativistic analog of eq. (12) which is derived by replacing the four-velocity perturbations in the energy conservation law using eq. (18). The resulting equation has the form

$$\nabla_a[(\rho + p)Q^{ab}\nabla_b \delta U] - Q^{ab}\nabla_a p \nabla_b \delta U + \Psi \delta U = \delta F(\delta g_{ab}) \tag{19}$$

where Ψ depends on the frequency of the mode and the unperturbed structure of the star, and δF depends on δg_{ab}. This equation is particularly useful when the dynamics of a mode is driven primarily by hydrodynamic rather than gravitational forces. Such is the case for the higher-order modes of stars (Lindblom and Splinter 1990), as well as the modes of objects like accretion disks where self gravitational effects are not important. Under these circumstances the metric perturbations may be ignored and the complete dynamics of the general relativistic mode is determined by eq. (19) with $\delta F = 0$. This equation is no harder to solve in the relativistic case than it is for Newtonian stellar models. The equation in this form has been used to determine the modes of relativistic accretion disks (Ipser 1996).

4   Dissipative Effects

Dissipation plays an important role in the stability of rotating stars. The general arguments outlined in Sect. 2 show that gravitational radiation tends to make all rotating stars unstable (Friedman and Schutz 1978, Friedman 1978) while internal fluid dissipation processes (e.g. viscosity) tend to suppress this instability and make sufficiently slowly rotating stars stable (Lindblom and Hiscock 1983, Lindblom and Detweiler 1977). In this section the techniques are described which have been used to evaluate the effects of dissipation on the stability of the normal modes of rotating stars. The principal tool that is used in this analysis is the equation that determines the evolution of the energy of the perturbation due to dissipative effects. For example, the evolution of $\tilde{E}$ defined in eq. (2) can be evaluated using the equations for a dissipative Newtonian fluid including the effects of gravitational radiation reaction forces (Ipser and Lindblom 1991b):

$$
\begin{aligned}
\frac{d\tilde{E}}{dt} = & -\int \left[2\eta \delta\sigma^{ab}\delta\sigma^*_{ab} + \zeta\delta\sigma\delta\sigma^* \right] d^3x \\
& -(\omega + m\Omega)\sum_l N_l \omega^{2l+1}\delta D_{lm}\delta D^*_{lm}.
\end{aligned}
\tag{20}
$$

The thermodynamic functions η and ζ that appear on the right side of eq. (20) represent the viscosities of the fluid. The viscous forces in a fluid are driven by the shear $\delta\sigma_{ab}$ and the expansion $\delta\sigma$ of the perturbation:

$$
\delta\sigma_{ab} = \tfrac{1}{2}\left(\nabla_a \delta v_b + \nabla_b \delta v_a - \tfrac{2}{3}\delta_{ab}\nabla_c \delta v^c \right),
\tag{21}
$$

$$
\delta\sigma = \nabla_c \delta v^c.
\tag{22}
$$

The gravitational radiation reaction force couples to the evolution of the fluid via the mass multipole moments of the perturbation δD_{lm},

$$
\delta D_{lm} = \int \delta\rho\, r^l Y^*_{lm} d^3x,
\tag{23}
$$

with coupling constant N_l:

$$
N_l = \frac{4\pi G}{c^{2l+1}}\frac{(l+1)(l+2)}{l(l-1)[(2l+1)!!]^2}.
\tag{24}
$$

Now consider the normal modes of a rotating star that is subject to dissipative effects. Assume that the time dependence of the mode is $e^{i\omega t - t/\tau}$, where ω is the real part of the frequency and $1/\tau$ is the imaginary part. A mode is stable if $1/\tau$ is positive and unstable if negative. Thus the problem of evaluating the stability of a mode is reduced to determining the sign of the imaginary part of its frequency. Equation (20) provides a means of evaluating this quantity. The functional $\tilde{E}$ is real and quadratic in the perturbations, so its time dependence is $e^{-2t/\tau}$. It follows that the imaginary part of the frequency is given by

$$
\frac{1}{\tau} = -\frac{1}{2\tilde{E}}\frac{d\tilde{E}}{dt}.
\tag{25}
$$

The right side of eq. (25) is, using eqs. (2) and (20), a functional of the eigenfunction of the mode. This is an exact identity which is not however particularly useful. If the exact dissipative eigenfunctions of the star were known, then the frequency of the mode could easily be evaluated in a number of ways. Equation (25) is nevertheless an extremely useful tool for evaluating $1/\tau$ approximately. Dissipation is a relatively weak force in stars: gravitational radiation and internal fluid dissipative processes effect the evolution of the fluid in a star on time scales that are much longer than the dynamical time scale. Thus the presence of dissipation has a relatively small effect on the evolution of the fluid in a star, and so the exact eigenfunctions of a mode (including the effects of dissipation) differ only slightly from the more easily evaluated eigenfunctions based on dissipation-free hydrodynamics. Thus, the functional on the right hand side of eq. (25) has essentially the same value whether evaluated using the exact or the dissipation-free eigenfunctions. This functional is straightforward to evaluate approximately, therefore, using the dissipation-free eigenfunctions as determined in Sect. 3. This approximation is expected to give values for the imaginary part of the frequency that have fractional errors of order $\tau\omega$, the ratio of the dissipative to the dynamical time scales. Studies have shown that this ratio is extremely small in neutron stars (Cutler et al. 1990).

The imaginary part of the frequency can be evaluated numerically using eq. (25). All that is needed is the dissipation-free eigenfunction of the mode, and the thermodynamic functions η and ζ that describe the viscous forces in the stellar fluid. The viscosity coefficients have been evaluated for neutron star matter (Sawyer 1989, Cutler and Lindblom 1987), and these quantities are given approximately by

$$\zeta = 6.0 \times 10^{-59} \left(\frac{\rho}{\omega}\right)^2 T^6, \tag{26}$$

$$\eta = 6.0 \times 10^6 \left(\frac{\rho}{T}\right)^2. \tag{27}$$

Note that these viscosities depend on the thermodynamic temperature T of the star. The bulk viscosity ζ is proportional to T^6 and becomes very large when the temperature of the star is high. The shear viscosity η is proportional to T^{-2} so it becomes large when the temperature is low. These two types of viscosity are comparable in neutron stars when $T \approx 10^9$K. Viscosity tends to suppress the gravitational radiation instability in rotating stars. Hence it is clear that these viscous forces will be very effective in suppressing this instability in very hot and very cool neutron stars.

To determine which rotating stars are unstable, the imaginary parts of the frequencies of their modes must be evaluated using eq. (25). The modes with the lowest values of l and m couple most strongly to gravitational radiation, while the viscous coupling increases as l and m increase. The viscous forces tend to suppress the gravitational radiation driven secular instability. Thus, the only modes that are likely to be unstable in these stars are those with relatively small values of l and m. In practice the viscous forces are always found to suppress the gravitational instability in modes with $m \geq 6$. In sufficiently slowly rotating stars all of the modes (that have been examined) are stable. It is useful therefore to define the critical angular velocity Ω_{crit} where some mode first becomes unstable, that is where

$$0 = \frac{1}{\tau(\Omega_{\text{crit}})}. \tag{28}$$

Figure 7 illustrates the critical angular velocities for a range of neutron star temperatures (Lindblom 1995). The critical angular velocity is displayed in units of Ω_{max} the maximum angular velocity for which there exists an equilibrium stellar model. Figure 7 reveals that in very cool neutron stars, $T < 10^7$K, the critical angular velocity is identical to Ω_{max}. Thus, the viscous forces completely suppress the gravitational radiation instability in these stars. Similarly in hot neutron stars, $T > 10^{10}$K, the bulk viscosity suppresses the

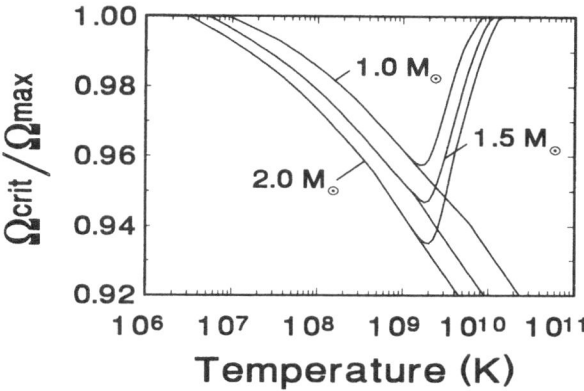

Figure 7 The temperature dependence of the critical angular velocities of neutron stars. The critical angular velocities Ω_{crit} are expressed in terms of Ω_{max} the maximum angular velocity for which there exists an equilibrium neutron star model.

instability. Only neutron stars with temperatures in the range $10^7 < T < 10^{10}$K are subject to the gravitational radiation driven secular instability. Further, this instability only occurs in the most rapidly rotating stars. Even for the most extreme case, $T \approx 2 \times 10^9$K, only those stars with angular velocities greater than about $0.96\Omega_{\text{max}}$ may be subject to the gravitational radiation driven secular instability. Figure 7 illustrates that there is only a moderate dependence of Ω_{crit} on the mass of the star. (More massive stars couple more strongly to gravitational radiation and hence have somewhat lower Ω_{crit}.) The discussion of the effects of dissipation up to this point has been based on Newtonian hydrodynamics, with the effects of gravitational radiation added as a small correction. Some work has been done, however, to estimate the effects of general relativistic dynamics on these results. Figure 8 illustrates the critical angular velocities based on a calculation

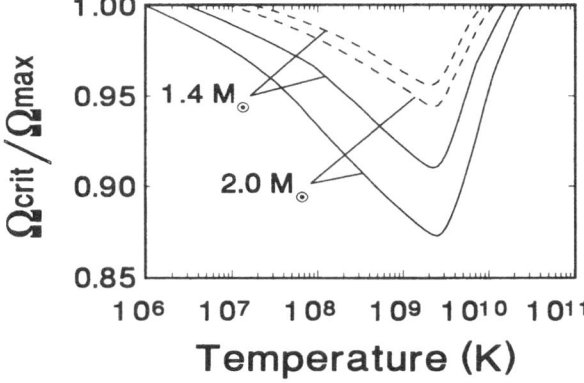

Figure 8 The temperature dependence of the critical angular velocities of neutron stars using Newtonian (dashed curves) and post-Newtonian (solid curves) gravitation and hydrodynamics.

that uses the post-Newtonian frequencies for the modes as described in Sect. 3 (Lindblom 1995). This calculation shows that post-Newtonian effects tend to enhance the gravitational radiation instability in these stars. This increases the range of temperatures where this instability may set in, and lowers the critical angular velocities to about $0.91\Omega_{max}$ in the most extreme case for $1.4M_\odot$ stars. The effects of post-Newtonian hydrodynamics on these stability results are quite striking. It illustrates the need for us to press on to a more accurate fully relativistic analysis of this problem.

The earliest studies of the secular instabilities of rotating stars were concerned with the viscosity driven instability (Thompson and Tait 1883), rather than the gravitational radiation driven instability discussed extensively here. The viscosity driven instability occurs in a different set of modes, but the formalism described here can easily be turned to study it. Such studies reveal that the viscosity driven secular instability probably does not play any role in neutron stars at all. The principle reason is that the viscosity driven secular instability only occurs in stars with very stiff equations of state. In stars with polytropic equations of state, $p = \kappa\rho^\gamma$, the adiabatic index γ must exceed 2.237 for a viscosity driven secular instability to exist at all (James 1964). The equation of state of real neutron star matter appears to be not quite stiff enough. Analysis has shown that the viscosity driven instability does not occur in any of thirteen realistic equations of state for $1.4M_\odot$ neutron star models (Skinner and Lindblom 1996, Bonazzola et al. 1996). These realistic equations of state become stiffer at higher densities, however. In a few of the stiffest equations of state, it has been found that the most massive neutron star models are subject to this instability in the most rapidly rotating models. It remains to be seen whether the actual equation of state in neutron stars is stiff enough to allow this viscosity driven instability, and whether this instability plays any role in the astrophysics of real neutron stars.

Acknowledgments: I thank James Ipser for several helpful discussions. This research was supported by NSF grant PHY-9796079 and NASA grant NAG5-4093.

References

Backer, D.C., Kulkarni, S.R., Heiles, C., Davis, M.M., and Goss, W.M. (1989): A Millisecond Pulsar. *Nature*, **300**, 615–618

Bonazzola, S., Frieben, J., and Gourgoulhon, E. (1996): Spontaneous Symmetry Breaking of Rapidly Rotating Stars in General Relativity. *Astrophys. J.*, **460**, 379–389

Bryan, G.H. (1889): The Waves on a Rotating Liquid Spheroid of Finite Ellipticity. *Phil. Trans. Roy. Soc. London*, **A180**, 187–219

Chandrasekhar, S. (1969): *Ellipsoidal Figures of Equilibrium.* Yale University Press, New Haven

Chandrasekhar, S. (1970a): Solutions of Two Problems in the Theory of Gravitational Radiation. *Phys. Rev. Letters*, **24**, 611–615

Chandrasekhar, S. (1970b): The Effect of Gravitational Radiation on the Secular Stability of the Maclaurin Spheroid. *Astrophys. J.*, **161**, 561–569

Clement, M.J. (1981): Normal Modes of Oscillation for Rotating Stars. I. The Effect of Rigid Rotation on Four Low-Order Pulsations. *Astrophys. J.*, **249**, 746–760

Cook, G.B., Shapiro, S.L., and Teukolsky, S.A. (1994): Rapidly Rotating Neutron Stars in General Relativity— Realistic Equations of State. *Astrophys. J.*, **424**, 823–845

Cutler, C. (1991): Post-Newtonian Effects on the Modes of Rotating Stars. *Astrophys. J.*, **374**, 248–254

Cutler, C. and Lindblom, L. (1987): The Effect of Viscosity on Neutron Star Oscillations. *Astrophys. J.*, **314**, 234–241

Cutler, C. and Lindblom, L. (1992): Post-Newtonian Frequencies for the Pulsations of Rapidly Rotating Neutron Stars. *Astrophys. J.*, **385**, 630–641

Cutler, C., Lindblom, L., and Splinter, R.J. (1990): Damping Times for Neutron Star Oscillations. *Astrophys. J.*, **363**, 603–611

Detweiler, S.L. and Ipser, J.R. (1973): A Variational Principle and a Stability Criterion for the Non-Radial Motions of Pulsation of Stellar Models in General Relativity. *Astrophys. J.*, **185**, 685–707

Detweiler, S.L. and Lindblom, L. (1977): On the Evolution of the Homogeneous Ellipsoidal Figures. *Astrophys. J.*, **213**, 193–199

Detweiler, S. and Lindblom, L. (1985): On the Nonradial Pulsations of General Relativistic Stellar Models. *Astrophys. J.*, **292**, 12–15

Friedman, J. (1978): Generic Instability of Rotating Relativistic Stars. *Comm. Math. Phys.*, **62**, 247–278

Friedman, J.L. and Schutz, B.F. (1975): On the Stability of Relativistic Systems. *Astrophys. J.*, **200**, 204–220

Friedman, J.L. and Schutz, B.F. (1978): Secular Instability of Rotating Newtonian Stars. *Astrophys. J.*, **222**, 281–296

Friedman, J.L., Ipser, J.R., and Parker, L. (1986): Rapidly Rotating Neutron Star Models. *Astrophys. J.*, **304**, 115–139

Ipser, J.R. (1996): Relativistic Accretion Disks: Low Frequency Modes and Frame Dragging. *Astrophys. J.*, **458** 508–513

Ipser, J.R. and Lindblom, L. (1989): Oscillations and Stability of Rapidly Rotating Neutron Stars. *Phys. Rev. Letters*, **62**, 2777–2780; Errata, **63**, 1327

Ipser, J.R. and Lindblom, L. (1990): The Oscillations of Rapidly Rotating Newtonian Stellar Models. *Astrophys. J.*, **355**, 226–240

Ipser, J.R. and Lindblom, L. (1991a): On the Adiabatic Pulsations of Accretion Disks and Rotating Stars. *Astrophys. J.*, **379**, 285–289

Ipser, J.R. and Lindblom, L. (1991b): The Oscillations of Rapidly Rotating Newtonian Stellar Models II. Dissipative Effects. *Astrophys. J.*, **373**, 213–221

Ipser, J.R. and Lindblom, L. (1992): On the Pulsations of Relativistic Accretion Disks and Rotating Stars: The Cowling Approximation. *Astrophys. J.*, **389**, 392–399

Ipser, J.R. and Managan, R.A. (1985): An Eulerian Variational Principle and a Criterion for the Occurrence of Nonaxisymmetric Neutral Modes Along Rotating Axisymmetric Sequences. *Astrophys. J.*, **292**, 517–521

James, R.A. (1964): The Structure and Stability of Rotating Gas Masses. *Astrophys. J.*, **140**, 552–582

Lebovitz, N.L. (1965): On Schwarzschild's Criterion for the Stability of Gaseous Masses. *Astrophys. J.*, **142**, 229–242

Lindblom, L. (1979): On the Secular Instability from Thermal Conductivity in Rotating Stars. *Astrophys. J.*, **233**, 974–980

Lindblom, L. (1983): Necessary Conditions for the Stability of Rotating Newtonian Stellar Models. *Astrophys. J.*, **267**, 402–408

Lindblom, L. (1992): Determining the Nuclear Equation of State from Neutron Star Masses and Radii. *Astrophys. J.*, **398**, 569–573

Lindblom, L. (1995): Critical Angular Velocities of Rotating Neutron Stars. *Astrophys. J.*, **438**, 265–268

Lindblom, L. (1996): Stellar Stability According to Newtonian Theory and General Relativity. In *Proceedings of the 14th International Conference on General Relativity and Gravitation* (eds. M. Francavagilia, G. Longhi, L. Lussana, and E. Sorace). World Scientific, Singapore, in press

Lindblom, L. and Detweiler, S.L. (1977): On the Secular Instabilities of the Maclaurin Spheroids. *Astrophys. J.*, **211**, 565–567

Lindblom, L. and Detweiler, S.L. (1983): The Quadrupole Oscillations of Neutron Stars. *Astrophys. J. Supp.*, **53**, 73–92

Lindblom, L. and Hiscock, W.A. (1983): On the Stability of Rotating Stellar Models in General Relativity Theory. *Astrophys. J.*, **267**, 384–401

Lindblom, L. and Splinter, R.J. (1990): The Accuracy of the Relativistic Cowling Approximation. *Astrophys. J.*, **348**, 198–202

Lynden-Bell, D. and Ostriker, J. (1967): On the Stability of Differentially Rotating Bodies. *Mon. Not. Roy. Astron. Soc.*, **136**, 293–310

Managan, R. (1986): Nonaxisymmetric Normal Modes and Secular Instabilities of Rotating Stars. *Astrophys. J.*, **309**, 598–608

Press, W.H. and Teukolsky, S.A. (1973): On the Evolution of the Secularly Unstable, Viscous Maclaurin Spheroids. *Astrophys. J.*, **181**, 513–517

Roberts, P.H. and Stewartson, K. (1963): On the Stability of a Maclaurin Spheroid of Small Viscosity. *Astrophys. J.*, **137**, 777–790

Sawyer, R.F. (1989): Bulk Viscosity of Hot Neutron-Star Matter and the Maximum Rotation Rates of Neutron Stars. *Phys. Rev. D*, **39**, 3804–3806

Schwarzschild, M. (1958): *Structure and Evolution of the Stars.* Dover Publications, New York

Skinner, D. and Lindblom, L. (1996): On the Viscosity Driven Secular Instability in Rotating Neutron Stars. *Astrophys. J.*, **461**, 920–926

Thompson, W. and Tait, P.G. (1883): *Principles of Natural Philosophy.* Clarendon Press, Oxford

Thorne, K.S. (1969): Nonradial Pulsation of General-Relativistic Stellar Models. III. Analytic and Numerical Results for Neutron Stars. *Astrophys. J.*, **158**, 1–16

Thorsett, S.E., Arzoumanian, Z., McKinnon, M.M., and Taylor, J.H. (1993): The Masses of Two Binary Neutron Star Systems. *Astrophys. J. Letters*, **405**, L29–L32

Yoshida, S. and Eriguchi, Y. (1995): Gravitational Radiation Driven Secular Instability of Rotating Polytropes. *Astrophys. J.*, **438**, 830–840

The Fate of Stars in the Vicinity of Supermassive Black Holes

Pablo Laguna

Department of Astronomy and Center for Gravitational Physics & Geometry
525 Davey Lab, Pennsylvania State University, University Park, PA 16802, USA
pablo@astro.psu.edu

Abstract

When a star approaches the neighborhood of a supermassive black hole, the tidal forces acting on the star trigger a vast range of physical effects that could yield unique observational signatures. Of particular interest are ultraclose passages of stars by massive black holes. These encounters could not only unveil general relativistic effects but also provide favorable conditions for the detonation of the star. This article presents a general review of the tidal stellar disruption phenomenon.

1   Introduction

The first studies of tidal disruption of stars by supermassive black holes ($M_h \geq 10^6 \, M_\odot$) were motivated by the need to find a mechanism for fuelling AGNs (Hills 1975, Frank and Rees 1976). In brief, the breakup of a star caused by the tidal forces from the black hole was suggested to be the source of the gas that feeds the hole itself. Other early studies, however, also indicated that, in order to explain the observed luminosity of AGNs (Rees 1990), a high stellar density is required in the galactic core to generate enough gas. The end result is a situation in which the gas production is dominated by star-star collisions instead of by tidal disruptions (Hills 1978, Young 1977, Young et al. 1977, Frank 1978). Even if tidally disrupted stars do not provide the main feeding mechanism of AGNs, the environment in an active galactic core containing a putative supermassive black hole is suitable for the capture and breakup of stars on nearly radial orbits, causing potentially observational effects (Rees 1988). For instance, the bound debris from the destruction of each star could yield a brief accretion flare as the material is swallowed by the hole. On the other hand, for ultraclose encounters, the dynamics of the star, as well as the ejecta, will be subject to relativistic effects and could also generate a high compression phase suitable for effectively detonating the star.

The main goal of this contribution is to provide a general picture of the different aspects of the problem of the disruption of stars by massive black holes. For details or specific results, I have made an attempt to provide the reader with a comprehensive bibliography.

2   The Tidal Stellar Disruption Problem

In general terms, the tidal breakup of a star of mass m_* and radius r_* by massive black holes can be chronologically divided into three stages or phases. The first phase involves the gravitational capture of a star by the black hole. The second stage consists of processes taking place inside the tidal or Roche radius, during which the star gets disrupted. Finally, the problem reduces to dynamics of stellar debris. The tidal or Roche radius, r_t, is essentially defined as the point at which the surface gravity of the star, Gm_*/r_*^2, equals the tidal force from the black hole, $GM_h r_*/r_t^3$. Thus,

$$r_t \simeq r_* \left(\frac{M_h}{m_*}\right)^{1/3} \simeq 23.3\, r_g \left(\frac{M_h}{10^6 M_\odot}\right)^{-2/3} \tag{1}$$

where $r_g \equiv 2GM_h/c^2$ is the Schwarzschild radius of the hole. Therefore, for black holes with masses $M_h \geq 10^8 M_\odot$, the tidal radius is inside the Schwarzschild radius of the hole, meaning that the star is swallowed by the hole without being disrupted. The strength of the tidal disruption is basically characterized by the penetration factor $\beta \equiv r_t/r_p$, with r_p the pericentric distance. Disruption occurs then for $\beta \geq 1$.

2.1   Capture of Stars

The presence of a supermassive black hole in the core of a galaxy has a fundamental influence on the orbits and spatial distribution of stars in the galaxy's inner regions (Frank and Rees 1976, Bahcall and Wolf 1976, Lightman and Shapiro 1977, Young 1980, Sigurdsson and Rees 1997). The radius of influence of a black hole of mass M_h is $r_h = GM_h/\sigma_c^2$, where $\sigma_c \simeq (Gm_* n_c r_c^2)^{1/2}$ is the characteristic (one-dimensional) velocity dispersion within the stellar cluster of number density n_c and radius r_c. Since typically one expects that $M_h \ll n_c r_c^3 m_*$, the black hole will only dominate the dynamics of the star within a radius $r_h = (M_h/m_* n_c r_c^3)r_c$.

Inside r_h, the dynamics of stars can still be perfectly described ignoring the internal structure of the star (point particle approximation) unless the star approaches the neighborhood of the black hole. At about r_t, tidal effects become relevant, and the system becomes a full hydrodynamical problem. The study of the consumption rates of stars by massive objects at the center of stellar clusters consists then on examining the manner in which stars diffuse inward from r_h and are consumed at r_t. First studies addressing consumption rates treated this diffusion problem only in energy space $-GM_h/r_t < \epsilon < -GM_h/r_h$ (Bahcall and Wolf 1976, Peebles 1972). These studies ignored the removal of stars with $\epsilon \gg -GM_h/r_t$ and low angular momentum $J \leq [2(\epsilon + GM_h/r_t)]^{1/2}$. These stars have a pericentric distance $r_p < r_t$. When these large energy small angular momentum stars are included, the problem becomes a two-dimensional problem in energy and momentum. Furthermore, viewed in terms of velocity distribution of ingoing stars with fixed energy, one can represent the capture flux of stars as scattering into a *loss-cone* in J-space (Lightman and Shapiro 1977). For estimates of the capturing rate of stars see (Frank and Rees 1976, Frank 1978, Lightman and Shapiro 1977).

2.2   Disruption of Stars

The disruption of a star flying by a massive black hole can be stated as the trade of orbital kinetic energy to overcome the star's self-binding energy. The star at this point ceases to act as a point particle, and its finite size plays an essential role.

As a star approaches the radius r_t, tidal forces will tend to stretch the star along the direction of the black hole and compress the star in directions perpendicular to it; that is, the star in principle acquires a cigar shape pointing towards the hole. However, this cigar shape never fully develops because the star's orbital motion causes a misalignment of the deformation with respect to the black hole direction. As a consequence, a tidal torque on the star $\tau \simeq GM_h r_*^2/r_p^3$ acts over the encounter duration $t_o \simeq (GM_h/r_p^3)^{-1/2}$. The specific spin angular momentum l_s is then $l_s = \tau\, t_o \simeq (r_*/r_p)^2 l_o \simeq (m_*/M_h)^{2/3}\beta^2 l_o$, with $l_o = (2GM_h r_p)^{1/2}$ the specific orbital angular momentum. The rotation due to the tidal torque induces a surface velocity $v_s = l_s/r_* \simeq \beta^{3/2} v_e$, where $v_e = (2Gm_*/r_*)^{1/2}$ is the star's surface escape velocity (Evans and Kochanek 1989). Since $\beta \geq 1$, the disruption of a star is, hence, due to the combined effect of tidal stretching and large surface velocities. Another effect due to the star's finite size is the spreading of specific binding energy to the hole of debris elements; this is because each mass element in the star probes different regions of the potential well of the black hole. This energy spread, $\Delta\epsilon$, is approximately given by the differences of the hole's potential well across the star at pericentric distance, namely $\Delta\epsilon \simeq GM_h r_*/r_p^2$. Comparison of $\Delta\epsilon$ with the specific self-binding energy of the star, $\epsilon_b = Gm_*/r_*$, and the specific orbital kinetic energy at pericentric, $\epsilon_k = GM_h/r_p$, yields

$$\Delta\epsilon \simeq \left(\frac{M_h}{m_*}\right)^{1/3}\beta^2\, \epsilon_b \simeq \left(\frac{M_h}{m_*}\right)^{-1/3}\beta\, \epsilon_k\,. \tag{2}$$

Notice that since $\beta \geq 1$ and $m_* \ll M_h$, then $\epsilon_b < \Delta\epsilon < \epsilon_k$.

2.3   Fate of Stellar Debris

Assuming an initially parabolic orbit, numerical experiments show that the ratio of bound to unbound debris after disruption is roughly unity. I will first consider the infall rate dm/dt of stellar debris bound to the hole.

The working assumption is that the differential mass distribution with respect to specific energy of the debris, $dm/d\epsilon$, is roughly constant (Rees 1988); that is, $dm/d\epsilon \approx m_*/2\Delta\epsilon$. Numerical experiments show that this is a good approximation only for mild encounters ($\beta \leq 3$) (Evans and Kochanek 1989, Laguna et al. 1993b). The rainfall rate of debris is then derived from $dm/dt = (dm/d\epsilon)(d\epsilon/dt) \approx (m_*/2\Delta\epsilon)(d\epsilon/dt)$. Using $t/2\pi = GM_h(2|\epsilon|)^{-3/2}$, one obtains an infall rate

$$\frac{dm}{dt} \approx \frac{1}{3}\frac{m_*}{t_m}\left(\frac{t}{t_m}\right)^{-5/3} \approx 3\beta^3\left(\frac{t}{t_m}\right)^{-5/3}\left(\frac{M_h}{10^6 m_\odot}\right)^{-1/2} M_\odot\,\mathrm{yr}^{-1}, \tag{3}$$

where t_m is the period of the most tightly bound gas given by

$$t_m \equiv t(-\Delta\epsilon) \approx 0.32\,\beta^{-3}\left(\frac{M_h}{10^6 m_\odot}\right)^{1/2}\mathrm{yr}. \tag{4}$$

The rainfall of material is then expected to be super-Eddington,

$$\dot{M}_{Edd} = 0.02\,\eta_{0.1}^{-1}\left(\frac{M_h}{10^6 m_\odot}\right)M_\odot \text{yr}^{-1}\,, \tag{5}$$

for radiative efficiency $\eta = 0.1\eta_{0.1}$, with this flare lasting during a period

$$\Delta t \simeq 20\,\beta^{9/5}\,\eta_{0.1}^{3/5}\left(\frac{M_h}{10^6 M_\odot}\right)^{-9/10} t_m\,. \tag{6}$$

The conversion of the infall rate (3) into a black hole swallowing rate requires considering a series of post-disruption effects (Roos 1992, Rees 1994). Among those effects are the return of debris after first passage, relativistic precession leading to self intersection of debris with different orbital periods, as well as the circularization and formation of an accretion torus. Little work has been done attempting to numerically simulate this regime (Kochanek 1994, Lee and Monaghan 1994). Because of the vast range of timescales involved (orbital circularization, radiative cooling, viscosity, etc.), the problem quickly becomes numerically expensive.

The unbound debris undergoes a less spectacular fate. The maximum specific kinetic energy of the unbound material is $\sim \Delta\epsilon$, expanding adiabatically with velocity $v_{max} = (2\Delta\epsilon)^{1/2} \simeq 6200\,\beta(M_h/10^6 M_\odot)^{1/6}$km s^{-1}. Rees (1988) pointed out that it is likely that the only direct observational effect arising from the unbound debris could occur from the interaction with the external medium. Although recently, the unbound stellar debris from tidally disrupted stars has been suggested to provide a natural explanation for the broad emission lines and UV bump observed in AGN spectra (Roos 1992).

3   Ultraclose Encounters

Stars penetrating deep inside the tidal radius ($\beta \geq 3$) are subject to extreme tidal compression that it could trigger the explosive release of energy. In addition, if the encounter has a pericentric separation $r_p \leq \text{few}\,r_g$, with $r_g \equiv 2GM_h/c^2$ the Schwarzschild radius, general relativistic effects start playing a fundamental role. I will deal with these two types of effects separately.

3.1   Detonation of Stars

Carter and Luminet (1982) were the first to notice that tidal forces compressing the star in the direction perpendicular to its orbital plane could render a short-lived, pancake star configuration such that, if the density and temperature of the star rise enough, the star undergoes detonation. This process has been also suggested as a mechanism for generating cosmic gamma-ray bursts (Carter 1992).

The original order of magnitude argument to investigate whether the conditions are favorable for thermonuclear energy release started by assuming that, at some point within the tidal radius, tidal forces on the star completely dominate over internal pressure and self-gravitational forces. During this regime, the mass elements in the star can be considered to free fall in the potential well of the black hole, with their trajectories confined to planes all intersecting at the line joining the black hole with the pericentric point of the star's center of mass. The opening angle of these planes is $\alpha \approx r_*(r_p r_t)^{-1/2} = \beta^{1/2}(M_h/m_*)^{-1/3}$. The maximum velocity, u, acquired by a mass element, viewed from the center of mass frame of the star, is then $u \approx \alpha v$, with $v = (GM_h/r_p)^{1/2} = \beta^{1/2}(M_h/m_*)^{1/3}v_*$ the orbital velocity at pericentric and $v_* = (Gm_*/r_*)^{1/2}$ the characteristic sound speed of the star. Thus, $u \approx \beta v_*$.

As the mass elements converge, the density and temperature of the star rise. The collapse is stopped eventually by the buildup of pressure forces. The problem translates into estimating how much of the kinetic energy of the mass elements is converted into internal energy of the star. Carter and Luminet assumed that all of the kinetic energy goes into internal energy of the gas. Thus, the maximum temperature, Θ_m, achieved in the core of the star is

$$\frac{\Theta_m}{\Theta_*} \approx \left(\frac{u}{v_*}\right)^2 \approx \beta^2 , \tag{7}$$

with Θ_* the unperturbed core temperature of the star. Similarly, if the gas in the star is non-relativistic ($\Theta \sim \rho^{2/3}$), the maximum density, ρ_m, at the core of the star relative to its initial core density, ρ_*, is given by

$$\frac{\rho_m}{\rho_*} \approx \left(\frac{\Theta_m}{\Theta_*}\right)^{3/2} \approx \beta^3 . \tag{8}$$

The duration of the compression phase is $\tau_m \approx \beta^{-4}(G\rho_*)^{-1/2}$, with $(G\rho_*)^{-1/2}$ the characteristic time scale of the star. Therefore, for encounters of $\beta \geq 10$, enough increase in the temperature and density is produced at the core of the star, in a sufficiently short timescale, for thermonuclear detonation to be induced (Carter and Luminet 1983, Luminet and Pichon 1989, Pichon 1985).

The above order of magnitude estimates were corroborated by Carter and Luminet themselves using an affine model of a star (Carter and Luminet 1983, 1985). Under the affine model, a star is treated as a collection of constant density ellipsoids. Deformations of these ellipsoids are constrained to be linear; thus, the star can not be followed much longer beyond its total disruption.

A debate emerged regarding the occurrence of tidal detonation. The main point of discussion is of course the conversion of collapse kinetic energy into internal energy of the star. In other words, the pancake collapse of the star is halted by a buildup of the pressure, with the eventual development of a shock. Depending on when the shock forms during the compression, shock heating could prevent the rise of density and temperature required for detonating the star. Addressing the onset of shocks and their impact on constraining the temperature and density at the core of a star requires a full numerical hydrodynamic treatment. Bicknell and Gingold (1983) first consider studying this problem using a smoothed particle hydrodynamic technique (Benz 1990). Unfortunately because of the coarse resolution of their simulations (~ 500 particles), artificial viscosity effects prevented the buildup of the density and temperature to the levels predicted by Carter and Luminet's arguments. With Miller and Zurek, I later studied the same effect, using also smooth particle techniques, but with a larger number of particles (~ 7000 at that time and more recently with ~ 80000 particles) and a less diffusive implementation of artificial viscosity (Laguna et al. 1993b). Our results were in close agreement with those of Carter and Luminet. Similar results have been obtained by Marck et al. (1996) using powerful pseudo-spectral methods. More recently, Fulbright et al. (1995) addressed the same problem applying a smooth particle hydrodynamic method based on spheroidal kernels and also found a buildup of density and temperature, though not as high as Carter and Luminet's predictions.

3.2   Relativistic Encounters

General relativistic effects will become important ($> 5\%$) for encounters with pericentric separations $r_p \leq 10\, r_g$. Such situations do not not necessarily imply deep penetration inside the tidal radius (large β). From the definition of tidal radius, Eq. (1), we have that $r_p/r_g \simeq 23.3\, \beta^{-1}\, (M_h/10^6 M_\odot)^{-2/3}$. For black hole masses $M_h > 10^7 M_\odot$, one gets that $r_p/r_g \leq 5\, \beta^{-1}$; thus even gentle encounters ($\beta \sim 1$) exhibit general relativistic effects.

In collaboration with Miller and Zurek, I developed a general relativistic version of smooth particle hydrodynamics suitable to simulate fluid flows in fixed, curved spacetime backgrounds (Laguna et al. 1993a). Based on this technique, we constructed a numerical code to study the relativistic tidal disruption problem for the case of a Schwarzschild black hole (Laguna et al. 1993b). Indeed, we found that without the inclusion of relativistic effects, not only the central density and temperature of the star at maximum compression were underestimated, but also the structure of the ejecta. Because each mass element in the star undergoes its own relativistic precession, the ejected debris acquires the shape of a radially thin, expanding, circular band. This configuration is completely missed if a Newtonian calculation is performed instead.

Another important feature only present if relativistic effects are included is the occurrence of multiple compressions. This effect was first noticed by Luminet and Mark (1985). They found, using the affine star model, that if, again because of relativistic precession, the center of mass trajectory of the star self-intersects inside the tidal radius, the dynamics of the mass elements of the star admit multiple phases of expansion and compression. Later, this effect was demonstrated using full hydrodynamic simulations (Laguna et al. 1993b).

Only recently, the generalization of relativistic encounters to rotating black holes has been considered (Diener et al. 1997, Laguna et al. 1997). The main differences in this case, with respect to the non-rotating one, are the changes of the central density and temperatures depending on the relative orientation between the orbital angular momentum of the star and the spin angular momentum of the hole. An interesting extension to these studies would be to investigate the Lense-Thirring precession of the debris.

4   Conclusions

In spite of the considerable amount of the work done on the problem of the tidal disruption of stars by supermassive black holes, there are still unanswered questions. Of particular interest is a detailed study of the fate of bound debris. Such work should provide a better understanding of the duration and spectral properties of the flare produced when the bound material rains down into the black hole. Other point requiring detailed attention is the development of shocks during ultraclose encounters. The question is not whether they are present or not, but when they form. Early shock production will invalidate Carter and Luminet's predictions. Unfortunately, both the late stages of tidal disruption and tidal detonation require a considerable investment of computer resources, in addition to development of more accurate numerical techniques. I will end with a most likely incomplete list of the numerical, hydrodynamical studies of the tidal stellar disruption problem (Evans and Kochanek 1989, Laguna et al. 1993b, Lee and Monaghan 1994, Bicknell and Gingold 1983, Marck et al. 1996, Fulbright et al. 1995, Regev and Portnoy 1987, Khokhlov et al. 1993a,b, Novikov et al. 1992).

Acknowledgment: It is a pleasure to thank my collaborators Warner Miller, Mike Warren and Wojciech Zurek. I would also like to thank H. Ruder, H.-P. Nollert, H. Riffert and F.W. Hehl for their kind invitation to the WE-Heraeus-Seminar. This research has been supported in part by DOE through Los Alamos National Laboratory and NSF through grants PHY-9309834, PHY-9357219 (NYI).

References

Bahcall, J.N. and Wolf, R.A. (1976): Star Distribution around a Massive Black Hole in a Globular Cluster. *Astrophys. J.*, **209**, 214

Benz, W. (1990): Numerical Modeling of Stellar Pulsation: Problems & Prospects (ed. J.R. Buchler). Kluwer, Dordrecht

Bicknell, G.V. and Gingold, R.A. (1983): On Tidal Detonation of Stars by Massive Black Holes. *Astrophys. J.*, **273**, 749

Carter, B. (1992): Cosmic Gamma-Ray Bursts from Black Hole Tidal Disruption of Stars? *Astrophys. J.*, **391**, L67

Carter, B. and Luminet, J. P. (1982): Pancake detonation of stars by black holes in galactic nuclei. *Nature*, **296**, 211

Carter, B. and Luminet, J.P. (1983): Tidal Compression of a Star by a Large Black Hole. *Astron. Astrophys.*, **121**, 97

Carter, B. and Luminet, J.P. (1985): Mechanics of the Affine Star Model. *Mon. Not. Roy. Astron. Soc.*, **212**, 23

Diener, P., Frolov, V.P., Khokhlov, A.M., Novikov, I.D., and Pethick, C.J. (1997): Relativistic Tidal Interactions of Stars with a Rotating Black Hole. *Astrophys. J.*, **479**, 164

Evans, C.R. and Kochanek, C.S. (1989): The Tidal Disruption of a Star by a Massive Black Hole. *Astrophys. J.*, **346**, L13

Frank, J. (1978): Effects of Massive Central Black Holes on Dense Stellar Systems. *Mon. Not. Roy. Astron. Soc.*, **184**, 375

Frank, J. and Rees, M.J. (1976): Effects of massive central black holes on dense stellar systems. *Mon. Not. Roy. Astron. Soc.*, **176**, 633

Fulbright, M.S., Benz, W., and Davies, M.B. (1995): A Method of Smoothed Particle Hydrodynamics Using Spheroidal Kernels. *Astrophys. J.*, **440**, 254

Hills, J.G. (1975): Possible Power Source of Seyfert Galaxies and QSO's. *Nature*, **254**, 295

Hills, J.G. (1978): Stellar Debris Clouds in Quasars and Related Objects. *Mon. Not. Roy. Astron. Soc.*, **182**, 517

Khokhlov, A., Novikov, I.D., and Pethick, C.J. (1993a): Weak Tidal Encounters of a Star with a Massive Black Hole. *Astrophys. J.*, **418**, 163

Khokhlov, A., Novikov, I.D., and Pethick, C.J. (1993b): Strong Effects during Close Encounters of a Star with a Massive Black Hole. *Astrophys. J.*, **418**, 181

Kochanek, C.S. (1994): The Aftermath of Tidal Disruption: The Dynamics of Thin Gas Streams. *Astrophys. J.*, **422**, 508

Laguna, P., Miller, W. A., and Zurek, W.H. (1993): Smoothed Particle Hydrodynamics near a Black Hole. *Astrophys. J.*, **404**, 678

Laguna, P., Miller, W.A., Warren, M.S., and Zurek, W.H. (1997): Disruption of Stars by Rotating Black Holes. *Astrophys. J.*, submitted

Laguna, P., Miller, W.A., Zurek, W.H., and Davies, M.B. (1993): Tidal Disruptions by Supermassive Black Holes: Hydrodynamic Evolution of Stars on a Schwarzschild Background. *Astrophys. J.*, **410**, L83

Lee, H.M. and Monaghan, J.J. (1994): The Nuclei of Normal Galaxies (eds. R. Genzel and A.I. Harris). Kluwer, Dordrecht

Lightman, A.P. and Shapiro, S.L. (1977): The Distribution and Consumption Rate of Stars around a Massive Collapsed Object. *Astrophys. J.*, **211**, 244

Luminet, J.P. and Marck, J.A. (1985): Tidal squeezing of Stars by Schwarzschild Black Holes. *Mon. Not. Roy. Astron. Soc.*, **212**, 57

Luminet, J.P. and Pichon, B. (1989): Tidal Pinching of White Dwarfs. *Astron. Astrophys.*, **209**, 103

Marck, J.A., Lioure, S., and Bonazzola, S. (1996): Numerical Study of the Tidal Interaction of a Star and a Massive Black Hole. *Astron. Astrophys.*, **306**, 666

Novikov, I.D., Pethick, C.J., and Polnarev, A.G. (1992): Tidal Capture of Stars by a Massive Black Hole. *Mon. Not. Roy. Astron. Soc.*, **255**, 276

Peebles, P.J.E. (1972): Star Distribution near a Collapsed Object. *Astrophys. J.*, **178**, 371

Pichon, B. (1985): Helium Detonation in Pancake Stars. *Astron. Astrophys.*, **145**, 387

Rees, M. (1988): Tidal Disruption of Stars by Black Holes of $10^6 - 10^8$ Solar Masses in Nearby Galaxies. *Nature*, **333**, 523

Rees, M. (1990): "Dead Quasars" in Nearby Galaxies? *Science*, **247**, 817

Rees, M.J. (1994): The Nuclei of Normal Galaxies (eds. R. Genzel and A.I. Harris). Kluwer, Dordrecht

Regev, O. and Portnoy, D. (1987): Tidal disruption of stars by a massive black hole. *Ap. Space Sci.*, **132**, 249

Roos, N. (1992): Gas Clouds from Tidally Disrupted Stars in Active Galactic Nuclei. *Astrophys. J.*, **385**, 108

Sigurdsson, S. and Rees, M.J. (1997): Capture of stellar-mass compact objects by massive black holes in galactic cusps. *Mon. Not. Roy. Astron. Soc.*, **284**, 318

Young, P.J. (1977): The Black Tidal Model of QSO's. II. Destruction in an Isothermal Sphere. *Astrophys. J.*, **215**, 36

Young, P.J. (1980): Numerical Models of Star Clusters with a Central Black Hole. I. Adiabatic Models. *Astrophys. J.*, **242**, 1232

Young, P.J., Shields, G.A., and Wheeler, J.C. (1977): The Black Tide Model of QSO's. *Astrophys. J.*, **212**, 367

Newtonian and Post-Newtonian Calculations of Coalescing Compact Binaries

Frederic A. Rasio

Department of Physics
Massachusetts Institute of Technology, 77 Massachusetts Ave., Cambridge, MA 02139, USA
rasio@mit.edu

Abstract

Coalescing binary neutron stars are important sources of gravitational waves that should become detectable with the laser interferometers now being built as part of LIGO, VIRGO and GEO. Post-Newtonian (PN) approximation methods have been used to calculate waveform templates in the low-frequency, slow-inspiral phase of the binary evolution. These theoretical templates can be used to extract parameters such as the neutron star (NS) masses and spins. In the slow-inspiral phase the two stars are still well separated and can be treated essentially as point masses. Near the end of the coalescence, however, hydrodynamic effects and the interior structure of the stars play an increasingly important role. Hydrodynamics becomes dominant when the two stars finally merge together into a single object. The shape of the corresponding burst of gravitational waves provides a direct probe into the interior structure of a NS and the nuclear equation of state (EOS). The interpretation of the gravitational waveform data will require detailed theoretical models of the complicated 3D hydrodynamic processes involved. This review summarizes recent work on the hydrodynamic aspects of NS binary coalescence. Newtonian and, more recently, relativistic calculations have been performed. The methods include both approximate quasi-analytic techniques and large-scale numerical hydrodynamics calculations on supercomputers. Also included here is a brief discussion of coalescing white dwarf (WD) binaries, which are important sources of very low-frequency gravitational waves, potentially detectable by LISA.

1 Introduction

The coalescence and merging of two stars into a single object is the almost inevitable endpoint of compact binary evolution. Dissipation mechanisms such as friction in common envelopes and the emission of gravitational radiation are always present and cause the binary orbit to decay. The terminal stage of this orbital decay is generally hydrodynamic in nature, with the final merging of the two stars taking place on a time scale comparable to the orbital period. In addition to the angular momentum loss to gravitational radiation, *global hydrodynamic instabilities* can drive the binary system to rapid coalescence once the tidal interaction between the two stars becomes sufficiently strong (Rasio and Shapiro 1992–1997, Lai et al. 1993a,b, 1994a,b,c, Lai and Shapiro 1995, New and Tohline 1997). The existence of these global instabilities for binary systems containing a compressible fluid was demonstrated for the first time in Rasio and Shapiro (1992) using numerical hydrodynamic calculations. In addition, the classical *analytic* work for binaries containing an *incompressible* fluid (Chandrasekhar 1969) was extended to compressible fluids in the work of Lai et al. This new analytic study confirmed the existence of dynamical and secular instabilities for sufficiently close binary systems containing polytropes. Although the simplified analytic studies have given us much physical insight into difficult questions of global fluid instabilities, fully numerical calculations remain essential for establishing the stability limits of close binaries accurately and for following the nonlinear evolution of unstable systems all the way to complete coalescence. Given the absence of any underlying symmetry in the problem, these calculations must be done in 3D and therefore require supercomputers. A number of different groups have now performed such calculations, using a variety of numerical methods and focusing on different aspects of the problem. Nakamura and collaborators [see Nakamura (1994) and references therein] were the first to perform 3D hydrodynamic calculations of binary NS coalescence, using a traditional Eulerian finite-difference code. Rasio and Shapiro have been using the Lagrangian method SPH (Smoothed Particle Hydrodynamics) and have focused on determining the stability properties of initial binary models in strict hydrostatic equilibrium and calculating the emission of gravitational waves from the coalescence of unstable binaries. Many of the results of Rasio and Shapiro have now been independently confirmed in the work of New and Tohline (1997), who used completely different numerical methods but also focused on stability questions. Zhuge et al. (1994) have also used SPH and studied the dependence of the gravitational waveforms on the initial NS spins. Davies et al. (1994) and Ruffert et al. (1996, 1997) have incorporated a treatment of the nuclear physics in their hydrodynamic calculations (done using SPH and PPM codes, respectively) but discussed their results primarily in the context of gamma-ray burst models.

For compact binaries, relativistic effects combine nonlinearly with Newtonian tidal effects so that close binary configurations can become dynamically unstable earlier during the spiral-in phase (i.e., at larger binary separation and lower orbital frequency) than predicted by Newtonian hydrodynamics alone. The combined effects of relativity and hydrodynamics on the stability of close compact binaries have only very recently begun to be studied. Preliminary results have been obtained using both analytic approximations [basically, PN generalizations of Lai et al.; see Lai (1996), Taniguchi and Nakamura (1996), Lai and Wiseman (1997), Lombardi et al. (1997)] as well as numerical hydrodynamics calculations in 3D incorporating simplified treatments of relativistic effects (Wilson and Mathews 1995, Shibata 1996, Baumgarte et al. 1997, Mathews and Wilson 1997). A NASA Grand Challenge project is under way (E. Seidel in this volume) that will ultimately attempt a fully relativistic calculation of the final coalescence, combining the techniques of numerical relativity and numerical hydrodynamics in 3D.

This review will concentrate on the coalescence of compact binaries, containing either two NS (§2) or two WD (§3). Although relativity plays a less important role during the final merging of two WD, the very low-frequency gravitational waves emitted during the inspiral could be easily detected by space-based laser interferometers such as those planned for the LISA project (see the article by K. Danzmann in this volume). Many of the results obtained for WD binaries are also relevant to low-mass main-sequence stars in contact binaries and the important related problem of blue straggler formation in star clusters (Rasio and Shapiro 1995, Rasio 1995, Lombardi et al. 1996).

2 Coalescing Neutron Star Binaries

2.1 Astrophysical Motivation

Coalescing compact binaries are the most promising known sources of gravitational radiation that could be detected by the new generation of laser interferometers: the Caltech-MIT LIGO (Abramovici et al. 1992, Cutler et al. 1993) and the European projects VIRGO (Bradaschia et al. 1990) and GEO (Danzmann, this volume). In addition to providing a major new confirmation of Einstein's theory of general relativity, including the first direct proof of the existence of black holes (Flanagan and Hughes 1997, Lipunov et al. 1997), the detection of gravitational waves from coalescing binaries at cosmological distances could provide accurate measurements of the Hubble constant and mean density of the Universe (Schutz 1986, Chernoff and Finn 1993, Marković 1993). For recent reviews on the detection and sources of gravitational radiation, see Thorne (1995, 1996).

Expected rates of binary coalescence in the Universe, as well as expected event rates in forthcoming laser interferometers, have now been calculated by many groups. Although there is some disparity between various published results, the estimated rates are generally encouraging. Statistical arguments based on the observed local population of binary radio pulsars with probable NS companions lead to an estimate of the rate of NS binary coalescence in the Universe of order $10^{-7}\,\mathrm{yr}^{-1}\,\mathrm{Mpc}^{-3}$ (Narayan et al. 1991, Phinney 1991). Using this estimate, Finn and Chernoff (1993) predict that an advanced LIGO detector could observe as many as 70 events per year. These numbers are based on a Galactic merger rate $R \simeq 10^{-6}\,\mathrm{yr}^{-1}$ derived from radio pulsar surveys. More recently, however, van den Heuvel and Lorimer (1996) revised this number to $R \simeq 0.8 \times 10^{-5}\,\mathrm{yr}^{-1}$, using the latest galactic pulsar population model of Curran and Lorimer (1995). This value is consistent with the upper limit of $10^{-5}\,\mathrm{yr}^{-1}$ for the Galactic binary NS birth rate derived by Bailes (1996) on the basis of very general statistical considerations about pulsars. In addition, theoretical models of the binary star population in our Galaxy also suggest that the NS binary coalescence rate may be as high as $\gtrsim 10^{-6}\,\mathrm{yr}^{-1}\,\mathrm{Mpc}^{-3}$ [Tutukov and Yungelson (1993), see also the more recent studies by Portegies Zwart and Spreeuw (1996) and Lipunov et al. (1997)].

Most recent calculations of the gravitational radiation waveforms from coalescing binaries have focused on the signal emitted during the last few thousand orbits, as the frequency sweeps upward from about 10 Hz to 1000 Hz. The waveforms in this regime can be calculated fairly accurately by performing high-order PN expansions of the equations of motion for two *point masses* (Lincoln and Will 1990, Junker and Schäfer 1992, Kidder et al. 1992, Wiseman 1993, Will 1994, Blanchet et al. 1996). High accuracy is essential here because the observed signals will be matched against theoretical templates. Since the templates must cover $\gtrsim 10^3$ orbits, a phase error as small as $\sim 10^{-3}$ can prevent detection (Cutler et al. 1993, Cutler and Flanagan 1994, Finn and Chernoff 1993).

Near the end of the inspiral, when the binary separation becomes comparable to the stellar radii, hydrodynamic effects become important and the character of the waveforms will change. Special purpose narrow-band detectors that can sweep up frequency in real time will be used to try to catch the corresponding final few cycles of gravitational waves (Meers 1988, Strain and Meers 1991, Danzmann, this volume). In this terminal phase of the coalescence, the waveforms contain information not just about the effects of general relativity, but also about the internal structure of the stars and the nuclear EOS at high density. Extracting this information from observed waveforms, however, requires detailed theoretical knowledge about all relevant hydrodynamic processes.

Many models of gamma-ray bursts at cosmological distances are also based on coalescing NS-NS systems (Paczyński 1986, Eichler et al. 1989, Narayan et al. 1992, Mészáros, this volume). The isotropic angular distribution of the bursts detected by the BATSE experiment on the Compton GRO satellite (Meegan et al. 1992) strongly suggests a cosmological origin, and the rate of gamma-ray bursts detected by BATSE, of order one per day, is in rough agreement with theoretical predictions for the rate of NS binary coalescence in the Universe (cf. above). However, the complete hydrodynamic and nuclear evolution during final merging, especially in the outermost, low-density regions of the merger, must be understood in details before realistic models can be constructed for the gamma-ray emission. Numerical calculations of binary coalescence including some treatment of the nuclear physics have been performed by Davies et al. (1994) and Ruffert et al. (1996, 1997). The most recent results from these calculations indicate that, even under the most favorable conditions, the energy provided by $\nu\bar{\nu}$ annihilation is too small by at least an order of magnitude, and more probably two or three orders of magnitude, to power typical gamma-ray bursts at cosmological distances (Janka and Ruffert 1996).

2.2 Hydrodynamic Instabilities

Hydrostatic equilibrium configurations for binary systems with sufficiently close components can become *dynamically unstable* (Chandrasekhar 1975, Tassoul 1975). The physical nature of this instability is common to all binary interaction potentials that are sufficiently steeper than $1/r$ [see, e.g., Goldstein (1980), §3.6]. It is analogous to the familiar instability of circular orbits sufficiently close to a black hole (BH) (Shapiro and Teukolsky 1983, §12.4). Here, however, it is the *tidal interaction* that is responsible for the steepening of the effective interaction potential between the two stars and for the destabilization of the circular orbit (Lai et al. 1994a). The tidal interaction exists of course already in Newtonian gravity and the instability is therefore present even in the absence of relativistic effects. For sufficiently compact binaries, however, the combined effects of relativity and hydrodynamics lead to an even stronger tendency toward dynamical instability (see below).

The stability properties depend sensitively on the NS EOS. Close binaries containing NS with stiff EOS (adiabatic exponent $\Gamma \gtrsim 2$) are particularly susceptible to the dynamical instability. This is because tidal effects are stronger for stars containing a less compressible fluid. As the dynamical stability limit is approached, the secular orbital decay driven by gravitational wave emission can be dramatically accelerated (Lai et al. 1993b, 1994a). The two stars then plunge rapidly toward each other, and merge together into a single object in just a few rotation periods. This dynamical instability was first identified in Rasio and Shapiro (1992), where the evolution of Newtonian binary equilibrium configurations was calculated for two identical polytropes with $\Gamma = 2$. It was found that when $r \lesssim 3R$ (r is the binary separation and R the radius of an unperturbed NS), the orbit becomes unstable to radial perturbations and the two stars undergo rapid coalescence. For $r \gtrsim 3R$, the system could be evolved dynamically for many orbital periods without showing any sign of orbital evolution (in the absence of dissipation). Many of the results derived in Rasio and Shapiro and Lai et al. concerning the stability properties of NS binaries have been confirmed recently in completely independent work by New and Tohline (1997), using very different numerical methods (a combination of a 3-D self-consistent field code for constructing equilibrium configurations and a finite-difference code for following the dynamical evolution of the binaries).

The dynamical evolution of an unstable, initially synchronized (i.e., rigidly rotating) binary can be described typically as follows (Rasio and Shapiro 1992, 1994). During the initial, linear stage of the instability, the two stars approach each other and come into contact after about one orbital revolution. In the corotating frame of the binary, the relative velocity remains very subsonic, so that the evolution is adiabatic at this stage. This is in sharp contrast to the case of a head-on collision between two stars on a free-fall, radial orbit, where shocks are very important for the dynamics (Rasio and Shapiro 1992). Here the stars are constantly being held back by a (slowly receding) centrifugal barrier, and the merging, although dynamical, is much more gentle. After typically two orbital revolutions the innermost cores of the two stars have merged and the system resembles a single, very elongated ellipsoid. At this point a secondary instability occurs: *mass shedding* sets in rather abruptly. Material is ejected through the outer Lagrangian points of the effective potential and spirals out rapidly. In the final stage, the spiral arms widen and merge together. The relative radial velocities of neighboring arms as they merge are supersonic, leading to some shock-heating and dissipation. As a result, a hot, nearly axisymmetric rotating halo forms around the central dense core. No measurable amount of mass escapes from the system. The halo contains about 20% of the total mass and has a pseudo-barotropic structure (Tassoul 1978, §4.3), with the angular velocity decreasing as a power-law $\Omega \propto \varpi^{-\nu}$ where $\nu \lesssim 2$ and ϖ is the distance to the rotation axis (Rasio and Shapiro 1992). The core is rotating uniformly near breakup speed and contains about 80% of the mass still in a cold, degenerate state.

The emission of gravitational radiation during dynamical coalescence can be calculated perturbatively using the quadrupole approximation (Rasio and Shapiro 1992). Both the frequency and amplitude of the emission peak somewhere during the final dynamical coalescence, typically just before the onset of mass shedding. Immediately after the peak, the amplitude drops abruptly as the system evolves towards a more axially symmetric state. For an initially synchronized binary containing two identical polytropes, the properties of the waves near the end of the coalescence depend very sensitively on the stiffness of the EOS. When $\Gamma < \Gamma_{crit}$, with $\Gamma_{crit} \approx 2.3$, the final merged configuration is perfectly axisymmetric[1] and the amplitude of the waves drops to zero in just a few periods (Rasio and Shapiro 1992). In contrast, when $\Gamma > \Gamma_{crit}$, the dense central core of the final configuration remains *triaxial* (its structure is basically that of a compressible Jacobi ellipsoid; cf. Lai et al. 1993a) and therefore it continues to radiate gravitational waves. The amplitude of the waves first drops quickly to a nonzero value and then decays more slowly as gravitational waves continue to carry angular momentum away from the central core (Rasio and Shapiro 1994). Because realistic NS models give effective Γ values precisely in the range 2–3 (Lai et al. 1994a), i.e., close to $\Gamma_{crit} \approx 2.3$, a simple determination of the absence or presence of persisting gravitational radiation after the coalescence (i.e., after the peak in the emission) could place a strong constraint on the stiffness of the EOS.

2.3   Mass Transfer and the Dependence on the Mass Ratio

Clark and Eardley (1977) suggested that secular, *stable* mass transfer from one NS to another could last for hundreds of orbital revolutions before the lighter star is tidally disrupted. Such an episode of stable mass transfer would be accompanied by a secular *increase* of the orbital separation. Thus if stable mass transfer could indeed occur, a characteristic "reversed chirp" would be observed in the gravitational wave signal at the end of the inspiral phase (Jaranowski and Krolak 1992).

The question was reexamined recently by Kochanek (1992) and Bildsten and Cutler (1992), who both argued against the possibility of stable mass transfer on the basis that very large mass transfer rates and extreme mass ratios would be required. Moreover, in Lai et al. (1994a) it was pointed out that mass transfer has in fact little importance for most NS binaries (except perhaps those containing a very low-mass NS). This is because for $\Gamma \gtrsim 2$, *dynamical instability always arises before the Roche limit* along a sequence of binary configurations with decreasing r. Therefore, by the time mass transfer begins, the system is already in a state of dynamical coalescence and it can no longer remain in a nearly circular orbit. Thus stable mass transfer from one NS to another appears impossible.

In Rasio and Shapiro (1994) a complete dynamical calculation was presented for a system containing two polytropes with $\Gamma = 3$ and a mass ratio $q = 0.85^2$. For this system it is found that the dynamical stability limit is at $r/R \approx 2.95$, whereas the Roche limit is at $r/R \approx 2.85$. The dynamical evolution turns out to be quite different from that of a system with $q = 1$. The Roche limit is quickly reached while the system is still in the linear stage of growth of the instability. Dynamical mass transfer from the less massive to the more massive star begins within the first orbital revolution. Because of the proximity of the two components, the fluid acquires very little velocity as it slides down from the inner Lagrangian point to the surface of the other star. As a result, relative velocities of fluid particles remain largely subsonic and the coalescence proceeds quasi-adiabatically, just as in the $q = 1$ case. In fact, the mass transfer appears to have essentially no effect on the dynamical evolution. After about two orbital revolutions the smaller-mass star undergoes complete tidal disruption. Most of its material is quickly spread on top of the more massive star, while a small fraction of the mass is ejected from the outermost Lagrangian point and forms a single-arm spiral outflow. The more massive star, however, remains little perturbed during the entire evolution and simply becomes the inner core of the merged configuration.

The dependence of the peak amplitude $h_{\max}$ of gravitational waves on the mass ratio q appears to be very strong, and nontrivial. In Rasio and Shapiro (1994) an approximate scaling $h_{\max} \propto q^2$ was derived. This is very different from the scaling obtained for a detached binary system with a given binary separation. In particular, for two point masses in a circular orbit with separation r the result would be $h \propto \Omega^2 \mu r^2$, where $\Omega^2 = G(M + M')/r^3$ and $\mu = MM'/(M + M')$. At constant r, this gives $h \propto q$. This linear scaling is obeyed (only approximately, because of finite-size effects) by the wave amplitudes of the various systems at the *onset* of dynamical instability. For determining the *maximum* amplitude, however, hydrodynamics plays an essential role. In a system with $q \neq 1$, the more massive star tends to play a far less active role in the hydrodynamics and, as a result, *there is a rapid suppression of the radiation efficiency as q departs even slightly from unity*. For the peak luminosity of gravitational radiation Rasio and Shapiro found approximately $L_{\max} \propto q^6$. Again, this is a much steeper dependence than one would expect based on a simple point-mass estimate, which gives $L \propto q^2(1 + q)$ at constant r.

---

[2] This is the most probable value of the mass ratio in the binary pulsar PSR 2303+46 (Thorsett et al. 1993) and represents the largest observed departure from $q = 1$ in any observed binary pulsar with likely NS companion. For comparison, $q = 1.386/1.442 = 0.96$ in PSR 1913+16 (Taylor and Weisberg 1989) and $q = 1.32/1.36 = 0.97$ in PSR 1534+12 (Wolszczan 1991).

2.4   Measuring the Radius of a Neutron Star with LIGO/VIRGO

The most important parameter that enters into quantitative estimates of the gravitational wave emission during the final coalescence is the relativistic parameter M/R for a NS (here we take $G = c = 1$). In particular, for two identical point masses we know that the wave amplitude obeys $(r_O/M)h \propto (M/R)$, where r_O is the distance to the observer, and the total luminosity $L \propto (M/R)^5$. Thus one expects that any quantitative measurement of the emission near maximum should lead to a direct determination of the radius R, assuming that the mass M has already been determined from the low-frequency inspiral waveform (Cutler and Flanagan 1994). Most current NS EOS give $M/R \sim 0.1$, with $R \sim 10 \, \mathrm{km}$ nearly independent of the mass in the range $0.8 M_\odot \lesssim M \lesssim 1.5 M_\odot$ [see, e.g., Baym (1991), Cook et al. (1994), Lai et al. (1994a)].

However, the details of the hydrodynamics also enter into this determination. The importance of hydrodynamic effects introduces an explicit dependence of all wave properties on the internal structure of the stars (which we represent here by a single dimensionless parameter Γ), and on the mass ratio q. If relativistic effects were taken into account for the hydrodynamics itself, an additional, nontrivial dependence on M/R would also be present. This can be written conceptually as

$$\left(\frac{r_O}{M}\right) h_{\mathrm{max}} \;\equiv\; \mathcal{H}(q,\Gamma,M/R) \times \left(\frac{M}{R}\right) \tag{1}$$

$$\frac{L_{\mathrm{max}}}{L_o} \;\equiv\; \mathcal{L}(q,\Gamma,M/R) \times \left(\frac{M}{R}\right)^5 \tag{2}$$

Combining all the results of Rasio and Shapiro, we can write, in the limit where $M/R \to 0$ and for q not too far from unity,

$$\mathcal{H}(q,\Gamma,M/R) \approx 2.2\, q^2 \qquad \mathcal{L}(q,\Gamma,M/R) \approx 0.5\, q^6, \tag{3}$$

essentially independent of Γ in the range $\Gamma \approx 2$–3 (Rasio and Shapiro 1994). This is in the case of synchronized spins. For nonsynchronized configurations, the spin frequency of the stars must be considered as additional parameters.

2.5   Nonsynchronized Binaries

Recent theoretical work suggests that the synchronization time in close NS binaries remains always longer than the orbital decay time due to gravitational radiation (Kochanek 1992, Bildsten and Cutler 1992). In particular, Bildsten and Cutler (1992) show with simple dimensional arguments that one would need an implausibly small value of the effective viscous time, $t_{\mathrm{visc}} \sim R/c$, in order to reach complete synchronization just before final merging.

In the opposite limiting regime where viscosity is completely negligible, the fluid circulation in the binary system is conserved during the orbital decay and the stars behave approximately as Darwin-Riemann ellipsoids (Kochanek 1992, Lai et al. 1994a).

Of particular importance are the irrotational Darwin-Riemann configurations, obtained when two initially nonspinning (or, in practice, slowly spinning) NS evolve in the absence of significant viscosity. Compared to synchronized systems, these irrotational configurations exhibit smaller deviations from point-mass Keplerian behavior at small r. However, as shown in Lai et al. (1994a) and Rasio and Shapiro (in prep.), irrotational configurations for binary NS with $\Gamma \gtrsim 2$ can nevertheless become dynamically unstable near contact. Thus the final coalescence of two NS in a nonsynchronized binary system must still be driven by hydrodynamic instabilities.

The details of the hydrodynamics are very different, however (Rasio and Shapiro, in prep.). Because the two stars appear to be counter-spinning in the corotating frame of the binary, a vortex sheet with $\Delta v = |v_+ - v_-| \approx \Omega r$ appears when the surfaces come into contact. Such a vortex sheet is Kelvin-Helmholtz unstable on all wavelengths and the hydrodynamics is therefore rather difficult to model accurately given the limited spatial resolution of 3D calculations. The breaking of the vortex sheet generates a large turbulent viscosity so that the final configuration may no longer be irrotational. In numerical simulations, however, vorticity is generated mostly through spurious shear viscosity introduced by the spatial discretization. An additional difficulty is that nonsynchronized configurations evolving rapidly by gravitational radiation emission tend to develop significant tidal lags, with the long axes of the two components becoming misaligned (Lai et al. 1994c). This is a purely dynamical effect, present even if the viscosity is zero, but its magnitude depends on the entire previous evolution of the system. Thus the construction of initial conditions for hydrodynamic calculations of nonsynchronized binary coalescence must incorporate the gravitational radiation reaction *self-consistently*. Instead, previous studies of nonsynchronized, equal-mass binary coalescence by Shibata et al. (1992), Davies et al. (1994), and Zhuge et al. (1994) used very approximate initial conditions consisting of two identical *spheres* (polytropes with $\Gamma \approx 2$) placed on an inspiral trajectory calculated for two point masses.

2.6 Relativistic Effects on the Stability of Compact Binaries

Most of the results discussed so far in this section are based on purely Newtonian calculations of NS binaries. Over the last year or so, various efforts have started to calculate the stability limits for NS binaries including both hydrodynamic finite-size (tidal) effects and relativistic effects. Note that, strictly speaking, equilibrium circular orbits do not exist in general relativity because of the emission of gravitational waves. However, the stability of quasi-circular orbits can still be studied in the framework of general relativity by truncating the radiation-reaction terms in a PN expansion of the equations of motion (Lincoln and Will 1990, Kidder et al. 1992, Will 1994). Alternatively, one can solve the full Einstein equations numerically in the $3 + 1$ formalism (see the article by Seidel in this volume) on time slices with a spatial 3-metric chosen to be conformally flat (Wilson and Mathews 1989, 1995, Wilson et al. 1996, Baumgarte et al. 1997). This effectively minimizes the gravitational wave content of space-time. The field equations then reduce to a set of coupled elliptic equations for the lapse and shift functions and the conformal factor.

Several groups are now working on PN generalizations of the semi-analytic Newtonian treatment of Lai et al. based on ellipsoids. Taniguchi and Nakamura (1996) consider NS-BH binaries and adopt a modified version of the pseudo-Newtonian potential of Paczyński and Wiita (1980) to mimic general relativistic effects near the BH. Lai and Wiseman (1997) concentrate on NS-NS binaries and the dependence of the results on the NS EOS. They add a restricted set of PN orbital terms to the dynamical equations given in Lai and Shapiro (1995) for a binary system containing two NS modeled as Riemann-S ellipsoids (cf. Lai et al. 1993a,b, 1994a,b,c), but neglect relativistic corrections to the fluid motion, self-gravity and tidal interaction. Lombardi et al. (1997) include PN corrections affecting both the orbital motion and the interior structure of the stars and explore the consequences not only for orbital stability but also for the stability of each NS against collapse. The most important result, on which these various studies all seem to agree, is that neither the relativistic effects nor the Newtonian tidal effects can be neglected if one wants to obtain a quantitatively accurate determination of the stability limits. In particular, the critical frequency corresponding to the onset of dynamical instability can be much lower than the value obtained when only one of the two effects is included. This critical frequency for the "last stable circular orbit" is a measurable quantity (with LIGO/VIRGO) and can provide direct information on the NS EOS.

A surprising result coming from the numerical $3 + 1$ relativistic calculations of Wilson and Mathews (1995, Mathews and Wilson 1997) is the appearance of a binary-induced collapse instability of the NS. This must be a purely relativistic effect, since the Newtonian tidal effects in fact tend to *stabilize* the NS against collapse (Lai 1996). In effect, the maximum stable mass of a NS in a relativistic close binary system could be slightly lower than that of a NS in isolation. Initially stable NS close to the maximum mass could then collapse to BH before getting to the final phase of binary coalescence. It should be noted, however, that the numerical results of Wilson and Mathews have yet to be confirmed independently by other studies. Even if it is real, the effect would be of importance only if the NS EOS is very soft, and the maximum stable mass for a NS in isolation is not much larger than $1.4 M_\odot$.

3   Coalescing White Dwarf Binaries

3.1   Astrophysical Motivation

Close WD binaries are expected to be extremely abundant in our Galaxy. Iben and Tutukov (1984, 1986) predict that $\sim 20\%$ of all binary stars produce close pairs of WD at the end of their stellar evolution. The most common systems should be those containing two low-mass helium WD. Their final coalescence can produce an object massive enough to start helium burning. Bailyn (1993) suggests that extreme horizontal branch stars in globular clusters may be such helium-burning stars formed by the coalescence of two WD. Paczyński (1990) has proposed that the peculiar X-ray pulsar 1E 2259+586 may be the product of a recent WD-WD merger. Planets in orbit around a massive WD may also form following a merger (Livio et al. 1992).

Coalescing WD binaries may also be progenitors for Type Ia supernovae (Iben and Tutukov 1984, Webbink 1984, Paczyński 1985, Mochkovitch and Livio 1989, Yungelson et al. 1994). To produce a supernova, the total mass of the system must be above the Chandrasekhar mass. Given evolutionary considerations, this requires two C-O or O-Ne-Mg WD. Yungelson et al. (1994) show that the expected merger rate for close pairs of WD with total mass exceeding the Chandrasekhar mass is consistent with the rate of type Ia supernovae deduced from observations. Alternatively, a massive enough merger may collapse to form a rapidly rotating NS (Nomoto and Iben 1985, Colgate 1990). Chen and Leonard (1993) have discussed the possibility that most millisecond pulsars in globular clusters may have formed in this way. In some cases planets may form in the disk of material ejected during the coalescence and left in orbit around the central pulsar (Podsiadlowski et al. 1991). Indeed the first extrasolar planets have been discovered in orbit around a millisecond pulsar, PSR B1257+12 (Wolszczan 1994). A merger of two highly magnetized WD might lead to the formation of a NS with extremely high magnetic field, and this scenario has been proposed as a source of gamma-ray bursts (Usov 1992).

Coalescing WD binaries are also important sources of very low-frequency gravitational waves that should be easily detectable by future space-based interferometers such as LISA (Danzmann, this volume). Evans et al. (1987) estimate a WD merger rate of order one every 5 yr in our own Galaxy. Coalescing systems closest to Earth should produce quasi-periodic gravitational waves of amplitude $h \sim 10^{-21}$ in the frequency range ~ 10–100 mHz. In addition, the total number ($\sim 10^4$) of close WD binaries in our Galaxy emitting at lower frequencies ~ 0.1–1 mHz (the emission lasting for $\sim 10^2$–10^4 yr before final coalescence) should provide a continuum background signal of amplitude $h_c \sim 10^{-20}$–10^{-21}. Individual sources should be detectable by LISA above this background when frequency becomes $\gtrsim 10$ mHz. The detection of the final burst of gravitational waves emitted during the actual merging would provide a unique opportunity to observe in "real time" the hydrodynamic interaction between the two WD, possibly followed immediately by a supernova explosion, nuclear outburst, or some other type of electromagnetic signal.

3.2   Hydrodynamics of Coalescing White Dwarf Binaries

The results of Rasio and Shapiro (1995) for polytropes with $\Gamma = 5/3$ show that hydrodynamics also plays an important role in the coalescence of two WD, either because of dynamical instabilities of the equilibrium configuration, or following the onset of mass transfer. Systems with $q \approx 1$ must evolve into deep contact before they become dynamically unstable and merge. Instead, equilibrium configurations for binaries with q sufficiently far from unity never become dynamically unstable. However, when these binaries reach the Roche limit, *dynamically unstable mass transfer* occurs and the less massive star is completely disrupted after a small number (< 10) of orbital periods [see also Benz et al. (1990)]. In both cases, the final merged configuration is an axisymmetric, rapidly rotating object with a core-halo structure similar to that obtained for coalescing NS binaries [Rasio and Shapiro (1994, 1995); see also Mochkovitch and Livio (1989)].

For two massive enough WD, the merger product may be well above the Chandrasekhar mass $M_{\rm Ch}$. The object may therefore explode as a (type Ia) supernova, or perhaps collapse to a NS. The rapid rotation and possibly high mass (up to $2M_{\rm Ch}$) of the object must be taken into account for determining its final fate. Unfortunately, this is not done in current theoretical calculations of accretion induced collapse (AIC), which always consider a nonrotating WD just below the Chandrasekhar limit accreting matter slowly and quasi-spherically (Canal et al. 1990, Nomoto and Kondo 1991, Nomoto et al. 1995). Under these assumptions it is found that collapse to a NS is possible only for a narrow range of initial conditions. In most cases, a supernova explosion follows the ignition of the nuclear fuel in the degenerate core. However, the fate of a much more massive object with substantial rotational support and large deviations from spherical symmetry (as would be formed by dynamical coalescence) may be very different.

References

Abramovici, A., Althouse, W.E., Drever, R.W.P., Gürsel, Y., Kawamura, S., Raab, F., Shoe-maker, D., Sievers, L., Spero, R.E., Thorne, K.S., Vogt, R.E., Weiss, R., Whitcomb, S.E., and Zucker, M.E. (1992): LIGO: The Laser Interferometer Gravitational-Wave Observatory. *Science*, **256**, 325

Bailes, M. (1996): In: *Compact Stars in Binaries* (eds. J. van Paradijs et al.). Kluwer, Dordrecht, 213

Bailyn, C.D. (1993): In: *Structure and Dynamics of Globular Clusters* (eds. S.G. Djorgovski and G. Meylan). San Francisco: ASP Conf. Series, **50**, 191

Baumgarte, T.W., Shapiro, S.L., Cook, G.B., Scheel, M.A., and Teukolsky, S.A. (1997): In: Proceedings of the 18th Texas Symposium on Relativistic Astrophysics (eds. A. Olinto et al.). World Scientific, in press

Baym, G. (1991): In: *Neutron Stars: Theory and Observation* (eds. J. Ventura and D. Pines). Kluwer, Dordrecht, 21

Benz, W., Bowers, R.L., Cameron, A.G.W., and Press, W.H. (1990): Dynamic Mass Exchange in Doubly Degenerated Binaries. I. 0.9 and 1.2 $M\odot$ Stars. *Astrophys. J.*, **348**, 647

Bildsten, L. and Cutler, C. (1992): Tidal Interactions of Inspiraling Compact Binaries. *Astrophys. J.*, **400**, 175

Blanchet, L., Iyer, B.R., Will, C.M., and Wiseman, A.G. (1996): Gravitational waveforms from inspiralling compact binaries to second-post-Newtonian order. *Class. Quant. Grav.*, **13**, 575

Bradaschia, C., Del Fabro, R., Di Virgilio, A., Giazotto, A., Kautzky, H., Montelatici, V., Pas-suello, D., Brillet, A., Cregut, O., Hello, P., Man, C.N., Manh, P.T., Marraud, A., Shoe-maker, D., Vinet, J.Y., Barone, F., Di Fiore, L., Milano, L., Russo, G., Aguirregabiria, J.M., Bel, H., Duruisseau, J.P., Le Denmat, G., Tourrenc, Ph., Capozzi, M., Longo, M., Lops, M., Pinto, I., Rotoli, G., Damour, T., Bonazzola, S., Marck, J.A., Gourghoulon, Y., Holloway, L.E., Fuligni, F., Iafolla, V., and Natale, G. (1990): The Virgo Project: A Wide Band Antenna for Gravitational Wave Detection. *Nucl. Instr. Methods A*, **289**, 518

Canal, R., García, D., Isern, J., and Labay, J. (1990): Can C+O White Dwarfs Form Neutron Stars? *Astrophys. J.*, **356**, L51

Chandrasekhar, S. (1969): *Ellipsoidal Figures of Equilibrium*. Yale University Press, New Haven; Revised Dover edition 1987

Chandrasekhar, S. (1975): On Coupled Second-Harmonic Oscillations of the Congruent Darwin Ellipsoids. *Astrophys. J.*, **202**, 80

Chen, K. and Leonard, P.J.T. (1993): Does the Coalescence of White Dwarfs Produce Millisecond Pulsars in Globular Clusters? *Astrophys. J.*, **411**, L75

Chernoff, D.F. and Finn, L.S. (1993): Gravitational Radiation, Inspiring Binaries, and Cosmol-ogy. *Astrophys. J.*, **411**, L5

Clark, J.P.A. and Eardley, D.M. (1977): Evolution of Close Neutron Star Binaries. *Astrophys. J.*, **251**, 311

Colgate, S.A. (1990): In: *Supernovae* (ed. S.E. Woosley). Springer-Verlag, New York, 585

Cook, G.B., Shapiro, S.L., and Teukolsky, S.A. (1994): Rapidly Rotating Neutron Stars in General Relativity: Realistic Equations of State. *Astrophys. J.*, **424**, 823

Curran, S.J. and Lorimer, D.R. (1995): Pulsar statistics. III. Neutron star binaries. *Mon. Not. Roy. Astron. Soc.*, **276**, 347

Cutler, C. and Flanagan, E.E. (1994): Gravitational Waves from Merging Compact Binaries: How Accurately Can One Extract the Binary's Parameters from the Inspiral Waveform? *Phys. Rev. D*, **49**, 2658

Cutler, C., Apostolatos, T.A., Bildsten, L., Finn, L.S., Flanagan, E.E., Kennefick, D., Markovic, D.M., Ori, A., Poisson, E., Sussman, G.J., and Thorne, K.S. (1993): The Last Three Minutes: Issues in Gravitational-Wave Measurements of Coalescing Compact Binaries. *Phys. Rev. Lett.*, **70**, 2984

Davies, M.B., Benz, W., Piran, T., and Thielemann, F.K. (1994): Merging Neutron Stars. *Astrophys. J.*, **431**, 742

Eichler, D., Livio, M., Piran, T., and Schramm, D.N. (1989): Nucleosynthesis, Neutrino Bursts and γ-Rays from Coalescing Neutron Stars. *Nature*, **340**, 126

Evans, C.R., Iben, I., and Smarr, L. (1987): Degenerate Dwarf Binaries as Promising, Detectable Sources of Gravitational Radiation. *Astrophys. J.*, **323**, 129

Finn, L.S. and Chernoff, D. (1993): Observing Binary Inspiral in Gravitational Radiation: One Interferometer. *Phys. Rev. D*, **47**, 2198

Flanagan, E.E. and Hughes, S.A. (1997): Measuring gravitational waves from binary black hole coalescences: I. Signal to noise for inspiral, merger, and ringdown. *Phys. Rev. D*, submitted

Goldstein, H. (1980): *Classical Mechanics*. Addison-Wesley, Reading

Iben, I., Jr., and Tutukov, A.V. (1984): Supernovae of Type I as end products of the evolution of binaries with components of moderate initial mass $(M \lesssim 9 M_\odot)$. *Astrophys. J. Suppl.*, **54**, 335

Iben, I., Jr., and Tutukov, A.V. (1986): On the Number-Mass Distribution of Degenerate Dwarfs Produced by Interacting Binaries and Evidence for Mergers of Low-Mass Helium Dwarfs. *Astrophys. J.*, **311**, 753

Janka, H.-T. and Ruffert, M. (1996): Can Neutrinos from Neutron Star Merges Power γ-Ray Bursts? *Astron. Astrophys.*, **307**, L33

Jaranowski, P. and Krolak, A. (1992): Detectability of the Gravitational Wave Signal from a Close Neutron Star Binary with Mass Transfer. *Astrophys. J.*, **394**, 586

Junker, W. and Schäfer, G. (1992): Binary Systems: Higher Order Gravitational Radiation Damping and Wave Emission. *Mon. Not. Roy. Astron. Soc.*, **254**, 146

Kidder, L.E., Will, C.M., and Wiseman, A.G. (1992): Innermost stable orbits for coalescing binary systems of compact objects. *Class. Quantum Grav.*, **9**, L125

Kochanek, C.S. (1992): Coalescing Binary Neutron Stars. *Astrophys. J.*, **398**, 234

Lai, D. (1996): Tidal Stabilization of Neutron Stars and White Dwarfs. *Phys. Rev. Lett.*, **76**, 4878

Lai, D. and Shapiro, S.L. (1995): Hydrodynamics of Coalescing Binary Neutron Stars: Ellipsoidal Treatment. *Astrophys. J.*, **443**, 705

Lai, D. and Wiseman, A.G. (1997): Innermost Stable Circular Orbit of Inspiraling Neutron-Star Binaries: Tidal Effects, Post-Newtonian Effects, and the Neutron-Star Equation of State. *Phys. Rev. D.*, **54**, 3958

Lai, D., Rasio, F.A., and Shapiro, S.L. (1993a): Hydrodynamic instability and coalescence of close binary systems. *Astrophys. J. Suppl.*, **88**, 205

Lai, D., Rasio, F.A., and Shapiro, S.L. (1993b): Hydrodynamic Instability and Coalescence of Close Binary Systems. *Astrophys. J.*, **406**, L63

Lai, D., Rasio, F.A., and Shapiro, S.L. (1994a): Hydrodynamic Instability and Coalescence of Binary Neutron Stars. *Astrophys. J.*, **420**, 811

Lai, D., Rasio, F.A., and Shapiro, S.L. (1994b): Equilibrium, Stability, and Orbital Evolution of Close Binary Systems. *Astrophys. J.*, **423**, 344

Lai, D., Rasio, F.A., and Shapiro, S.L. (1994c): Hydrodynamics of Rotating Stars and Close Binary Interactions: Compressible Ellipsoid Models. *Astrophys. J.*, **437**, 742

Lincoln, C.W. and Will, C.M. (1990): Coalescing Binary Systems of Compact Objects to (Post)$^{5/2}$-Newtonian Order: Late-Time Evolution and Gravitational-Radiation Emission. *Phys. Rev. D*, **42**, 1123

Lipunov, V.M., Postnov, K.A., and Prokhorov, M.E. (1997): Black holes and gravitational waves: simultaneous discovery by initial laser interferometers. *Astron. Lett.*, in press

Livio, M., Pringle, J.E., and Saffer, R.A. (1992): Planets around Massive White Dwarfs. *Mon. Not. Roy. Astron. Soc.*, **257**, 15

Lombardi, J.C., Rasio, F.A., and Shapiro, S.L. (1996): Collisions of Main-Sequence Stars and the Formation of Blue Stragglers in Globular Clusters. *Astrophys. J.*, **468**, 797

Lombardi, J.C., Rasio, F.A., and Shapiro, S.L. (1997): Post-Newtonian Models of Binary Neutron Stars. *Phys. Rev. D*, in press

Marković, D. (1993): Possibility of Determining Cosmological Parameters from Measurements of Gravitational Waves Emitted by Coalescing, Compact Binaries. *Phys. Rev. D*, **48**, 4738

Mathews, G.J. and Wilson, J.R. (1997): Binary Induced Neutron-Star Compression, Heating, and Collapse. *Astrophys. J.*, in press

Meegan, C.A., Fishman, G.J., Wilson, R.B., Paciesas, W.S., Pendleton, G.N., Horack, J.M., Brock, M.N., and Kouveliotou, C. (1992): Spatial Distribution of γ-Ray Bursts Observed by BATSE. *Nature*, **355**, 143

Meers, B.J. (1988): Recycling in Laser-Interferometric Gravitational-Wave Detectors. *Phys. Rev. D*, **38**, 2317

Mochkovitch, R. and Livio, M. (1989): The Coalescence of White Dwarfs and Type I Supernovae. *Astron. Astrophys.*, **209**, 111

Nakamura, T. (1994): In: *Relativistic Cosmology* (ed. M. Sasaki). Universal Academy Press, Tokyo, 155

Narayan, R., Paczyński, B., and Piran, T. (1992): Gamma-Ray Bursts as the Death Throes of Massive Binary Stars. *Astrophys. J.*, **395**, L83

Narayan, R., Piran, T., and Shemi, A. (1991): Neutron Star and Black Hole Binaries in the Galaxy. *Astrophys. J.*, **379**, L17

New, K.C.B. and Tohline, J.E. (1997): The Relative Stability against Merger of Close, Compact Binaries. *Astrophys. J.*, in press

Nomoto, K. and Iben, I., Jr. (1985): Carbon Ignition in a Rapidly Accreting Degenerate Dwarf: A Clue to the Nature of the Merging Process in Close Binaries. *Astrophys. J.*, **297**, 531

Nomoto, K. and Kondo, Y. (1991): Conditions for Accretion-Induced Collapse of White Dwarfs. *Astrophys. J.*, **367**, L19

Nomoto, K., Yamaoka, H., Shigeyama, T., and Iwamoto, K. (1995): In: *Supernovae and Supernova Remnants* (eds. R.A. McCray et al.). Cambridge University Press, Cambridge

Paczyński, B. (1985): In: *Cataclysmic Variables and Low-mass X-ray Binaries* (eds. D.Q. Lamb and J. Patterson). Reidel, Dordrecht, 1

Paczyński, B. (1986): Gamma-Ray Bursters at Cosmological Distances. *Astrophys. J.*, **308**, L43

Paczyński, B. (1990): X-Ray Pulsar IE 2259+586: A Merged White Dwarf with a 7 Second Rotation Period? *Astrophys. J.*, **365**, L9

Paczyński, B. and Wiita, P.J. (1980): Thick Accretion Disks and Supercritical Luminosities. *Astron. Astrophys.*, **88**, 23

Phinney, E.S. (1991): The Rate of Neutron Star Binary Mergers in the Universe: Minimal Predictions for Gravity Wave Detectors. *Astrophys. J.*, **380**, L17

Podsiadlowski, P., Pringle, J.E., and Rees, M.J. (1991): The Origin of the Planet Orbiting PSR 1829-10. *Nature*, **352**, 783

Portegies Zwart, S.F. and Spreeuw, H.N. (1996): The Galactic Merger Rate of (NS, NS) Binaries. I. Perspective for Gravity-Wave Detectors. *Astron. Astrophys.*, **312**, 670

Rasio, F.A. (1995): The Minimum Mass Ratio of W Ursae Majoris Binaries. *Astrophys. J.*, **444**, L41

Rasio, F.A. and Shapiro, S.L. (1992): Hydrodynamical Evolution of Coalescing Binary Neutron Stars. *Astrophys. J.*, **401**, 226

Rasio, F.A. and Shapiro, S.L. (1994): Hydrodynamics of Binary Coalescence. I. Polytropes with Stiff Equations of State. *Astrophys. J.*, **432**, 242

Rasio, F.A. and Shapiro, S.L. (1995): Hydrodynamics of Binary Coalescence. II. Polytropes with $\Gamma = 5/3$. *Astrophys. J.*, **438**, 887

Ruffert, M., Janka, H.-T., and Schäfer, G. (1996): Coalescing Neutron Stars – A Step towards Physical Models. *Astron. Astrophys.*, **311**, 532

Ruffert, M., Janka, H.-T., Takahashi, K., and Schäfer, G. (1997): Coalescing neutron stars – a step towards physical models: II. Neutrino emission, neutron tori, and gamma-ray bursts. *Astron. Astrophys.*, in press

Schutz, B.F. (1986): Determining the Hubble Constant from Gravitational Wave Observations. *Nature*, **323**, 310

Shapiro, S.L. and Teukolsky, S.A. (1983): *Black Holes, White Dwarfs, and Neutron Stars*. Wiley, New York

Shibata, M. (1996): Instability of Synchronized Neutron Stars in the First Post-Newtonian Approximation of General Relativity. *Prog. Theor. Phys.*, **96**, 317

Shibata, M., Nakamura, T., and Oohara, K. (1992): Coalescence of Spinning Binary Neutron Stars of Equal Mass. *Prog. Theor. Phys.*, **88**, 1079

Strain, K.A. and Meers, B.J. (1991): Experimental Demonstration of Dual Recycling for Interferometric Gravitational-Wave Detectors. *Phys. Rev. Lett.*, **66**, 1391

Taniguchi, K. and Nakamura, T. (1996): Innermost stable circular orbit of coalescing neutron star – black hole binary. *Prog. Theor. Phys.*, **96**, 693

Tassoul, M. (1975): On the Stability of Congruent Darwin Ellipsoids. *Astrophys. J.*, **202**, 803

Tassoul, J.-L. (1978): *Theory of Rotating Stars*. Princeton University Press, Princeton

Taylor, J.H. and Weisberg, J.M. (1989): Further Experimental Tests of Relativistic Gravity Using the Binary Pulsar PSR 1913+16. *Astrophys. J.*, **345**, 434

Thorne, K.S. (1995): Gravitational Waves. In: *Proceedings of the Snowmass 95 Summer Study on Particle and Nuclear Astrophysics* (eds. E.W. Kolb and R. Peccei). World Scientific, Singapore

Thorne, K.S. (1996): In: *Compact Stars in Binaries*, IAU Symp. 165 (eds. J. van Paradijs et al.). Kluwer, Dordrecht, 153

Thorsett, S.E., Arzoumanian, Z., McKinnon, M.M., and Taylor, J.H. (1993): The Masses of Two Binary Neutron Star Systems. *Astrophys. J.*, **405**, L29

Tutukov, A.V. and Yungelson, L.R. (1993): The Merger Rate of Neutron Star and Black Hole Binaries. *Mon. Not. Roy. Astron. Soc.*, **260**, 675

Usov, V.V. (1992): Millisecond Pulsars with Extremely Strong Magnetic Fields as a Cosmological Source of γ-Ray Bursts. *Nature*, **357**, 472

van den Heuvel, E.P.J. and Lorimer, D.R. (1996): On the Galactic and Cosmic Merger Rate of Double Neutron Stars. *Mon. Not. Roy. Astron. Soc.*, **283**, L37

Webbink, R.F. (1984): Double White Dwarfs as Progenitors of R Coronae Borealis Stars and Type I Supernovae. *Astrophys. J.*, **277**, 355

Will, C.M. (1994): In: *Relativistic Cosmology* (ed. M. Sasaki). Universal Academy Press, Tokyo, 83

Wilson, J.R. and Mathews, G.J. (1989): In: *Frontiers in Numerical Relativity* (eds. C.R. Evans et al.). Cambridge University Press, Cambridge, 306

Wilson, J.R. and Mathews, G.J. (1995): Instabilities in Close Neutron Star Binaries. *Phys. Rev. Lett.*, **75**, 4161

Wilson, J.R., Mathews, G.J., and Marronetti, P. (1996): Relativistic numerical model for close neutron-star binaries. *Phys. Rev. D*, **54**, 1317

Wiseman, A.G. (1993): Coalescing Binary Systems of Compact Objects to (post)$^{5/2}$-Newtonian Order. IV. The Gravitational Wave Tail. *Phys. Rev. D*, **48**, 4757

Wolszczan, A. (1991): A nearby 37,9-ms Radio Pusar in a Relativistic Binary System. *Nature*, **350**, 688

Wolszczan, A. (1994): Confirmation of Earth-Mass Planets Orbiting the Millisecond Pulsar PSR B 1257+12. *Science*, **264**, 538

Yungelson, L.R., Livio, M., Tutukov, A.V., and Saffer, R.A. (1994): Are the Observed Frequencies of Double Dengerates and SN Ia Contradictory? *Astrophys. J.*, **420**, 336

Zhuge, X., Centrella, J.M., and McMillan, S.L.W. (1994): Gravitational Radiation from Coalescing Binary Neutron Stars. *Phys. Rev. D*, **50**, 6247

Quasinormal Ringdown: The Late Stage of Neutron Star Mergers

Hans-Peter Nollert

Institut für Theoretische Astrophysik
Universität Tübingen, Auf der Morgenstelle 10, D-72076 Tübingen
nollert@tat.physik.uni-tuebingen.de

Abstract

Perturbation techniques have been applied successfully to obtain the total radiated energy as well as wave forms for the collision of two black holes. Therefore, it appears promising to extend these methods to the case of colliding neutron stars as well. *Stationary* perturbations (quasi-normal modes) both of black holes and of neutron stars have been investigated extensively. *Time-dependent* perturbations, yielding radiated energy and wave function for a given initial configuration, have mostly been studied for black holes. We review the basic techniques for treating perturbations of black holes and neutron stars, summarize the results, and discuss the problems that have to be solved if a time-dependent perturbation treatment is to be applied to the collision of neutron stars. We will limit this discussion to final objects which are spherically symmetric; however, the technique can be extended to other cases, such as final black holes or neutron stars carrying spin.

1 Introduction

The coalescence of massive bodies, in particular binary neutron stars or black holes, is one of the most fascinating topics in general relativity, and it carries considerable astrophysical significance. Such an event is expected to produce significant gravitational radiation, carrying a wealth of information about the process as well as about the objects involved in it. Several large interferometric gravitational wave detectors are currently under construction (Danzmann 1997). The availability of templates for the expected waveforms is vital for interpreting the data of these detectors, adding urgency to efforts to complete theoretical studies of such mergers.

The Grand Challenge alliance attempts to solve this problem for the coalescence of black holes by developing a fully relativistic, three-dimensional computer code (Finn 1997). More recently, projects are getting underway to do similar calculations for neutron stars. Figure 1 shows a schematic representation of the coalescence of a binary system. We can distinguish three major phases:

I: A semi-stable phase, where the two compact objects are in a close orbit. The orbit slowly decays due to energy lost by gravitational radiation. This phase has been treated in post-Newtonian approximations to various orders.

II: After the orbit has decayed past a certain limit, it becomes unstable, and the actual merger of the two objects occurs. Newtonian calculations of the neutron star mergers have been carried out [Rasio and Shapiro (1992), Rasio (1997) and references therein, Ruffert et al. (1996), Zhuge et al. (1994, 1996)]. However, owing to the strength of the gravitational field, this phase will generally require a relativistic treatment. Calculations involving post-Newtonian or other approximations have been done, e.g., by Taniguchi and Nakamura (1996), Baumgarte et al. (1997), Wilson et al. (1996), and Shibata (1997). It remains to be seen if these results prove reliable, or if a fully relativistic numerical treatment is necessary to obtain correct results.

III: After the merger, the final object has essentially formed, but has not yet reached its final, stationary state. The deviations from this final state will be radiated away during the ringdown phase. This phase can be treated using perturbation techniques.

The supercomputer calculations for binary black hole coalescence are extremely demanding, both in terms of techniques and algorithms to be developed, and in terms of the required computer resources. For neutron stars, the absence of a horizon eliminates some problems; on the other hand, the hydrodynamics of the stars' matter must now be included.

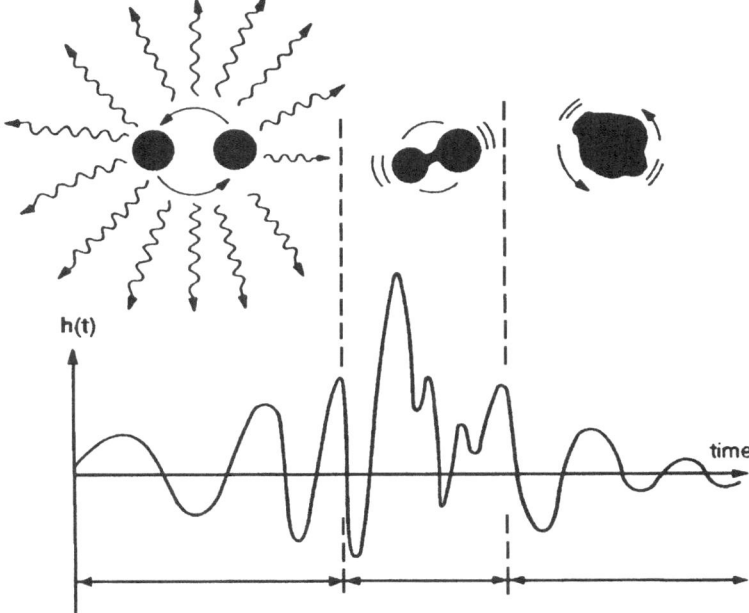

Figure 1 From Vogt (1992): Schematic representation of the coalescence of a binary system.

It turns out, though, that perturbation methods can actually be applied to a much broader range than what we have labelled as "phase III" above. Price and Pullin (1994) first explored this for a head-on collision of initially static black holes. They assume that the two black holes are so close to each other at the beginning of the collision that they have already formed a common horizon, and treat this initial configuration as a perturbation of a single black hole (Fig. 2). Comparison with the results of supercomputer simulations shows that this approach yields realistic results for larger initial separations than one would expect a priori. It has further been shown (Abrahams and Cook 1994, Baker and Li 1997) that for head-on collisions, connecting Newtonian calculations for large separations directly to the perturbation treatment (in effect, eliminating phase II) is a viable and robust procedure.

The inclusion of second-order perturbation terms improves these results further, and also provides an independent error estimate for the first-order results (Gleiser et al. 1996).

It is not clear yet whether the perturbation approach will give similarly excellent results for more complicated initial configurations, such as including spinning black holes or non-axisymmetric configurations. A further complication arises if the initial configuration — and therefore the final black hole as well — contains net angular momentum, which is obviously the case for the inspiral coalescence of binary black holes.

In any case, it is clear that the perturbation approach provides a very interesting alternative to computing the radiated energy and the wave forms of black hole collisions. It can serve as an alternative route in its own right, and also as a benchmark for testing fully relativistic numerical codes.

The question thus arises naturally whether perturbation techniques can be used to study collisions of neutron stars as well. Not much work has been done in this direction. Stationary perturbations of neutron stars (quasinormal modes) have been studied extensively. However, if we want to specify an initial configuration, and then determine the total energy and the wave form of the gravitational radiation that is emitted in the ensuing collision, then we need to treat time-dependent perturbations. We will also have to find a way to derive initial data for the perturbations from the physical configuration that we wish to study.

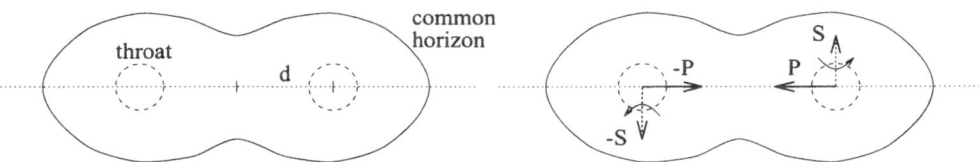

Figure 2 $1+1 \rightarrow 1$: Treating a close collision of compact object as a perturbation of the final object.

2   Perturbations of Compact Objects

2.1   Perturbations of Black Holes

Spherically Symmetric Equilibrium State

The spherically symmetric equilibrium state of a black hole is, of course, the well-known Schwarzschild spacetime:

$$\left(ds^2\right)_0 = -\left(1 - \frac{2M}{r}\right)dt^2 + \left(1 - \frac{2M}{r}\right)^{-1}dr^2 + r^2\left(d\vartheta^2 + \sin^2\vartheta d\phi^2\right). \tag{1}$$

Perturbation of the Equilibrium

We will write a perturbed Schwarzschild metric in the usual way:

$$ds^2 = \left(ds^2\right)_0 + h_{\mu\nu}dx^\mu dx^\nu, \tag{2}$$

where h is assumed to be small in an appropriate sense. At this point, we have not chosen a particular gauge for the perturbation. Using a complete set of tensor spherical harmonics $T_{\mu\nu}^{lmp}(\vartheta,\phi)$ (Zerilli 1970a, Moncrief 1974), we may represent the perturbation as the sum

$$h_{\mu\nu}(t,r,\vartheta,\phi) = \sum_{l=0}^{\infty}\sum_{m=-l}^{l}\sum_{n=1}^{10} C^{lmn}(t,r)T_{\mu\nu}^{lmn}(\vartheta,\phi). \tag{3}$$

Just as their scalar counterparts, the tensor spherical harmonics possess distinct values of l and m. In addition, they can be classified into even and odd parity. For each value of l and m, there are 10 tensors; we use the letter n as an additional label to distinguish them.

This kind of decomposition makes sense here: Due to the spherical background, perturbations belonging to different values of l and m or to different parities will not mix. Therefore, we can examine the contribution of each (l,m,P) mode separately, adding up the final result.

From the coefficients $C^{lmn}(t,r)$ in Eq. (3), we can construct a scalar function (usually called the Zerilli function) (Detweiler 1979), separately for each l and m and for even and odd parity contributions. This construction can be found in (Abrahams and Price 1996). Here we only note that it is linear, and write it symbolically as

$$Q^+(t,r) = \mathcal{L}^+(C^{n,\text{even}}) \tag{4}$$
$$Q^\times(t,r) = \mathcal{L}^\times(C^{n,\text{odd}}). \tag{5}$$

From now on, we are omitting the indices l,m.

After considerable analytical acrobatics (Regge and Wheeler 1957, Zerilli 1970b, Chandrasekhar 1975, Chandrasekhar 1983), the linearized field equations lead to a differential equation for the Zerilli function which has the form of a wave equation with a potential:

$$\left[\frac{\partial^2}{\partial t^2} - \frac{\partial^2}{\partial r_*^2} + V(r_*)\right]Q(t,r) = 0, \tag{6}$$

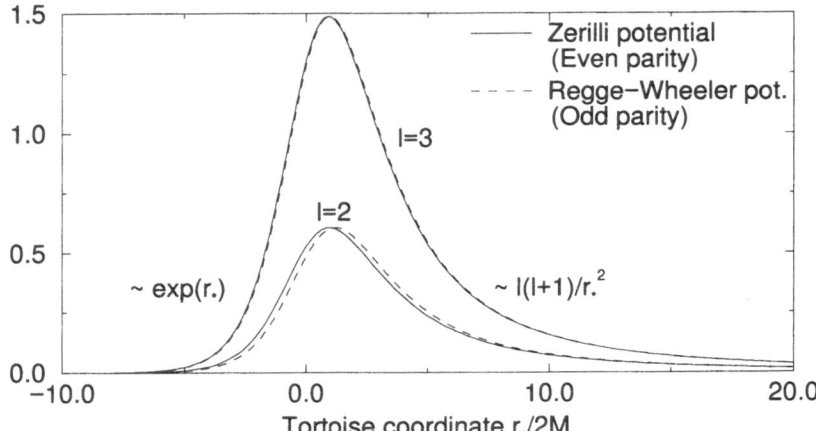

Figure 3   Potentials in the wave equation for time dependent perturbations of Schwarzschild black holes.

where M is the mass of the background black hole, and r_* is the tortoise coordinate $r_* = r + 2M \ln(r/2M - 1)$. While the form of Eq. (6) is the same for odd and even parity, the potentials are different:

Even parity (Zerilli potential):

$$V(r) = \left(1 - \frac{2M}{r}\right) \left[\frac{72M^3}{r^5\lambda^2} - \frac{12M}{r^3\lambda^2}(l-1)(l+2)\left(1 - \frac{3M}{r}\right) + \frac{(l-1)\,l\,(l+2)(l+1)}{r^2\lambda}\right],$$
$$(7)$$

$$\lambda = l(l+1) - 2 + \frac{6M}{r}.$$

Odd parity (Regge-Wheeler potential):

$$V(r) = \left(1 - \frac{2M}{r}\right) \left[\frac{l(l+1)}{r^2} - \frac{6M}{r^3}\right].$$
$$(8)$$

Note that the potentials, and therefore the wave equation, depend on l, but not on m. While the functional forms of the Zerilli and the Regge-Wheeler potential appear to be quite different, Fig. 3 shows that their values differ only slightly from each other.

2.2   Perturbations of Neutron Stars

Spherically Symmetric Equilibrium State

Due to the extra freedom associated with the material of the star, the determination of the equilibrium state of a neutron star is somewhat more involved than for a black hole. The following is only a brief summary; extensive discussions can be found in just about any textbook on general relativity, e.g. Misner et al. (1973) and Weinberg (1972).

We start by writing the metric (inside as well as outside the star) in the form

$$ds^2 = (ds^2)_0 = -e^{\nu(r)}dt^2 + e^{\lambda(r)}dr^2 + r^2\left(d\vartheta^2 + \sin^2\vartheta d\phi^2\right).$$
$$(9)$$

Assuming that the material of the neutron star behaves like an ideal fluid, we have for the energy-momentum tensor

$$T_{\mu\nu} = pg_{\mu\nu} + (p + \rho)u_\mu u_\nu, \tag{10}$$

where $\rho(r)$ is the total energy density and $p(r)$ the total pressure. Conservation of energy-momentum, i.e., the condition $T^\mu_{\nu;\mu} = 0$, yields

$$\frac{d\nu}{dr} = -2(\rho + p)^{-1}\frac{dp}{dr}. \tag{11}$$

Using the mass function $m(r)$, defined as

$$m(r) = \int_0^r 4\pi r'^2 \rho dr', \tag{12}$$

in connection with the field equations, we obtain

$$e^{-\lambda} = 1 - \frac{2m}{r}. \tag{13}$$

Again using the field equations, as well as Eq. (11), we obtain the relativistic generalization for the condition of hydrostatic equilibrium:

$$\frac{dp}{dr} = -\frac{(\rho + p)(m + 4\pi r^3 p)}{r(r - 2m)}. \tag{14}$$

Equations (14) and (11), together with Eq. (12), are usually referred to as the Oppenheimer-Volkov (OV) or Tolman-Oppenheimer-Volkov (TOV) equations (Weinberg 1972, Misner et al. 1973, Shapiro and Teukolsky 1983).
The equation of state for the star's material,

$$p = p(\rho), \tag{15}$$

completes the required equations for the determination of the unknown functions. Realistic equations of state are usually given numerically; either in the form of tables which must be interpolated, or in the form of subroutines which compute the required values "on the fly". Polytropic equations of state are popular as analytic approximations.
We can use the equation of state (15) to eliminate $p(r)$ from Eqs. (14) and (12), providing us with a system of two ordinary differential equations for the two unknown functions $\rho(r)$ and $m(r)$. One initial condition is given by $m(r = 0) = 0$. This leaves us with a choice for one additional parameter, i.e., the central energy density $\rho(0)$. Spherically symmetric stars consisting of an ideal fluid (with given equation of state) therefore form a one-parameter family.
Once $\rho(r)$ and $m(r)$ are determined, we can use the equations above to find $p(r)$ and $\lambda(r)$. For the other function in the metric, it turns out that

$$\nu(r) = -\lambda(r). \tag{16}$$

Using $m(r = 0) = 0$ and $\rho(0)$, we integrate outward until we reach a radius $r = R_{\text{star}}$ such that $p(r = R_{\text{star}}) = 0$, which defines the radius of the star. Outside the star, we obtain, of course, a Schwarzschild metric with the mass parameter given by the total mass of the star:

$$M = \int_0^{R_{\text{star}}} 4\pi r^2 \rho dr. \tag{17}$$

Perturbation of the Star's Equilibrium

We will again denote the perturbation of the metric with $h_{\mu\nu}(t,r,\vartheta,\phi)$. In addition to the metric perturbations, we now have to deal with the perturbations of the energy density, $\delta\epsilon$, and of the pressure, δp. The Lagrangian displacement of the fluid elements is called δu_μ. Since $\delta\epsilon$ and δp are scalar quantities, the usual scalar spherical harmonics can be used for the separation of the angular variables ϑ and ϕ:

$$\delta\epsilon = \rho^{lm}(t,r)Y_{lm}(\vartheta,\varphi), \tag{18}$$

where a sum over l and m is implied. The pressure perturbation δp is determined from $\delta\epsilon$ via the equation of state. Using vector spherical harmonics for δu_μ, we have

$$
\begin{aligned}
\delta u_0 &= u_0^{lm}(t,r)Y_{lm}(\vartheta,\varphi) = -S_1^{lm}(t,r)Y_{lm}(\vartheta,\varphi) \\
\delta u_1 &= u_1^{lm}(t,r)Y_{lm}(\vartheta,\varphi) \\
\delta u_2 &= u_2^{lm}(t,r)\frac{\partial}{\partial\vartheta}Y_{lm}(\vartheta,\varphi) - u_3^{lm}(t,r)\frac{1}{\sin\vartheta}\frac{\partial}{\partial\varphi}Y_{lm}(\vartheta,\varphi) \\
\delta u_3 &= u_2^{lm}(t,r)\frac{\partial}{\partial\varphi}Y_{lm}(\vartheta,\varphi) + u_3^{lm}(t,r)\sin\vartheta\frac{\partial}{\partial\vartheta}Y_{lm}(\vartheta,\varphi).
\end{aligned}
\tag{19}
$$

The normalization of the four-velocity has been used to determine u_0 in terms of S_1, which is related to the (tt) component of the metric perturbation. Owing to the form of the vector spherical harmonics, the expansion coefficients do not have a one-to-one relationship to the components of the displacement velocity.

Instead of writing out the complete expansion of the metric perturbation in terms of tensor spherical harmonics, we will just indicate which components of $h_{\mu\nu}$ give rise to which expansion coefficients:

$$
h_{\mu\nu} \to
\begin{array}{ccc}
t & r & \vartheta,\varphi
\end{array}
\left(
\begin{array}{ccc}
S_1 & S_2 & V_1, V_2 \\
* & S_3 & V_3, V_4 \\
* & * & T_1, T_2, T_3
\end{array}
\right)
\begin{array}{c}
t \\ r \\ \vartheta,\varphi
\end{array}
. \tag{20}
$$

The perturbation of the energy-momentum tensor can be expressed in terms of these quantities with the help of Eq. (10); we only need it to first order in the perturbations. We will use the field equations in their first-order form [see App., Eqs. (59) and (60)]. We therefore need an expansion for the extrinsic curvature as well:

$$
k_{ij} \to
\begin{array}{cc}
r & \vartheta,\varphi
\end{array}
\left(
\begin{array}{cc}
K_1 & K_2, K_3 \\
* & K_4, K_5, K_6
\end{array}
\right)
\begin{array}{c}
r \\ \vartheta,\varphi
\end{array}
. \tag{21}
$$

Due to the static nature of the background metric, the perturbation of the extrinsic curvature is identical to the total extrinsic curvature. Linearizing the field equations around the spherically symmetric background yields a system of first order equations for the coefficients of these quantities. It can be written schematically as

$$\frac{\partial}{\partial t}\vec{Q}(t,r) = \underline{A}(r)\frac{\partial^2}{\partial r^2}\vec{P}(t,r) + \underline{B}(r)\frac{\partial}{\partial r}\vec{P}(t,r) + \underline{C}(r)\,\vec{P}(t,r). \tag{22}$$

These are only the time evolution equations; in addition, the field equations yield 4 constraint equations. Just as in the case of black holes, these equations depend on l, but not on m. Again, they separate into a set for even perturbations and one for odd perturbations. The "vectors" $\vec{Q}$ and $\vec{P}$ for these are

$$
\begin{aligned}
\vec{Q}^{\text{even}} &= (\quad\quad S_3, \quad V_3, T_1, T_2, K_1, K_2, K_4, K_5, \rho, u_1, u_2) \\
\vec{P}^{\text{even}} &= (S_1, S_2, S_3, V_1, V_3, T_1, T_2, K_1, K_2, K_4, K_5, \rho, u_1, u_2)
\end{aligned}
\tag{23}
$$

$$
\begin{aligned}
\vec{Q}^{\text{odd}} &= (\quad V_4, T_3, K_3, K_6, u_3) \\
\vec{P}^{\text{odd}} &= (V_2, V_4, T_3, K_3, K_6, u_3) \, .
\end{aligned}
\tag{24}
$$

We obviously have four quantities, S_1, S_2, V_1, V_2, which appear only on the right-hand side of the equation; we do not have any evolution equation for them. They are the expansion coefficients associated with the lapse and the shift vector.

We might think of using the constraint equations to determine these coefficients using the values of the others. However, if the constraint equations are satisfied on an initial slice, then the time evolution equations guarantee that they will remain satisfied for all time. Therefore, the constraint equations do not provide any information that is not already contained in the time evolution equations. The answer to this problem is that we have to impose gauge conditions. This may seem inappropriate, since we have already fixed the gauge by using the specific form (9) for the background metric of the spherically symmetric neutron star. However, we are still free to specify conditions for the metric in terms of the first order perturbation quantities. This will only affect the total metric to first order in the perturbation quantities, and leave the background metric intact.

A gauge commonly used for neutron star perturbations is the Regge-Wheeler gauge. It is characterized by the conditions

$$
\begin{aligned}
\frac{\partial}{\partial \vartheta}(h_{02} \sin \vartheta) &= \frac{\partial}{\partial \varphi}\left(\frac{h_{03}}{\sin \vartheta}\right) \\
\frac{\partial}{\partial \vartheta}(h_{12} \sin \vartheta) &= \frac{\partial}{\partial \varphi}\left(\frac{h_{13}}{\sin \vartheta}\right) \\
h_{23} = 0 \quad\quad\quad & h_{33} = \sin^2 \vartheta h_{22}.
\end{aligned}
\tag{25}
$$

There is no particularly intuitive justification for this choice; its main advantage is that it allows a considerable simplification of Eq. (22). As expected, the coefficients S_1, S_2, V_1, V_2 are fixed by these conditions. In addition, T_1, T_3, V_3, K_4 either have trivial solutions or are identical to other coefficients. We are now left with 8 equations for 8 unknowns for even perturbations, and 4 equations for 4 unknowns for odd perturbations.

Odd Perturbations

In the case of odd perturbations, the equation for u_3, the only matter coefficient involved, has a very simple solution

$$
u_3(t,r) = f(r).
\tag{26}
$$

This solution represents stationary, differential rotations. Since we assume the star to consist of an ideal fluid, there are no restoring forces between adjacent fluid shells which could lead to oscillations or to braking the rotation. Furthermore, this motion does not give rise to gravitational radiation. Therefore, metric perturbations are decoupled from matter perturbations in the odd case.

The equations for the matter perturbation coefficients can be reduced to a wave equation for a single coefficient. This equation depends on the background metric, but not on other perturbation quantities. The situation is therefore much the same as for black holes: The neutron star simply provides the background for the evolution of the metric perturbations, but, at least in first order, it is not influenced by them.

Odd perturbations have been studied by Chandrasekhar and Ferrari (1991) and by Kokkotas (1994). Due to their similarity with black hole perturbations, we will not discuss them any further. In the following, whenever we mention neutron star perturbations, we will mean specifically even parity perturbations.

Even Perturbations

Equations (22) and (24) represent 8 partial differential equations which are of first order in t, but their right-hand sides still contain second derivatives with respect to r. Introducing further auxiliary functions results in a hyperbolic system of 12 partial differential equations which are of first order in space and time.

This system has six eigenvalues which are different from zero, they come in three pairs having equal absolute value and opposite sign. They represent three dynamical degrees of freedom, two of which are associated with gravitational radiation, and one with matter oscillations.

The equations governing even-parity perturbations cannot be simplified further by analytical means. They must be integrated numerically, using conditions for regularity at the origin, for continuity at the star's surface, and for outgoing radiation at spatial infinity as boundary conditions.

3   Quasi-Stationary Perturbations

Since the coefficients in the differential equations for perturbations both of black holes and neutron stars are not explicitly time dependent, we can study single-frequency solutions with a harmonic time dependence:

$$Q(t,r) = \phi(\omega,r)e^{i\omega t}. \tag{27}$$

3.1   Quasinormal Modes of Black Holes

Inserting Eq. (27) into the time-dependent wave equation (6), we obtain a time-independent, ordinary differential equation for $\phi(\omega,r_*)$. In the case of black holes, this equation becomes

$$\phi''(\omega,r_*) + \left(\omega^2 - V(r_*)\right)\phi(\omega,r_*) = 0. \tag{28}$$

This equation resembles a time-independent Schrödinger equation with potential $V(r_*)$. When perturbations of black holes were first studied numerically, oscillations at characteristic frequencies were found to dominate the response at relatively late times (Press 1971, Cunningham et al. 1978). This suggested that there are oscillatory modes which are equivalent to normal modes of an oscillating system, which could be obtained by subjecting Eq. (6) to a normal mode analysis. However, since the potential $V(r_*)$ is positive everywhere, we cannot impose on Eq. (28) the usual boundary conditions for bound systems, i.e., vanishing of the wave function for large values of the radial coordinate r_*. Instead, boundary conditions were derived ad hoc by requiring that there should be no radiation coming in from infinity, nor emerging from the black hole. This leads to the requirement

$$\phi(\omega,r_*) \sim e^{i\omega r_*} \quad \text{as} \quad r_* \to -\infty \quad \text{and} \quad \phi(\omega,r_*) \sim e^{-i\omega r_*} \quad \text{as} \quad r_* \to \infty. \quad (29)$$

More precisely, in order to obtain the harmonic solution and the time-independent version of Eq. (6), we have to subject Eq. (6) and its solution to a Laplace transform, rather than a Fourier transform (Nollert and Schmidt 1992). This leads to an *inhomogeneous* time-independent equation,

$$\phi''(\omega,r_*) + \left(\omega^2 - V(r_*)\right)\phi(\omega,r_*) = -\dot{Q}_0(r_*) - i\omega Q_0(r_*) \equiv J(\omega,r_*) \quad (30)$$

which explicitly contains the initial data for the time-dependent solution on its right-hand side. The complete time-dependent solution is then obtained via the inverse Laplace transform:

$$Q(t,r_*) = \frac{1}{2\pi} \int_{-\infty+i\epsilon}^{\infty+i\epsilon} \phi(\omega,r_*) e^{i\omega t} d\omega, \quad (31)$$

where ϵ is an arbitrarily small positive number. The solution for the time-independent equation (30) is obtained via a Green's function

$$\phi(\omega,r_*) = \int_{x_2}^{x_1} G(\omega,r_*,r_*') J(\omega,r_*') dr_*', \quad (32)$$

with the integral running over a range $x_2 \leq r_* \leq x_1$ where the initial data is non-zero. The Green's function $G(\omega,r_*,r_*')$ is constructed in the usual way from solutions of the homogeneous version of Eq. (30) that satisfy a boundary condition at only *one* boundary each (Nollert and Schmidt 1992). It then turns out that this Green's function has poles in the complex frequency plane which lead to the modes at the characteristic frequencies found in the numerical experiments. Analyzing the properties of the Green's function further allows to identify other characteristic contributions to the time-dependent perturbation response, such as the late-time power-law tail (Leaver 1986, Andersson 1997).

Numerical Results

It turns out that there can be no quasinormal frequencies with negative or vanishing imaginary part. This makes physical sense: Based on the time dependence of Eq. (27), a positive imaginary part yields a damped oscillation, the damping being due to energy being carried off as gravitational radiation. On the other hand, the boundary conditions of Eq. (29) imply that the quasinormal mode solutions diverge, as functions of r_*, both at the horizon and at spatial infinity. This is not as unphysical as it may seem: A contribution to the wave function very far from the source must have originated at the source a very long time ago, when the wave function had to be much larger than at a later time. This illustrates that the paradigm of a stationary situation — one that has existed for an arbitrarily long time and will continue without change for an arbitrarily long time — is not really appropriate here: An arbitrarily long time ago, the perturbations at the source would have had to be arbitrarily large. Instead, we have to take into account that the perturbations started at a definite time $t = 0$, and they will die exponentially with time. This is the reason why we talk here about a quasi-stationary, rather than a stationary, picture.

In fact, the Green's function approach described above takes this into account explicitly; the Green's function itself, and the solutions used to construct it, do not diverge along the integral path for the inverse Laplace transformation. Also, note that there is no divergence along the characteristics of the wave equation, i.e., towards past or future null infinity.

Nevertheless, the presence of exponentially growing versus exponentially dying solutions has been a stumbling block for the numerical determination of quasinormal frequencies with large imaginary parts. For a long time, this restricted the known frequencies to the fundamental (least damped) one, and maybe one or two higher modes (Chandrasekhar and Detweiler 1975, Ferrari and Mashhoon 1984). Questions that remained unsolved included the total number of modes (finite or infinite), or whether the frequencies were unconstrained or confined to a finite part of the complex plane. Adapting a continued fraction technique used in molecular physics, Leaver (1986) successfully determined frequencies with imaginary parts as high as $\sim 30 \times 2M$. Nollert and Schmidt (1992) found a way to express the Wronskian appearing in the construction of the Green's function (32) as a convergent series, confirming Leaver's numerical results. Figure 4 shows the first 11 (or 12) quasinormal frequencies for $l = 2$.

Since then, various other techniques have been introduced (e.g. Andersson and Linnæus 1992), providing essentially identical results. A somewhat lingering debate remains over the existence or non-existence of an imaginary frequency at $\omega = 4i$ (for $l = 2$). Bachelot and Motet-Bachelot (1993) have presented a proof that the number of quasinormal modes is indeed countably infinite. Extending the continued fraction method pioneered by Leaver (1986), Nollert (1993) found a way to compute quasinormal frequencies with arbitrarily large imaginary parts.

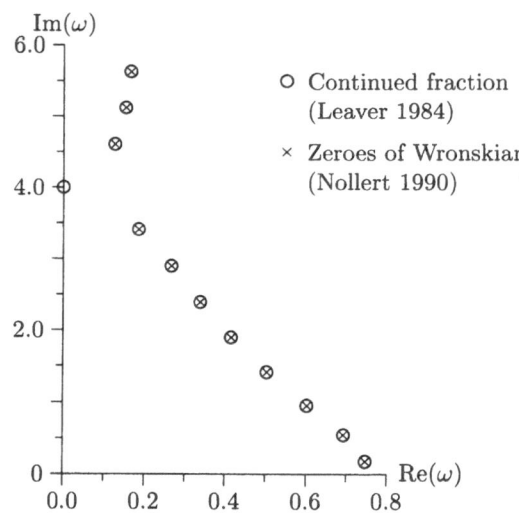

Figure 4   The first 11 (12) quasinormal frequencies of a Schwarzschild black hole for $l = 2$.

3.2   Quasinormal Modes of Neutron Stars

Using the harmonic time dependence of Eq. (27) for all perturbation quantities, we arrive at a system of 5 ordinary differential equations of first order. Detweiler and Lindblom (1985) have shown that this can be reduced to a system of 4 first-order equations. We do not have a horizon now, and the domain of integration therefore ranges from $r = 0$ to $r \to \infty$. Boundary conditions have to be imposed at $r = 0$, at the surface of the star, $r = R_{\text{star}}$, and at $r \to \infty$. The conditions and the restrictions they impose are:

- At the center ($r = 0$): The solutions have to be regular. This restricts 3 degrees of freedom.

- At the surface of the star: The Lagrangian change of density must vanish. However, this is not a real restriction: If the surface is defined by the pressure being zero in the equilibrium state, and if the density/pressure variation vanishes in the initial data, then the perturbation equations will guarantee that it continues to vanish at all times.

- At $r \to \infty$, the geometry of spacetime is identical to the black hole case, and we can impose the same boundary condition, resulting in purely outgoing radiation at spatial infinity. This restricts one degree of freedom.

Numerical Results

Neutron stars, unlike black holes, have a non-relativistic limit. Such a non-relativistic star will have a set of modes, associated with oscillations of the star's fluid. As long as we assume the star to consist of an ideal fluid and ignore internal dissipation, these oscillations will continue undamped forever, since there is no gravitational radiation to carry away energy. A relativistic star loses energy through gravitational radiation, but we expect this loss to be slow compared to the period of oscillation. We can therefore expect a set of modes with oscillation periods close to their Newtonian counterparts, and a damping rate small compared to the inverse oscillation period.

Such modes are indeed found, they are commonly called p-modes ('p' standing for pressure: the force driving the oscillation is the pressure of the fluid). Nobody has yet shown explicitly, though, that these relativistic modes will indeed approach the undamped modes of a Newtonian star if gravity is 'turned off' in some way.

Figure 5 shows the frequencies of all known types of quasinormal modes of relativistic neutron stars. In addition to these slowly damped p-modes, Kokkotas and Schutz (1992) found another set of modes which are much more strongly damped; their damping time is on the same order as the oscillation period. Inspired by a model system with very similar characteristics, Kokkotas and Schutz (1986) offer the interpretation that these modes are associated with the spacetime metric, rather than the material of the star. They call these modes w-modes; 'w' standing for (gravitational) wave. Since the metric perturbations contain less energy and are radiated away faster than the p-modes, the w-modes have to be much more strongly damped. In this respect, they should be analogous to quasinormal modes of a black hole, in particular since the metric of a spherically symmetric star outside the star itself is identical to the Schwarzschild metric. The w-modes, however, seem to be unbounded in their real parts, while the quasinormal modes of a black hole are bounded in the real parts and unbounded in the imaginary parts.

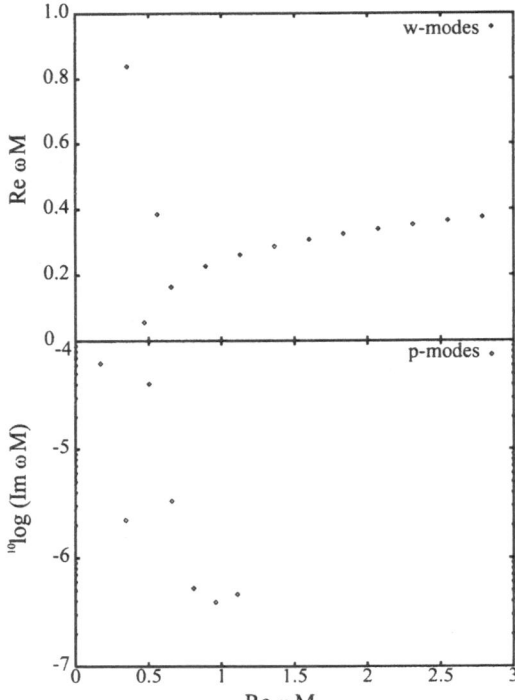

Figure 5   From Andersson et al. (1995): Some *p*-modes (bottom) and the two branches of *w*-modes of a relativistic neutron star.

Leins et al. (1993) found another branch of *w*-modes which appears to be more closely related to quasinormal modes of a black hole. For any given star, however, there is only a finite number of these *w*$_{\mathrm{II}}$-modes; this number increases if the star is made more compact. It turns out that neither the frequencies of the *w*-modes nor those of the *w*$_{\mathrm{II}}$-modes approach the quasinormal modes of a black hole if the (background) star is made as compact as general relativity allows (i.e., the star's Schwarzschild radius is 8/9 of its actual radius) (Leins et al. 1993).

3.3   Are Quasinormal Modes Physically Significant?

We have seen that quasinormal modes of black holes and neutron stars can be formally defined and numerically computed. At this point, the question arises whether they are meaningful and significant for the study of gravitational waves originating from perturbations of black holes and neutron stars, and ultimately for the study of close collisions. In fact, quasinormal modes were first studied because they were found in numerical studies of collapsing stars and black hole collisions (Press 1971, Vishveshwara 1970). Figure 6 shows gravitational waves being emitted from a perturbed Oppenheimer-Snyder dust collapse (Cunningham et al. 1979). At times between about $8 \times 4M$ and $30 \times 4M$, most of the radiation is made up of characteristic oscillations in the fundamental quasinormal mode. Subtracting the contribution of the fundamental mode from the total

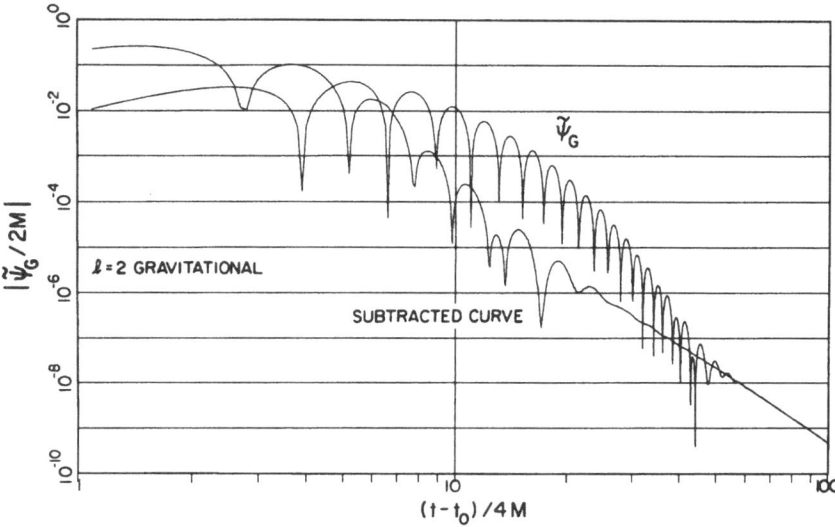

Figure 6 From Cunningham et al. (1979): Quasinormal ringing for quadrupole gravitational radiation, originating from a perturbed Oppenheimer-Snyder collaps.

signal, we find that the remainder is two orders of magnitude smaller than the total signal! However, at later time, the oscillations are swamped by a non-oscillatory power-law tail. This tail can obviously not be represented by a quasinormal mode, or any combination of them. Therefore, quasinormal modes do not contain the full information about the time evolution of perturbations. In addition to the tail, we notice in Fig. 4 that there are no quasinormal frequencies with large real parts, corresponding to fast oscillations. Therefore, perturbations containing fast oscillations cannot be represented by quasinormal modes.

The mathematical reason for this is that the differential equation (28), together with the boundary conditions (29), does not form a self-adjoint system. Linear bound systems of the kind that we usually study (such as a harmonic oscillator) form self-adjoint systems, guaranteeing that the set of normal modes will be orthogonal and complete. Therefore, studying the time evolution of normal modes will yield all the information one can obtain about the evolution of arbitrary initial data.

The situation for neutron stars is slightly different from black holes: The w-modes contain frequencies with large real parts. However, the time evolution of arbitrary initial data will still show a tail, which cannot be described by quasinormal ringing.

The 'missing contributions', such as the tail, high-frequency oscillations, and the response at early time, can be traced to the properties of the Green's function (32) for the solution of the time-independent equation. Extensive studies have been performed by Leaver (1986) and by Andersson (1997).

Figure 7 shows the extraction of waveforms from a fully non-linear computation of a head-on collision of two black holes. Again, quasinormal ringing is a prominent feature of the signal. Also shown is a fit with two modes $l = 2$. It turns out that they provide an excellent approximation to the original data, except for a short time after the onset of the signal. This result is rather surprising, since the quasinormal modes are based on a

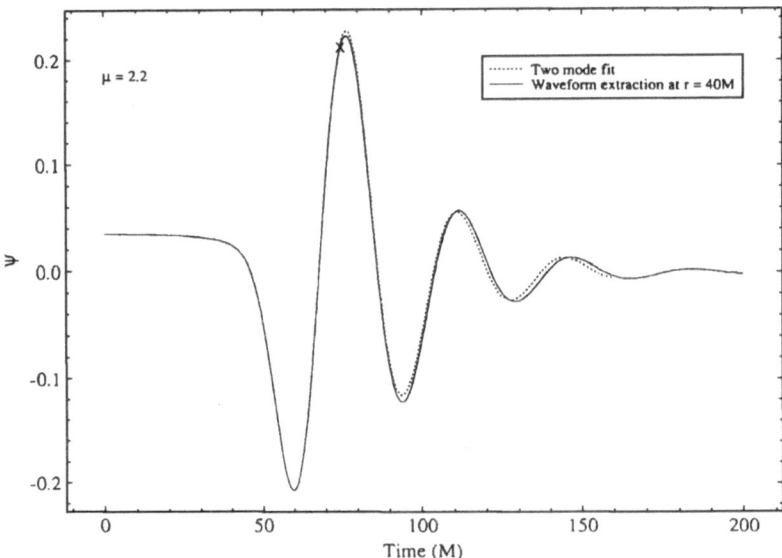

Figure 7   From (Anninos et al. 1993): Waveforms of gravitational radiation resulting from the head-on collision of two black holes, extracted at $R = 40M$ for $\ell = 2$. The cross marks the point where the quasinormal mode fit begins.

linearized perturbation treatment, while the data in Fig. 7 comes from a fully non-linear computation of an event which involves very strong curvature effects. Quasinormal ringing appears to be a very robust feature of processes which eventually form a stationary black hole, and probably of processes resulting in neutron stars as well.

4   Time-Dependent Perturbation Analysis

Why would we want to use time-dependent perturbation theory, even though the quasi-stationary picture of quasinormal modes seems to describe the time evolution of the perturbation so nicely?

We have already seen that the set of quasinormal modes does not yield a complete description of the time dependent wave function. As far as the power-law tail is concerned, we may choose to ignore this fact, since it contains only a tiny fraction of the total radiated energy. However, the figures in the last section also show that there is a time period from the onset of the signal, lasting about one or two full oscillations, where the quasinormal mode fits are completely inadequate. This time period contains a considerable part of the total energy and cannot be ignored.

Furthermore, we need to be able to quantify the strength of excitation of a given quasi-normal mode for a specific set of initial data. This can be achieved by fitting the modes to the time-dependent signal, but only if we have already determined this signal. An alternative is to deform the contour of integration in the inverse Laplace transform (31) to include the singularities that form the quasinormal modes (Sun and Price 1988). This procedure works well for initial data with compact support. Realistic astrophysical data, however, falls off far from the source, but it does not vanish completely. Together with the exponential growth of the quasinormal mode solution in space, this creates a problem when trying to determine the excitation coefficient. These problems have been studied (Sun and Price 1988, 1990), but it may be easier to do the time evolution explicitly.

4.1 Initial Data for Black Hole Collisions

We have seen that the time evolution for perturbations of black holes is exceedingly simple. Before we can do the time evolution, though, we need to find appropriate initial data. Specifically, we need to find the internal geometry and the extrinsic curvature of an initial slice of spacetime such that they satisfy the constraint equations, and that they can be regarded as describing two black holes which are close together and about to collide and merge. (See the appendix for the definition of the extrinsic curvature, its relationship to the internal geometry of the spatial slices, and the constraint equations.)

Constraint Equations, Conformally Flat Version

These four constraint equations are (Bowen and York 1980)

$$^3R - K_{ij}K^{ij} + K^2 \quad = \rho \qquad \text{Hamiltonian constraint} \qquad (33)$$
$$\nabla^i(K_{ij} - g_{ij}K) \quad = J_j \qquad \text{Momentum constraint,} \qquad (34)$$

where all indices i,j run from 1...3. g_{ij} is the interior geometry of the initial slice, ^3R the corresponding curvature scalar, and K_{ij} the extrinsic curvature, $K = Tr(K_{ij})$. ρ is the energy density and J_j the current density on the initial slice. In vacuum, we have

$$\rho = 0 \qquad J_j = 0. \qquad (35)$$

Equations (33) and (34) are four non-linear, coupled partial differential equations. Considerable simplification can be achieved if we assume that the initial geometry is conformally flat:

$$g_{ij} = \phi^4 \eta_{ij}, \qquad (36)$$

where η_{ij} is the metric of flat space. The extrinsic curvature then becomes

$$K_{ij} = \phi^{-2}\widehat{K}_{ij}. \qquad (37)$$

With this choice, the constraint equations are

$$\widehat{\nabla}^2\phi = -\tfrac{1}{8}\phi^{-7}\widehat{K}_{ij}\widehat{K}^{ij} \qquad \text{Hamiltonian constraint} \qquad (38)$$
$$\widehat{\nabla}^i\widehat{K}_{ij} = 0 \qquad \text{Momentum constraint.} \qquad (39)$$

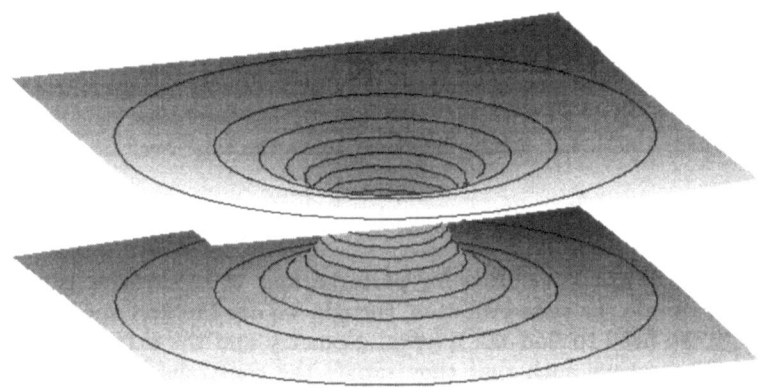

Figure 8   Schematic representation of a "wormhole" spatial slice of a Schwarzschild spacetime.

All quantities and operators in the conformal space are distinguished from their "physical" counterparts by a caret. Equations (38) and (39) are much simpler than Eqs. (33) and (34): The momentum constraint is linear in K, and it has now beome independent of ϕ. It can be solved first, leaving only the Hamiltonian constraint as one non-linear equation for the scalar function ϕ. Note that this is not an approximation: If ϕ and $\widehat{K}_{ij}$ satisfy Eqs. (38) and (39), then g_{ij} and K_{ij} will satisfy Eqs. (33) and (34) *exactly*. It is, however, a specific choice; and it is possible that not all solutions of the original Eqs. (33) and (34) will satisfy this pattern. In fact, this will generally be the case, since we can easily choose a slicing of a given spacetime in such a way that some spatial slice is not conformally flat.

Initial Deformation of Spacetime

To simplify things further, we will assume for now that the two black holes are initially at rest. The time derivative of the interior geometry of the spatial slices will then vanish initially. Setting the shift vector β^i to zero, and using the slicing of the Schwarzschild background spacetime, $\alpha = \sqrt{1 - 2M/r}$, we conclude that

$$K_{ij} = 0 = \widehat{K}_{ij} \qquad (40)$$

in this case. The momentum constraint is then satisfied trivially, and we are left with only the Hamiltonian constraint, which takes the simple form:

$$\widehat{\nabla}^2 \phi = 0. \qquad (41)$$

A simple solution, of course, is $\phi = 1$. This is not a very interesting solution, since it simply represents flat space; there are no black holes at all. There are many different solutions, characterized by the boundary conditions they satisfy. Misner (1963) has given an algorithm that allows the construction of solutions with an arbitrary number of momentarily static wormholes. A wormhole is a spatial slice of the Kruskal spacetime which contains two isometric parts (Fig. 8). The isometry condition at the throat defines the boundary condition at $r = a = m/2$. This boundary condition singles out the solution $\phi = 1 - a/r$ for $r > a$, and $\phi' = 1 - a/r'$, with $r' = (m/2)^2 1/r$, for $r < a$.

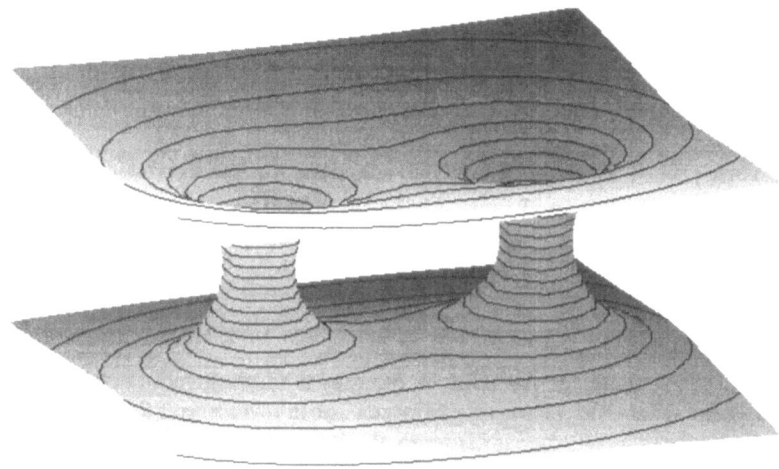

Figure 9 Schematic representation of a spatial slice containing two wormholes of equal size.

A solution describing two wormholes is more complicated; it consists of an infinite series of contributions. We now have isometry conditions across both throats (Fig. 9) with their associated boundary conditions. In either the upper or lower sheet, these wormholes look like two momentarily static black holes, which will then proceed to collide head-on. For several reasons, wormhole solutions are very popular in numerical relativity: They contain neither a singularity at the center, nor matter inside the horizon. They are given analytically and have well-known properties. They satisfy simple boundary conditions at the throats, which makes it possible to evolve only one sheet on the computer, saving memory and CPU time.

We should keep in mind, though, that wormhole solutions are simply one class of solutions of the constraint equations; there is no guarantee that they correspond to the physical situation of two black holes that are about to collide. It is clear that one wormhole will look from the outside like a Schwarzschild black hole. If we place two wormholes far enough apart that they are both in each other's asymptotic regime, it is safe to interpret them as two black holes far apart. However, it will generally be too expensive to follow the time evolution numerically from a point where the two black holes are asymptotically far apart. For the perturbation approach to be applicable, we also have to require them to be close initially, generally even closer than for numerical calculations.

It is not clear if the two-wormhole Misner solution is a good description of two close, initially static black holes. On the other hand, this configuration is not very realistic physically; there are probably no static black holes in the universe which are somehow held apart and are suddenly let go. We may therefore assume that the two-wormhole solution is as good an approximation for the initial data of an initially static black hole collision as any other, and feel justified in using it.

Initial Derivative of Metric Perturbation

If we want to allow the black holes to have momentum or spin initially, then we need to determine the extrinsic curvature, as a solution of the momentum constraint, for the initial spatial slice.
Bowen and York (1980) have found analytic solutions for the extrinsic curvature of one black hole carrying momentum or spin. These are simply

$$\widehat{K}_{ij} = \frac{3}{2r^2} \left[2P_{(i}n_{j)} - (\eta_{ij} - n_i n_j) P^c n_c \right] \tag{42}$$

$$\widehat{K}_{ij} = \frac{3}{r^3} \left(\epsilon_{kil} S^l n^k n_j + \epsilon_{kjl} S^l n^k n_i \right), \tag{43}$$

where P is the momentum, S the spin, n denotes a normal vector pointing from the center to the point where the extrinsic curvature is evaluated; and ϵ_{kjl} is the completely antisymmetric tensor.
Since the conformal form of the momentum constraint (39) is linear, we may simply add contributions from several black holes. In doing so, each contribution has to be computed with r and n relative to the center of the respective black hole. We generally use only the lowest order in L/r, which coincides with the lowest value of l, for the total extrinsic curvature.
While the Misner solution for the conformal factor is symmetrized to make the two sheets of the wormhole spacetime isometric, the expression (39) for the extrinsic curvature satisfy no such symmetry condition. It is possible to symmetrize tensors as well as scalar functions (Bowen and York 1980, Kulkarni 1984). However, we do not make use of this possibility, since the terms added for symmetrization are of higher order in L/r.
We need to multiply the conformal extrinsic curvature by ϕ^{-2} in order to obtain the extrinsic curvature in the original, physical space [Eq. (37)]. In this step, we simply use the conformal factor derived from the isotropic form of the Schwarzschild metric, rather than the correct solution for ϕ:

$$\phi^4 = \left(1 + \frac{M}{2\bar{r}} \right)^4, \tag{44}$$

where $\bar{r}$ is the isotropic Schwarzschild radial coordinate. Again, the difference is of higher order in L/r and needs not be considered in lowest order.
As an example, the extrinsic curvature of two initially static, counterrotating black holes (Cosmic Screw), in standard Schwarzschild coordinates, is given by

$$K = \frac{6SL}{r^2} \phi^2 \sin^2 \vartheta \left[\sin \vartheta \begin{pmatrix} 0 & 0 & 0 \\ 0 & 0 & 1 \\ 0 & 1 & 0 \end{pmatrix} + \frac{4}{r} \frac{2\bar{r} + M}{2\bar{r} - M} \cos \vartheta \begin{pmatrix} 0 & 0 & 1 \\ 0 & 0 & 0 \\ 1 & 0 & 0 \end{pmatrix} \right]. \tag{45}$$

Of course, the Hamiltonian constraint will not be as simple any more as it was in the initially static case. In general, we will not be able to find an analytic solution, and we will need to solve it numerically. However, if the momentum or spin is small, and if we are only interested in the lowest order, with respect to P or S, of the wave form or radiated energy, then we can use the fact that the right-hand side of the Hamiltonian constraint contains K quadratically:

$$\widehat{\nabla}^2 \phi \sim \phi^{-7} \widehat{K} \cdot \widehat{K} \approx 0. \tag{46}$$

This essentially tells us that initial momentum or spin will have a linear influence on the derivative of the spatial geometry, but it will deform the initial geometry itself only to higher order.

To summarize, the following approximations are made in the construction of initial data for black hole collisions which do not start out with an initially static configuration:

- The source term in the Hamiltonian constraint is ignored; the Misner solution for two wormholes is used.

- Only the lowest order in L/r, or l, is used in the extrinsic curvature.

- The extrinsic curvature is not symmetrized on the two sheets of the initial spatial slice.

- The conformal factor of a Schwarzschild spacetime is used when converting the conformal into the physical extrinsic curvature.

Each of these approximations is technical and used only to simplify the calculations. They could be relaxed if the use of more accurate initial data made a significant difference for the results to the order where the close approximation is valid. The strategy, then, is to use the Misner solution of the initially static case for the conformal factor ϕ, and the extrinsic curvature obtained from Eqs. (42) and (43) as a solution of Eq. (39) to determine the initial time derivative.

Initial Data for the Zerilli Equation

We now have to convert the initial spatial geometry and the extrinsic curvature into initial data for the Zerilli function. First, the spatial geometry has to be converted into a Schwarzschild background metric plus a perturbation:

$$g_{ij} = \phi\eta_{ij} = (g_0)_{ij} + h_{ij}, \tag{47}$$

where g_0 is a spatial slice of the Schwarzschild metric. This step has been worked out by Price and Pullin (1994). The (spatial) perturbation h_{ij} is then decomposed into tensor spherical harmonics as before, using only those tensors which do not have any non-zero time components:

$$h_{ij}(r,\vartheta,\phi) = \sum_{l=0}^{\infty} \sum_{m=-l}^{l} \sum_{n=5}^{10} C^{lmn}(r) T_{ij}^{lmn}(\vartheta,\phi). \tag{48}$$

The value of the Zerilli function itself is once again obtained by applying a linear operator to the resulting coefficients (Abrahams and Price 1996):

$$\text{Even parity:} \quad Q_0^+(r) \ = Q^+(t=0,r) = \bar{\mathcal{L}}^+(C^{n,\text{even}}) \tag{49}$$

$$\text{Odd parity:} \quad Q_0^\times(r) \ = Q^\times(t=0,r) = \bar{\mathcal{L}}^\times(C^{n,\text{odd}}). \tag{50}$$

The initial values for the derivative of the Zerilli function follow from the extrinsic curvature. Since the shift vector vanishes in the Schwarzschild spacetime, if using standard Schwarzschild coordinates, we simply have to apply the same procedure as above to the initial derivative of the spatial geometry (Abrahams and Price 1996), where α is the lapse function:

$$\partial_t g_{ij} = -2\alpha(r)K_{ij}(r,\vartheta,\phi) = \sum_{l=0}^{\infty} \sum_{m=-l}^{l} \sum_{n=5}^{10} \widetilde{C}^{lmn}(r)T_{ij}^{lmn}(\vartheta,\phi). \tag{51}$$

$$\text{Even parity:} \quad \dot{Q}^+{}_0(r) = \partial_t Q^+(t=0,r) = \bar{\mathcal{L}}^+(\tilde{C}^{p,\text{even}}), \tag{52}$$

$$\text{Odd parity:} \quad \dot{Q}^\times{}_0(r) = \partial_t Q^\times(t=0,r) = \bar{\mathcal{L}}^\times(\tilde{C}^{p,\text{odd}}). \tag{53}$$

4.2   Results for Black Hole Collisions

We have actually evolved only one scalar function which was derived from the full metric perturbation. It is possible to reconstruct the complete perturbed metric from the Zerilli function. However, we do not really need to do this, since we are mainly interested in the total energy radiated away as gravitational radiation, and in the wave form seen by an asymptotic observer. Therefore, all we need are certain components of the perturbed metric far away from the source.

Wave Forms and Total Radiated Energy

Taking the limit for large values of r, the power radiated into a solid angle element in a fixed direction is given by

$$\frac{dP}{d\Omega}(r \to \infty) = \frac{1}{16\pi}\left[\left(\frac{\partial}{\partial t}\frac{h_{\varphi\vartheta}}{\sin\vartheta}\right)^2 + \frac{1}{4}\left(\frac{\partial}{\partial t}\frac{h_{\varphi\varphi}}{\sin^2\vartheta} + \frac{\partial}{\partial t}h_{\vartheta\vartheta}\right)^2\right] \tag{54}$$

(Landau and Lifshitz 1975, Cunningham et al. 1978, Cunningham et al. 1979, Abrahams and Price 1996). Integrated over all angles, the total power radiated away towards spatial infinity is

$$P(r \to \infty) = \frac{1}{32\pi}\sum_{l,m}\left(\left(\frac{\partial}{\partial t}Q^+_{lm}\right)^2 + (Q^\times_{lm})^2\right). \tag{55}$$

Contributions from different values of l and m and from different parities can therefore simply be added for the total radiated energy. The wave form observed in a specific direction, however, will show interference between these different contributions; these interferences only cancel out upon integrating over all angles.

Different Configurations

The first configuration studied in the close approximation with the perturbation approach was the initially static collision (Price and Pullin 1994). The results are shown in Fig. 10, together with results of fully non-linear supercomputer calculations, and estimates from the two-phase approximation (Abrahams and Cook 1994, Baker and Li 1997b).

The most striking result is the excellent agreement between the close approximation and the full numerical calculations. The two black holes are surrounded by a common apparent horizon only for $\mu_0 < 1.36$. Even though the logarithmic scale of the energy de-emphasizes the errors to some extent, it is quite amazing that the result of the close approximation yields even anything reasonable for an initial configuration that barely forms a single black hole at all.

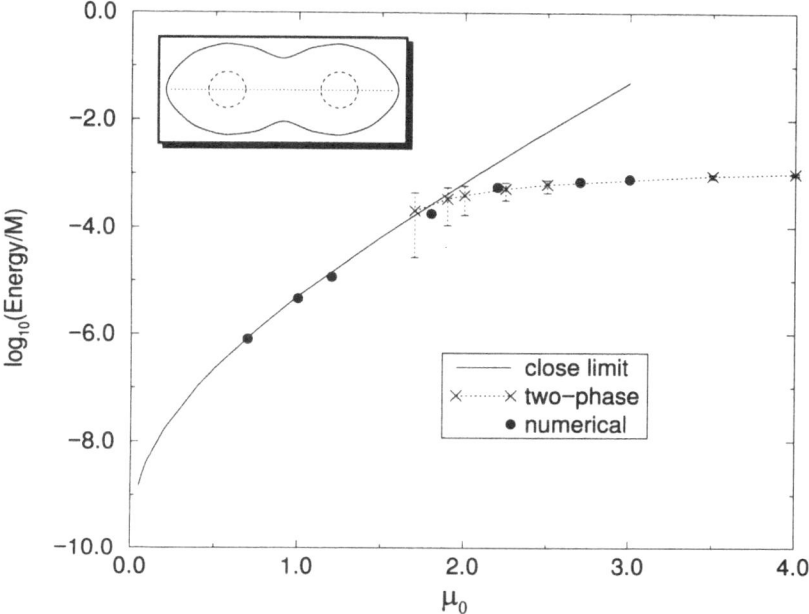

Figure 10   From Baker and Li (1997): Total radiated energy according to the close approximation, for an initially static collision of two black holes. The parameter μ_0 determines the initial separation.

One possible explanation for this surprisingly large range of validity of the close approximation is that most non-linear effects in the metric occur close to the horizon, and most of the gravitational radiation generated close to the horizon disappears into the black hole, anyway. A distant observer therefore never sees the differences between the correct metric distortions and those resulting from the close approximation.

This spectacular result encouraged the extension of this method to more complicated initial configurations. Allowing initial momentum such that the black holes approach each other can significantly increase the total radiated energy, as shown in Fig. 11. This situation is also somewhat more realistic, since it can be interpreted as the final stage of a collision that starts with the two black holes far apart.

It is somewhat surprising to see that the radiated energy actually decreases at first if the black holes are given some initial momentum. Only after increasing the initial momentum does the energy rise, eventually becoming much larger than the original, unboosted contribution. This dip can actually be understood as an interference between the contribution from the collision itself and from the initial momentum (Baker et al. 1997b). The perturbation approach provides additional insight in this case: Since it is linear, we can identify and follow the individual contributions through the whole process of generating initial data, computing the time evolution, and obtaining the radiated energy. This is not possible in the non-linear numerical calculation.

Table 1 lists the leading contributions to the energy emitted as gravitational radiation for the configurations we discuss. Both the contribution from the conformal factor itself, representing the initially static collision with $P = 0$, and the contribution from the extrinsic curvature due to the momentum, are characterized by $l = 2$, $m = 0$, and even

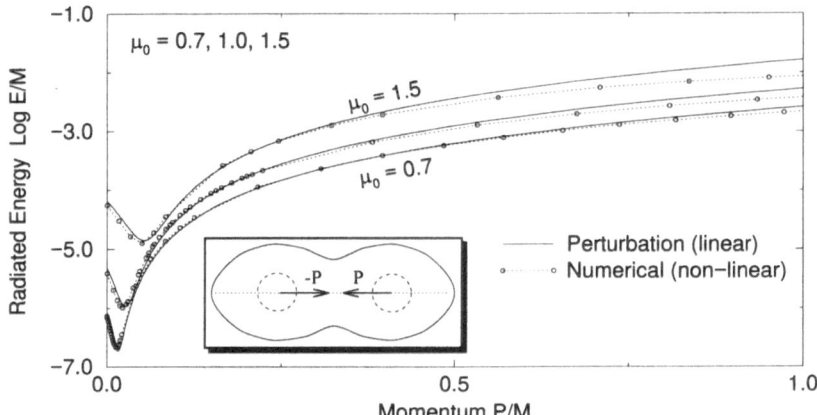

Figure 11 Total radiated energy according for black hole collisions, starting with the black holes moving towards each other. Three different initial separations are shown.

parity. Therefore, they combine to generate one Zerilli function. However, we can choose to determine the initial data, as well as the time evolved Zerilli function, separately for these two contributions. We then have to add them before computing the energy according to Eq. (55), leading to an interference and the dip in Fig. 11.

It turns out that the agreement between the linearized perturbation calculation and the non-linear, numerical results extends not only to rather large separations, as we have seen above, but also to rather large values of the initial momentum. This is again surprising, since we had explicitly used the assumption, in generating the initial data, that the initial momentum is small [cf. Eq. (46)].

		Contribution	Even	Odd
Initially Static	$l = 2$ $m = 0$	Misner	$Q = Q_{\mathrm{M}}$ $\dot{Q} = 0$	
Boosted Head-On	$l = 2$ $m = 0$	Misner Momentum	$Q = Q_{\mathrm{M}}$ $\dot{Q} = \dot{Q}_{\mathrm{P}}$	
Boosted Head-On	$l = 2$ $m = 0$	Misner	$Q = Q_{\mathrm{M}}$ $\dot{Q} = 0$	
		Momentum	$Q = 0$ $\dot{Q} = \dot{Q}_{\mathrm{P}}$	
Cosmic Screw	$l = 2$ $m = 0$	Misner	$Q = Q_{\mathrm{M}}$ $\dot{Q} = 0$	
		Spin		$Q = 0$ $\dot{Q} = \dot{Q}_{\mathrm{S}}$

Table 1 Leading contribution to the total energy emitted as gravitational radiation for the configurations discussed in the text.

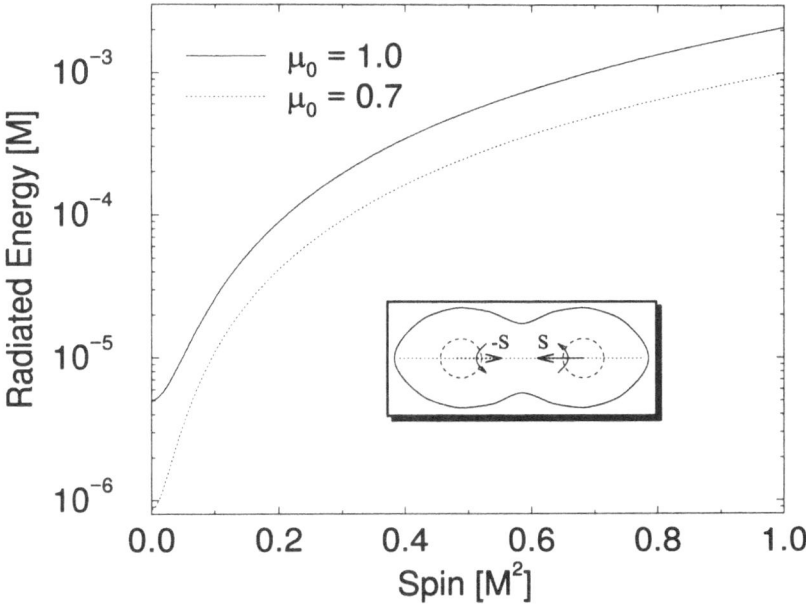

Figure 12   Total radiated energy for black hole collisions where the black holes carry spin, antiparallel and aligned with the direction of collision (cosmic screw). Two different initial separations are shown.

Another step towards astrophysically realistic collisions is to allow the black holes to carry spin. In order to be able to use the Schwarzschild metric as the background metric for the perturbation approach, we do not want overall spin in the initial (and therefore final) configuration. We will therefore assume that the spins of the two black holes are antiparallel and of equal magnitude, such that their sum vanishes. If the spins are furthermore aligned along the direction of collision, then the whole process will still be axisymmetric. Such a configuration is commonly referred to as a "cosmic screw".

Figure 12 shows that initial spin dramatically increases the total radiated energy, up to a factor of about 400 if $S = M^2$. Note that $S = 1M^2$ does not correspond to an extremal Kerr black hole; rather, S has to approach infinity for an extremal Kerr black hole [see Fig. 7 in Choptuik and Unruh (1986)]. If the two black holes are assumed to have collapsed from neutron stars which have been rotating as fast as millisecond pulsars, their spin will be much smaller, roughly $S = 0.05M^2$. However, the black holes might be spun up by accretion after collapsing, and black holes in the centers of galaxies could have spin parameters close to M^2 as well.

Contrary to the boosted case, there is no dip in the radiated energy as the spins are increased from zero. Table 1 shows that the conformal factor (Misner contribution) and the extrinsic curvature (spin contribution) generate different Zerilli functions. Therefore, we do not expect an interference when calculating the energy according to Eq. (55).

Further configurations have been studied, in particular non-axisymmetric collisions and pseudo-inspiral with small orbital angular momentum (Nollert 1997, Baker et al. 1997a).

4.3 Initial Data for Neutron Star Perturbations

Unlike the black hole case, there are no solutions readily available as initial data for neutron star collisions. The constraint equations now have to be solved including the matter source terms. Before we can even do that, we have to find the matter distribution and currents, which in turn depend on the prior evolution of the star, or on the (quasi-)equilibrium configuration that we want to start with.

Possible approaches to solving this problem include:

- Educated guess, just as in the black hole case. We need to find a solution of the constraint equations which represents both the two original stars that merged, and the one star that they merge into, which provides the background for the perturbation calculation. In the case of the black holes, the wormhole solutions have these properties: The individual throats, which continue to exist separately inside a potential common horizon, represent the two original black holes, while the common horizon characterizes the final black hole. No similar construction, as intuitively appealing as this, is available for the neutron star collision. It seems unlikely that analytic solutions can be found which retain this "double identity" for both the metric and the matter of the star.

- From fully relativistic numerical calculations of phase II. This may seem to defeat the purpose to some extent: If we have to start with a full numerical calculation, why not just keep running it into phase III, rather than switching to the perturbation technique? Numerical calculations are likely to be very expensive in terms of computer resources, it may therefore be desirable to stop the calculation as early as possible, and continue with a less demanding method. Even running numerical and perturbation calculations in parallel makes sense: It provides a check for the numerical code, and it may offer insights, such as in the black hole case, which would not be possible with a numerical calculation alone.

- From Newtonian or post-Newtonian calculations of phase I and II. Quasistable orbits of neutron star binaries can be computed to several post-Newtonian orders [e.g. Lincoln and Will (1990), Blanchet et al. (1996), and Shibata and Taniguchi (1997)]. It is also possible that hydrodynamic effects will dominate the actual merger phase (II) to such an extent that a treatment of the merger involving a post-Newtonian or other approximation [e.g. Taniguchi and Nakamura (1996), Baumgarte et al. (1997), Wilson et al. (1996), and Shibata (1997)], joined directly to a perturbation calculation, will actually provide a good approximation of the whole collision process, starting with a rather distant orbit and ending with the final equilibrium star. Comparison with fully relativistic numerical calculations will reveal just how important the effects of "full relativity" really are.

- For testing purposes, we may use initial data derived from the quasi-stationary results. If we chose a quasinormal mode of a particular frequency as initial data, then the time evolution will have to reproduce the known behavior of this quasinormal mode (Ruoff 1996). This has no astrophysical relevance, of course, since quasinormal modes do not correspond directly to any physical state of the system.

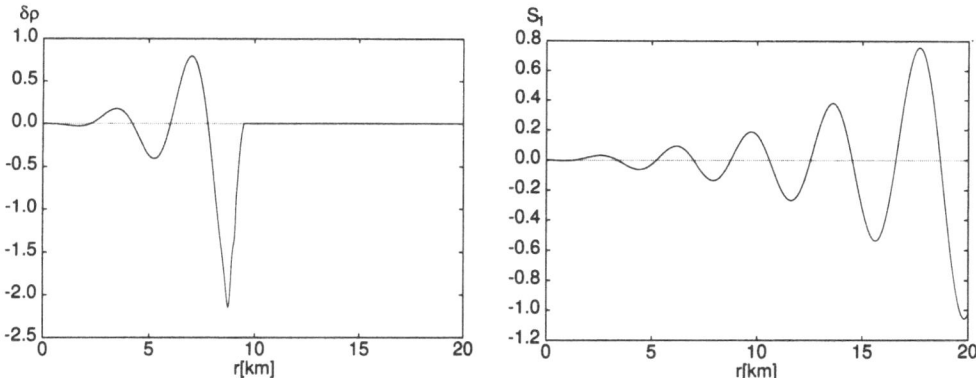

Figure 13   From Ruoff (1996): Initial data for the energy density perturbation (left) and one metric perturbation (right) in arbitrary units, using stationary solutions for the complex frequency $\omega = (1.3 + 0.1i)$ km^{-1}.

4.4   Results for Neutron Star Perturbations

No perturbation results are yet available for astrophysically interesting configurations. Codes exist for integrating the time-dependent perturbation equations (Ruoff 1996, Allen et al. 1997), but they have only been applied to test problems so far.

Ruoff (1996) has used results of stationary calculations performed by Leins et al. (1993). Instead of quasinormal frequencies, he used two arbitrary frequencies (one real, i.e., purely oscillatory without damping, and one complex, with damping). He has to omit the boundary condition at spatial infinity in order to obtain time-independent solutions for these frequencies. This implies that there is incoming radiation from infinity, but that does not present a problem for these test calculations. We simply have to make sure that the point of observation, the outer spatial boundary of the computational grid, and the time period of observation are chosen in such a way that the outer boundaries are not in the past light cone of the observation point at any moment during the observation. This also takes care of the problem that the time-independent solutions, whether they are quasinormal modes or not, do not vanish, but rather grow, outside the computational grid.

The equation of state for the background star is based on a Machleidt potential A (MPA) (Wu et al. 1991). The central pressure was chosen to be $p_c = 9.08 \times 10^{34}$ dyn/cm^2, corresponding to a central density of $\epsilon_c = 1.14 \times 10^{15}$ g/cm^3. This results in a total mass of $M = 0.802 M_\odot$ and a radius of $R = 10.09$ km.

Figure 13 shows part of the initial data for the complex frequency $\omega = (1.3 + 0.1i)$ km^{-1}. Shown are one of the metric perturbations, S_1, and the perturbation of the energy density ρ.

Figure 14 shows the evolution of two perturbation quantities as a function of the radial coordinate r and of time. In Fig. 15, the time dependence of S_1 at a fixed point inside the star, $r = 6.18$ km, is compared with the amplitude expected from the stationary calculation. The time-dependent integration is in excellent agreement with the expected behavior.

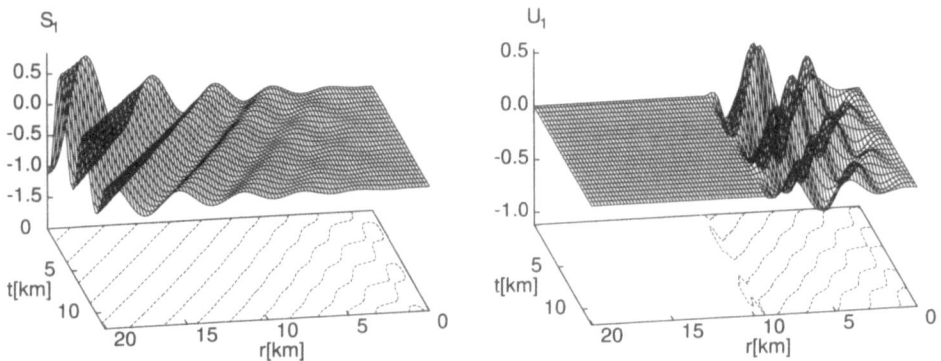

Figure 14   From Ruoff (1996): The perturbation of one metric coefficient S_1 and a component U_1 of the 4-velocity of the matter as functions of time and space, using the initial data shown in Fig. 13 (perturbations are in arbitrary units).

5   Discussion

The perturbation method for close collisions works surprisingly well for collisions of black holes. This could be due to the non-linear effects manifesting themselves mostly close to the horizon; most of the gravitational radiation carrying this information is therefore captured by the black hole and does not reach an asymptotic observer.

In fact, it turns out that a collision of two widely separated black holes can be treated using Newtonian calculations when they are far apart, directly joining them to a perturbation calculation when the black holes reach a certain separation. The results of this

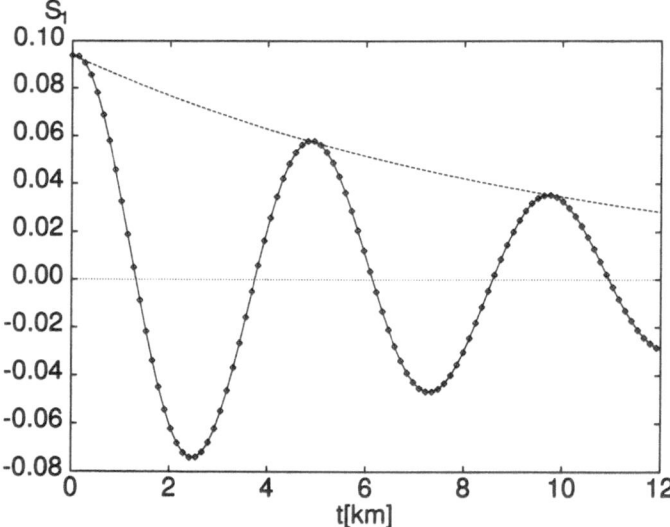

Figure 15   From Ruoff (1996): Comparison of the time dependence of the metric perturbation S_1 at a fixed point in space with the behavior expected from the stationary calculation.

procedure are very robust with respect to the choice of the point where the transition from Newtonian to perturbation approach is made (Abrahams and Cook 1994, Baker and Li 1997), and agree well with fully non-linear numerical calculations (Baker et al. 1997b).

It appears at first that the approach followed here should limit us not only to small separations, but also to small values of initial momentum or spin for the black holes. Again, it turns out that much larger values can be allowed than expected. The reason here is probably that the contributions of momentum or spin quickly become much larger than the one from the initial deformation of the spatial slice. This, in turn, depends on the term ϕ^{-7} on the right-hand side of the Hamiltonian constraint, which prevents the conformal factor from becoming large even if the extrinsic curvature is not small any more. The mistake we make in ignoring the influence of the extrinsic curvature K on the conformal factor via the Hamiltonian constraint can thus be ignored. This essentially means that details of the initial deformation, which are difficult to calculate, may not be important for the gravitational radiation any more when initial momentum or spin become fairly large.

All of these favorable coincidences do not occur in the neutron star case. There is no horizon which could swallow part of the gravitational radiation, destroying information about strong non-linear effects. The constraint equations now have source terms depending on the fluid of the star, making estimates about the behavior of their solutions for large separations and large momenta or spins more difficult.

Studying neutron star collisions in the perturbation approximation nevertheless makes sense: It will provide important checks for the much more complex numerical codes that are being developed. In addition, it is quite possible that the perturbation method will still work much better than expected, just as it did for black holes. In order to find out, we will have to follow this approach and examine the results we obtain.

The major problem in doing so lies in obtaining appropriate initial data. Other than for black holes, there are no analytical solutions available that we can use as an 'educated guess' for the correct initial data. We do not have the concept of two throats inside a common horizon any more, which allowed us to retain the signature of two individual black holes within the one black hole that formed in the collision. Therefore, we need to obtain initial data from numerical calculations specifically designed for this purpose, such as solving the constraint equations with source terms based on post-Newtonian results, or on results of a fully non-linear calculation of the merger phase.

Acknowledgments: We wish to thank J. Pullin, R. Price, P. Anninos, N. Andersson, LIGO, and J. Ruoff for permission to use figures and results from their work in this article. Discussions with, and comments from, J. Baker, R. Gleiser, P. Laguna, O. Nicasio, and J. Pullin have helped to improve this article. We acknowledge partial support through grant NSF PHY-9423950, by the Alfred P. Sloan foundation, and by the DFG through grant No 219/5-2.

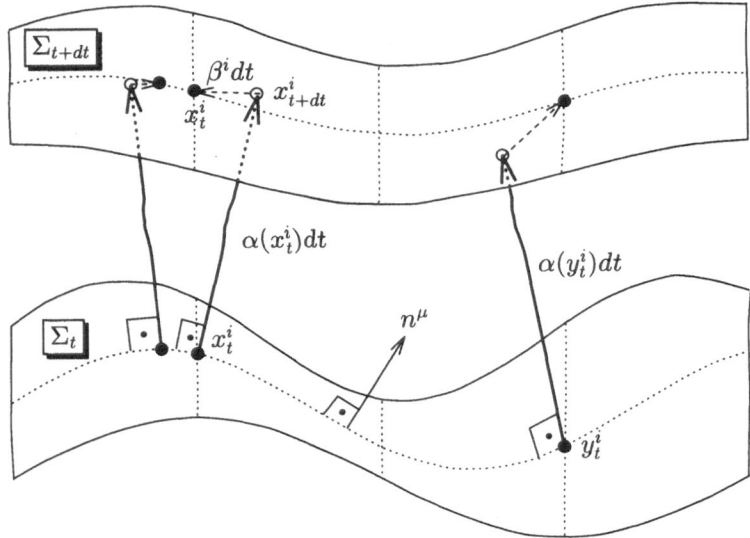

Figure 16 $3 + 1$ decomposition of spacetime into spatial slices (one spatial dimension suppressed).

Appendix

The 3+1 Decomposition of Spacetime

In order to describe the "time" evolution of a relativistic system, we may choose to represent the four-dimensional spacetime as a collection of three-dimensional space-like slices which are then "stacked" in a timelike direction.

Figure 16 shows the relationship between a slice at time t and one at time $t + dt$. The lapse function $\alpha(x^i)$ gives the distance between two points on adjacent slices along a normal direction. Conversely, choosing a lapse function $\alpha(x^i)$ will result in a specific slicing of a four-dimensional spacetime. The shift vector $\beta^i(x^i)$ represents the freedom to choose coordinates independently on each slice. Each of these slices has an interior, three-dimensional geometry, described by g_{ij}. The four-dimensional spacetime metric is directly related to this interior geometry, the lapse function, and the shift vector:

$$g_{\mu\nu} = \begin{pmatrix} \beta_k \beta^k - \alpha^2 & \beta_j \\ \beta_i & g_{ij} \end{pmatrix}.$$ (56)

As an example, we have for the Schwarzschild metric:

$$\alpha = \sqrt{1 - \frac{2M}{r}} \qquad \vec{\beta} = 0.$$ (57)

The three-dimensional extrinsic curvature K_{ij} describes how each slice is embedded in the four-dimensional spacetime:

$$K_{ij} = -n_{i;j}.$$ (58)

It is also related to the change of interior geometry from one slice to the next:

$$\partial_t g_{ij} = -2\alpha K_{ij} + g_{ij,k}\beta^k + g_{ki}\beta^k_{,j} + g_{kj}\beta^k_{,i}. \tag{59}$$

This dual role can be easily understood from Fig. 16: If we know the geometry in the neighborhood of a point x^i, we know how the normal vector "tilts" when we move from one point to another. This change of the normal vector defines the extrinsic curvature according to Eq. (58). On the other hand, if we know all normal vectors around x^i, as well as the lapse function $\alpha(x^i)$, then we can construct the surface at a later time $t + dt$ (assuming for a moment that $\beta^i = 0$), and then derive its internal geometry $g_{ij}(t+dt)$. Eventually, this leads to Eq. (59). The time derivative of the extrinsic curvature is given by:

$$\begin{aligned} \partial_t K_{ij} = \ & \alpha \ \left[R_{ij} + K K_{ij} - 2K_{ik}K^k{}_j - 8\pi \left(T_{ij} - \frac{1}{2}T g_{ij} \right) \right] \\ & - \ \alpha_{;j;i} + K_{ij,k}\beta^k + K_{ik}\beta^k_{,j} + K_{jk}\beta^k_{,i}, \end{aligned} \tag{60}$$

where $K = \mathrm{Tr}(K_{ij})$, $T = \mathrm{Tr}(T_{ij})$, and R_{ij} is the three-dimensional curvature tensor of the spatial slice. Choosing g_{ij} and K_{ij} as the unknown functions, we can regard Eqs. (59) and (60) as the first-order version of the gravitational field equations.

The interior geometry and the extrinsic curvature for any given spatial slice are not independent of each other. Four equations among the field equations are not related to the time evolution of g_{ij} and K_{ij}, but rather represent conditions which have to be satisfied in each spatial slice by itself. They are therefore called constraint equations. In the second order formulation of the field equations, the constraint equations contain at most first derivatives in time. In the first order version of Eqs. (59) and (60), they do not contain any time derivative at all. They do, however, always contain derivatives with respect to the spatial coordinates. The explicit form of the constraint equations is given by Eqs. (33) and (34).

References

Abrahams, A.M. and Cook, G.B. (1994): Collisions of boosted black holes: perturbation theory prediction of gravitational radiation. *Phys. Rev. D*, **50**, R2364

Abrahams, A.M. and Price, R.H. (1996): Applying black hole perturbation theory to numerically generated space times. *Phys. Rev. D*, **53**, 1963

Allen, G., Andersson, N., Kokkotas, K., and Schutz, B.F. (1997): Gravitational waves from pulsating stars: Evolving the perturbation equations for a relativistic star. gr-qc/9704023

Andersson, N. (1997): Evolving test-fields in a black-hole geometry. *Phys. Rev. D*, **55**, 468

Andersson, N. and Kokkotas, K.D. (1996): Gravitational waves and pulsating stars: What can we learn from future observations? *Phys. Rev. Lett.*, **77**, 4134

Andersson, N. and Linnæus, S. (1992): Quasinormal modes of a Schwarzschild black hole: improved phase-integral treatment. *Phys. Rev. D*, **46**, 4179

Andersson, N., Kokkotas, K.D., and Schutz, B.F. (1995): A new numerical approach to the oscillation modes of relativistic stars. *Mon. Not. Roy. Astron. Soc.*, **274**, 1039

Anninos, P., Hobill, D., Seidel, E., Smarr, L., and Suen, W.-M. (1993): Collision of Two Black Holes. *Phys. Rev. Lett.*, **71**, 2851

Bachelot, A. and Motet-Bachelot, A. (1993): Resonances of Schwarzschild Black Holes. In: *Proceedings of the "IV International Conference on Hyperbolic Problems"* (ed. A. Donato and F. Oliveri). Vieweg, Braunschweig

Baker, J. and Li, C.B. (1997): The two-phase approximation for black hole collisions: Is it robust? *Class. Quant. Grav.*, **14**, L77

Baker, J., Nollert, H.-P., and Nicasio, O. (1997a): Using the Close Limit Approximation for non-axisymmetric mergers of black holes. (in preparation)

Baker, J., Abrahams, A.M., Anninos, P., Brandt, S., Price, R.H., Pullin, J., and Seidel, E. (1997b): The collision of boosted black holes. *Phys. Rev. D*, **55**, 829

Baumgarte, T.W., Shapiro, S.L., Cook, G.B., Scheel, M.A., and Teukolsky, S.A. (1997): Binary Neutron Stars in Quasi-Equilibrium Circular Orbit: A Fully Relativistic Treatment. In: *Proceedings of the 18th Texas Symposium on Relativistic Astrophysics* (eds. A. Olinto et al.). World Scientific, in press

Blanchet, L., Iyer, B.R., Will, C.M., and Wiseman, A.G. (1996): Gravitational waveforms from inspiralling compact binaries to second-post-Newtonian order. *Class. Quant. Grav.*, **13**, 575

Bowen, J.M. and York Jr., J.W. (1980): Time-asymmetric initial data for black holes and black-hole collisions. *Phys. Rev. D*, **21**, 2047

Chandrasekhar, S. (1975): On the Equations Governing the Perturbations of the Schwarzschild Black Hole. *Proc. Roy. Soc. London A*, **343**, 289

Chandrasekhar, S. (1983): The Mathematical Theory of Black Holes. Clarendon Press, Oxford

Chandrasekhar, S. and Detweiler, S. (1975): The Quasi-Normal Modes of the Schwarzschild Black Hole. *Proc. Roy. Soc. London A*, **344**, 441

Chandrasekhar, S. and Ferrari, V. (1991): On the non-radial oscillations of a star III. A reconsideration of the axial modes. *Proc. Roy. Soc. Lond. A*, **434**, 449

Choptuik, M. and Unruh, W.G. (1986): An Introduction to the Multi-Grid Method for Numerical Relativists *Gen. Rel. Grav.*, **18**, 813

Cunningham, C., Price, R.H., and Moncrief, V. (1978): Radiation from Collapsing Relativistic Stars. I. Linearized Odd-Parity Radiation. *Astrophys. J.*, **224**, 643

Cunningham, C., Price, R.H., and Moncrief, V. (1979): Radiation from Collapsing Relativistic Stars. II. Linearized Even-Parity Radiation. *Astrophys. J.*, **230**, 870

Danzmann, K. (1997): Laser-Interferometric Gravitational Wave Detectors — on the Ground and in Deep Space. In: *This volume*

Detweiler, S.L. (1979): Black Holes and Gravitational Waves: Perturbation Analysis. In: *Sources of Gravitational Radiation* (ed. L. Smarr). Cambridge University Press, Cambridge

Detweiler, S.L. and Lindblom, L. (1985): On the nonradial pulsations of general relativistic stellar models. *Astrophys. J.*, **292**, 12

Ferrari, V. and Mashhoon, B. (1984): Oscillations of a Black Hole. *Phys. Rev. Lett.*, **52**, 1361

Finn, L.S. (1997): A numerical approach to binary black hole coalescence. In: *Proceedings of the 14th International Conference on General Relativity and Gravitation* (eds. M. Francaviglia, G. Longhi, L. Lusanna, and E. Sorace). World Scientific, Singapore

Gleiser, R., Nicasio, O., Price, R., and Pullin, J. (1996): Colliding black holes: how far can the close approximation go? *Phys. Rev. Lett.*, **77**, 4483

Kokkotas, K.D. (1994): Axial Modes for Relativistic Stars. *Mon. Not. Roy. Astron. Soc.*, **268**, 1015

Kokkotas, K.D. and Schutz, B.F. (1986): Normal Modes of a Model Radiating System. *Gen. Rel. Grav.*, **18**, 913

Kokkotas, K.D. and Schutz, B.F. (1992): W-Modes: A New Family of Normal Modes of Pulsating Relativistic Stars. *Mon. Not. Roy. Astron. Soc.*, **255**, 119

Kulkarni, A.D. (1984): Time-asymmetric initial data for the N black hole problem in general relativity. *J. Math. Phys.*, **25**, 1028

Landau, L.D. and Lifshitz, E.M. (1975): Eq. [107.12] in: *The Classical Theory of Fields*. Pergamon, London

Leaver, E. (1984): The Quasi-Normal Modes of Schwarzschild and Kerr Black Holes. *Ph.D. thesis*, Salt Lake City

Also published as: An Analytic Representation for the Quasi-Normal Modes of Kerr Black Holes. *Proc. Roy. Soc. Lond. A*, **402**, 285

Leaver, E. (1986): A spectral Decomposition of the Perturbation Response of the Schwarzschild Geometry. *Phys. Rev. D*, **34**, 384

Leins, M., Nollert, H.-P., and Soffel, M. (1993): Nonradial Oscillations of Neutron Stars: A New Branch of Normal Modes. *Phys. Rev. D*, **48**, 3467

Lincoln, C.W. and Will, C.M. (1990): Coalescing binary systems of compact objects to (post)5/2-Newtonian order: Late-time evolution and gravitational-radiation emission. *Phys. Rev. D*, **42**, 1123

Misner, C.W. (1963): The Method of Images in Geometrostatics. *Ann. Phys.*, **24**, 102

Misner, C.W., Thorne, K.S., and Wheeler, J.A. (1973): Gravitation. W.H. Freeman, New York

Moncrief, V. (1974): Gravitational Perturbations of Spherically Symmetric Systems. I. The Exterior Problem. *Annals of Physics*, **88**, 323

Nollert, H.-P. (1990): Astrophysik in der Schwarzschildmetrik am Beispiel von Quasi-Normalmoden schwarzer Löcher und Lichtablenkung bei Röntgenpulsaren. Ph.D. thesis, Universität Tübingen [see also: Nollert and Schmidt (1992)]

Nollert, H.-P. (1993): Quasinormal Modes of Schwarzschild Black Holes: The Determination of Quasinormal Frequencies with Very Large Imaginary Parts. *Phys. Rev. D*, **47**, 5253

Nollert, H.-P. (1997): A slightly less grand challenge: Colliding black holes using perturbation techniques. In: *Proceedings of the 18th Texas Symposium on Relativistic Astrophysics* (eds. A. Olinto et al.). World Scientific, Singapore, in press

Nollert, H.-P. and Schmidt, B.G. (1992): Quasi-Normal Modes of Schwarzschild Black Holes: Defined and Calculated via Laplace Transformation. *Phys. Rev. D*, **45**, 2617

Press, W.H. (1971): Long Wave Trains of Gravitational Waves from a Vibrating Black Hole. *Astrophys. J.*, **170**, L105

Price, R.H. and Pullin, J. (1994): Colliding Black Holes: The Close Limit. *Phys. Rev. Lett.*, **72**, 3297

Rasio, F.A. and Shapiro, S.L. (1992): Hydrodynamical Evolution of Coalescing Binary Neutron Stars. *Astrophys. J.*, **401**, 226

Rasio, F.A. (1997): Newtonian and Post-Newtonian Calculations of Coalescing Compact Binaries. *This volume*

Regge, T. and Wheeler, J.A. (1957): Stability of a Schwarzschild Singularity. *Phys. Rev.*, **108**, 1063

Ruffert, M., Janka, H.-Th., and Schäfer, G. (1996): Coalescing Neutron Stars: A Step towards Physical Models. I. Hydrodynamic Evolution and Gravitational Wave Emission. *Astron. Astrophys.*, **311**, 532

Ruoff, J. (1996): Schwingungen von Neutronensternen im Rahmen der (3+1)-Zerlegung. *Diplomarbeit*, Universität Tübingen

Shapiro, S.L. and Teukolsky, S.A. (1983): Black Holes, White Dwarfs, and Neutron Stars: The Physics of Compact Objects. Wiley, New York

Shibata, M. (1997): Numerical Study of Synchronized Binary Neutron Stars in the Post New-tonian Approximation of General Relativity. *Phys. Rev. D*, **55**, 6019

Shibata, M. and Taniguchi, K. (1997): Solving the Darwin problem in the first post-Newtonian approximation of general relativity: Compressible model. *Phys. Rev. D*, **56**, 811

Sun, Y. and Price, R. (1988): Excitation of Quasinormal Ringing of a Schwarzschild Black Hole. *Phys. Rev. D*, **38**, 1040

Sun, Y. and Price, R. (1990): Excitation of Schwarzschild Quasinormal Modes by Collaps. *Phys. Rev. D*, **41**, 2492

Taniguchi, K. and Nakamura, T. (1996): Innermost Stable Circular Orbit of Coalescing Neutron Star-Black Hole Binary: Generalized Pseudo-Newtonian Potential Approach. *Prog. Theor. Phys.*, **96**, 693

Vishveshwara, C.V. (1970): Scattering of Gravitational Radiation by a Schwarzschild Black-hole. *Nature*, **227**, 936

Vogt, R.E. (1992): The U.S. Ligo Project. In: *Proceedings of the 6th Marcel Grossmann meeting on general relativity* (ed. H. Sato and T. Nakamura). World Scientific, Singapore, 244

Weinberg, S. (1972): Gravitation and cosmology: principles and applications of the general theory of relativity. Wiley, New York

Wilson, J.R., Mathews, G.J., and Marronetti, P. (1996): Relativistic Numerical Model for Close Neutron Star Binaries. *Phys. Rev. D*, **54**, 1317

Wu, X., Müther, H., and Soffel, M. (1991): A new equation of state for dense matter and fast rotating pulsars. *Astron. Astrophys.*, **246**, 411

Zerilli, F. (1970a): Gravitational field of a Particle Falling in a Schwarzschild Geometry, Analyzed in Tensor Harmonics. *Phys. Rev. D*, **2**, 2141

Zerilli, F. (1970b): Effective Potential for Even-Parity Regge-Wheeler Gravitational Perturbation Equations. *Phys. Rev. Lett.*, **24**, 737

Zhuge, X., Centrella, J.M., and McMillan, S.L.W. (1994): Gravitational radiation from coalescing binary neutron stars. *Phys. Rev. D*, **50**, 6247

Zhuge, X., Centrella, J.M., and McMillan, S.L.W. (1996): Gravitational radiation from the coalescence of binary neutron stars: Effects due to the equation of state, spin, and mass ratio. *Phys. Rev. D*, **54**, 7261

A General Relativistic Approach to Neutron Star Binary Evolution

Edward Seidel

Max-Planck-Institut für Gravitationsphysik
Albert-Einstein-Institut, Schlaatzweg 1, D-14473 Potsdam
eseidel@aei-potdam.mpg.de

Abstract

In this paper I describe a new NASA project to develop a fully general relativistic approach to the important problem of neutron star coalescence. Unlike the other lectures in this volume, this contribution contains no results, only plans and motivation for work to be done! This project is just beginning, and its impact will be felt only after a couple of years of effort. My goal here is to outline plans to build on present approaches, including important physical effects that have not been possible until now. In addition to a deeper understanding of the neutron star coalescence, this project will also result in a fully general relativistic, parallel, hydro code that will be made available to the astrophysics community for a study of many problems in relativistic astrophysics. I will discuss some aspects of the difficulties one encounters in solving the fully coupled set of Einstein-Hydro equations.

1 Overview

Coalescing binary systems are thought to be common in the Universe. The most well known binary NS is the Hulse-Taylor binary pulsar PSR1913+16. This system's orbit is decaying slowly due to the emission of gravitational waves, and in about 10^8 years the stars will spiral rapidly together over a period of about 15 minutes. The gravitational waves will sweep up in frequency from about 10Hz to about 1500Hz; such signals should be seen by GEO, LIGO, and VIRGO, which should be operational around the year 2000. A number of independent calculations [see e.g. Finn (1996)] have been carried out to determine the final coalescence events which are detectable, yielding typical results such as NSNS coalescence observation rate of 29 yr^{-1} for $h = 0.5$ and 43 yr^{-1} for $h = 0.8$.

A new NASA "Grand Challenge" project has been funded recently to develop a high performance, parallel computer code to study the coalescence of NS binary systems in a fully general relativistic setting. The project, headed by computer scientist Paul Saylor at the University of Illinois, includes researchers at NCSA/University of Illinois, Washington University, SUNY-Stony Brook, Argonne, and Livermore. In the final coalescence stage, particularly through the last 5–10 orbits, general relativistic effects, hydrodynamic instabilities, the EOS, and neutrino transport may become extremely important governors of the gravitational wave signal that is generated and determine whether mass ejection occurs. It is during this merger phase, where fully relativistic effects are most important, that we intend to concentrate our efforts. After the completion of the project in a few years, the code developed for this project will also be made available to the astrophysics community, for application to many general relativistic astrophysics problems, ranging from AGN's to SN's.

Many groups have already studied the NS coalescence problem, using various different kinds of approximate schemes, and many important qualitative features of the coalescence processes have begun to emerge [see e.g. Rasio and Shapiro (1994), Oohara and Nakamura (1992), Kochanek (1992), Zhuge et al. (1996), Davies et al. (1994)]. In particular Rasio gave a review of recent approaches to this problem in his lectures at this meeting (see this volume). Although much has been learned about the coalescence process from such studies, an accurate quantitative description of the coalescence process, including many important physical processes, is still lacking. With so many simplifying assumptions of the essential physics, and especially the general relativistic effects, used to study this intrinsically general relativistic phenomena, the results to date are inconclusive. For example, the Japanese group (Oohara and Nakamura 1992) has shown that even the first order relativistic corrections may lead to drastic changes in the evolution. These differences in results from various groups, with various simplifications of the underlying physics, call for a full scale numerical treatment with the complete relativistic and hydrodynamic effects included within the framework of full general relativity. In particular, the goal of this NASA project is to provide a rigorous treatment of the gravitational waves and their backreaction, which is a driving mechanism for the decay of the binary orbit causing the coalescence in the first place. The treatment of the EOS of the NS's will be state-of-the-art (Lattimer and Swesty 1991), giving a more realistic model of the stars, with the matter evolution following the full relativistic equations of motion. It is our purpose to give a "complete" treatment of this very complex and fascinating problem in astrophysics.

2 Coupling Simulations to Observation: What Can we Learn from NS Coalescence?

We expect the simulation of neutron star coalescence events, when coupled with experimental data from NASA missions on high energy relativistic phenomena [such as the Compton Gamma Ray Observatory (GRO), Advanced X-ray Astronomy Facility (AXAF), the X-ray Timing Explorer (XTE)] and gravitational wave observatories being built, to yield important information ranging from the value of the Hubble parameter, to nuclear equation of state (EOS), and to gamma ray bursts. Here I discuss a few important aspects of the problem that will be investigated over the next few years.

2.1 Gravitational Wave Astronomy

As we have heard many times, a completely new window in observing the universe is about to open: gravitational wave astronomy. Gravitational waves are produced by coherent bulk motion of matter and travel nearly unscathed through space. We expect this new window to provide very different information about our universe that is either very difficult or impossible to obtain by traditional means.

The advent of broad band laser gravitational wave detectors such as GEO, LIGO, and VIRGO is making gravitational wave astronomy a reality. The GEO600 project is described by Danzmann in these lectures. Separate plans for a space detector (LISA) are underway at the European Space Agency. The LISA mission has recently been selected as one of the three "cornerstone missions" of the European agency. Also being developed are resonant detectors, which can be "tuned" to specific frequencies at which gravitational radiation detection from a particular source will be optimal. Coalescences of compact object binaries are considered to be one of the most promising sources of gravitational waves (GW) for these detectors.

The theoretical determination of the waveform is crucial for gravitational wave astronomy observations. The physical information in gravitational wave data is to be extracted through the standard template matching technique (Cutler et al. 1993). At present there is a large international collaboration using post-Newtonian (PN) techniques to determine the waveform templates for the earlier stage of the inspiral. However, PN techniques are not applicable for the late stage inspiral with the separation r of the two stars less than $10GM/c^2$ (M=total mass), when finite size effects and the internal structure of the stars become important. On the other hand, some of the most coveted information in the GW signal is coded in this late part of the signal: The bulk motion of matter upon coalescence may reveal the radius-mass relation, the structure of NS's, and the EOS of dense matter. Accurate waveform templates for the extraction of information from the final coalescence signal, which can be obtained only through detailed, fully relativistic hydrodynamics, will be crucial for filling in this important final step.

Fully relativistic calculations are also important for another reason. The coalescence of compact binaries is expected to lead often to the formation of BH's. In this, as well as other astrophysical processes involving BH's, a *unique* gravitational wave signal will be generated by the "ringing" modes of the BH. These BH fingerprints will give us the most direct confirmation of their existence, and their properties.

In short, in the emerging new field of gravitational wave astronomy, there are many important processes that are *impossible* to describe with Newtonian or PN approximations. Fully relativistic hydrodynamic calculations will be needed to treat these problems.

2.2   Determining the Hubble Parameter

Another very enticing reason for studying compact binary inspiral lies in the fact that observations of such inspirals by gravitational wave observatories may allow us to determine cosmological parameters like H_0 and q_0. It is a long standing quest of astronomers to determine accurately these parameters. There are several possible ways in which this data could be deduced from the GW signals, once the physical parameters of the systems are extracted through template matching. Schutz (1986) has pointed out that the signal from an inspiraling compact binary system by three LIGO type interferometers with different spatial orientations would be enough to determine the luminosity distance to the source. Methods [see, e.g. Finn (1996)] have been proposed for an accurate determination of the Hubble parameters without having to rely on the traditional distance ladder, the optical identification of the source, and is independent of the evolution of the source rate density with redshift. Computer simulations should greatly facilitate such a determination.

2.3   Nuclear Astrophysics

The study of NS-NS coalescence can lead to long sought-after information in nuclear astrophysics. Two particular areas of interest are determining the equation of state of dense matter and R-process nucleosynthesis.

Determining the NS Equation of State (EOS)

It is expected (Cutler et al. 1993) that GW signals from coalescing compact binaries, when combined with relativistic hydro simulations as proposed, may reveal information on the EOS of NS matter. Essential parameters include the nuclear compression modulus, the hadronic effective masses, the relative hyperon-nucleon and nucleon-nucleon coupling constants, possible kaon condensation and a quark/hadron phase transition. The gravitational waveform from the final coalescence may be sensitive to the overall stiffness of the EOS, which determines both the radius of single and coalesced NS's and the maximum mass. It will be important to investigate maximum masses for rapidly rotating configurations, and the sensitivity of waveforms to the equation of state. As mentioned above, the coalesced NS may form a BH. Some theoretical arguments suggest that the maximum neutron star mass is close to 1.5 $M_\odot$, and in this case, all binary mergers would result in BH formation. Besides the excitement of observing a BH in the process of formation, its signature could also be an important diagnostic for the EOS.

R-Process Nucleosynthesis

NS-NS coalescence simulations may help answer important questions in nuclear astrophysics independent of gravitational wave observations. NS-NS binary mergers may eject to infinity (Lattimer and Schramm 1976) extremely neutron-rich matter which decompresses, beta-decays and neutron captures, forming the classical r-process (Lattimer et al. 1977). The turn-on time of mergers, their low frequency, and their projected high yields seem to be consistent with abundances in metal-poor stars and thus with being a major contributant to the Galactic r-process (Mathews et al. 1992). However, shock heating might drive matter toward nuclear statistical equilibrium and ruin the r-process (Symbalisty and Schramm 1982, Davies et al. 1994). But at present, no calculations have been able to fully incorporate neutrino transport, a realistic EOS, and general relativity. One could couple multi-dimensional hydrodynamic calculations of mergers with dynamical nuclear network calculations to establish whether mass ejection occurs, and details of the nucleosynthesis, such as the sharpness of the r-process peaks and the relative yields of extremely heavy nuclei.

2.4   Gamma–Ray Bursts

Another important reason for studying NS-NS mergers is that they have been suggested as the origin of the gamma–ray bursts seen by the BATSE experiment on board the Compton GRO. There continues to be much debate as to whether the isotropy and inhomogeneity of the burst distributions indicate that they originate in the galactic halo or that they are of cosmological origin. Among those in favor of a cosmological component, one of the most popular models, but by no means the only one, has been NS-NS mergers (Paczynski 1986). However, to date there have only been a few numerical studies of the hydrodynamic process of such mergers in relation to this proposed gamma ray burst mechanism (Davies et al. 1994), often obtaining quite different results. The different conclusions could result from differences in the treatment of gravitational radiation in the two Newtonian calculations, and in any case fully relativistic effects have not been considered yet.

3   Difficulties of a Fully Relativistic Approach

There are many reasons why virtually all attempts to date to study neutron star coalescences have been made with basically Newtonian methods. Perhaps the most important reason is that the Einstein equations for the gravitational field are so complex that only now are crude 3D calculations becoming possible. They are among the most complex partial differential equations in all of physics, forming a system of dozens of coupled, nonlinear equations, with thousands of terms, of mixed hyperbolic, elliptic, and even parabolic types. Moreover, the evolution has elliptic constraints that should be satisfied at all times. These constraints are the relativistic generalization of the Poisson equation of Newtonian gravity, but instead of a single linear elliptic equation there are now four, coupled, highly nonlinear elliptic equations.

While the coupled Poisson-Hydro equations of Newtonian hydrodynamics have been solved for years, it is only over the last few years that serious 3D code development— coupled with adequate computer power—to solve the complete set of the vacuum part of the Einstein equations has begun. For the fully coupled Einstein-Hydro system, there are serious problems in virtually every aspect of the problem that must be solved. Here I describe several of these issues briefly to give an idea of the issues involved.

3.1   Initial Data

Einstein's equations in a 3+1 split (3 space and 1 time) naturally break up into two classes: The four initial value, or constraint equations, and the six evolution equations. The constraints are elliptic in nature; they are the relations between the 3–metric γ_{ab}, the extrinsic curvature K_{ab}, and the stress energy tensor T_{ab} on each constant t slice of the spacetime. They further split into the Hamiltonian constraint and the three momentum constraints.

These equations, which are generally considered as four coupled, nonlinear, elliptic equations, must be solved at the initial time to give the starting configuration of, say, two neutron stars in an unstable orbit. However, even in the vacuum case these equations have generally only been solved with certain simplifying assumptions, such as time symmetry (vanishing extrinsic curvature), or $trK = const$. Furthermore, as discussed by Cook et al. (1993), one does not have much control over the physical characteristics of the final data set produced by solving the constraints. Only the asymptotic mass and angular momentum can be specified as input parameters; other quantities, such as the gravitational wave content of the initial data, are difficult to control. The problem becomes even more complicated with matter present. Only in the last year have 3D data sets for binary neutron stars (Baumgarte et al. 1997), or for the general problem with arbitrary extrinsic curvature (Bernstein and Holst 1997), been attempted. There is much work to be done to get controlable, realistic, fully relativistic initial data for binary neutron stars that could be matched, say, to post-Newtonian orbit calculations of the early inspiral phase.

3.2   Evolution Codes

Several 3D codes have been developed by groups around the world. For our projects, we have developed two fundamentally different evolution codes, called the "G" and "H" codes, and applied them to a number of different 3D problems. The "G" (for General) code takes the standard "ADM" form of the Einstein equations, written in a very general 3D cartesian form, and can use different choices of explicit schemes. The "H" (for Hyperbolic) code uses a new formulation developed by our group and our collaborators (Bona et al. 1995). In this formulation, the full 3D Einstein equations are written in an explicitly hyperbolic, first order, flux-conservative form. This allows conservative finite difference techniques, developed after years of research in fluid dynamics, to be used on the Einstein equations for the first time. The use of operator splitting techniques allows one to treat the principal part of the system as a flux conservative first order system.

These codes severely strain the most advanced computing resources available at present. Even on the largest machines, with dozens of GBytes of memory, simulations of at most 300^3 zones can be attempted due to the huge number of variables that must be maintained. This resolution has proved to be rather crude for long term, accurate 3D calculations. Most of the work done to date with these 3D codes has been in vacuum studies of black holes (Anninos et al. 1995) and gravitational waves (Anninos et al. 1996).

3.3   GR Hydrodynamics

In order to solve the problem of coalescing neutron stars, these parallel relativity solvers must be combined with a parallel hydro code. Only a few attempts to couple fully selfconsistent, 3D matter fields to the Einstein equations have been made at the present time, and there is much to be learned about this coupling. The NCSA/WashU/Potsdam collaboration has experimented with self-gravitating scalar fields as a matter source to test coupling of matter to the gravitational field, and has recently begun developing full hydro codes for coupling to relativity solvers. At present only one such coupled 3D code, that evolves both hydrodynamics and the gravitational field, has been developed and tested [by Nakamura's group (Nakamura 1995)]. A number of 3D Newtonian hydrodynamics codes use standard artificial viscosity methods, or more advanced Godunov type methods, but Newtonian gravity is inadequate to treat properly many important problems in astrophysics. *Fixed spacetime* relativistic treatment of 3D hydrodynamics has been attempted by several groups (see e.g. Duncan et al.), but those codes do not currently allow for the dynamical evolution of the spacetime, and hence cannot treat situations involving strong dynamic gravitational fields. Clearly there is still much to be done before mature, coupled spacetime/hydro codes are applied to treat NS-NS coalescence and other problems in relativistic astrophysics.

3.4 Memory Requirements and Adaptive Mesh Refinement

As discussed above, the memory requirements for solving the coupled Relativity-Hydro equations are immense. As Einstein's theory is a tensor theory, there are many extra variables that need to be evolved when compared to a pure Newtonian treatment. The 3D relativity codes described above carry well over 100 3D arrays in permanent storage to efficiently evolve all the required fields. Our present 3D BH and NS calculations indicate that resolution equivalent to well over 2000^3 *fixed mesh* zones will be required for accurate strong field dynamics and radiation extraction for orbiting and coalescing binary systems, far exceeding the present capacity. This will not be possible in full GR, even on much larger machines, but by adding resolution *only where it is needed*, such calculations will be possible.

These space constraints have motivated the development of a software environment, called DAGH, which implements the adaptive mesh refinement (AMR) algorithms of Marsha Berger (1984) on parallel machines. DAGH has been developed by Manish Parashar, in collaboration with various members of the NSF funded binary black hole Grand Challenge Alliance. The use of AMR techniques will soon allow us to achieve extremely high spatial and temporal resolution in numerical simulations. This is a general parallel software environment which can be used in conjunction with any application-specific field solver to build an AMR application. The AMR system is based on MPI, and dynamically handles load balancing on parallel machines. DAGH will be used by the coupled Einstein-Hydro codes for treating 3D relativistic hydro problems, making it an indispensible tool to the astrophysical community when the code is released.

3.5 Gauge Conditions

One last area that I touch on is the implementation of gauge and slicing conditions in the Einstein-Hydro system. In standard Newtonian theory, one simply chooses a coordinate system in both space and time and marches forward. As we know, there is extra "gauge freedom" in Einstein's equations, so that in addition to the spacetime coordinate system, one must choose at every time slice (a) how time is advanced (by choosing the lapse function α), and (b) how the spatial coordinates move relative to the normal vector into the next slice (by choosing the shift vector β^i). This is detailed by York (1979) for those unfamiliar with this construction.

Although this formalism has been known for years, and a variety of lapse and shift conditions have been proposed, it is still very much an open research question to develop, and choose, appropriate conditions for 3D numerical relativity, with or without matter. Most shift conditions to date have been developed based on spherical or axisymmetry. Comparatively little 3D work has been done, and virtually no general relativistic calculations with orbital angular momentum have been attempted yet! Without appropriate shift conditions, coordinates can drift, twist, shear, etc., leading to coordinate pathologies and singularities that cause the code to crash or become extremely inaccurate. Similarly, without appropriate control of slicing conditions, time slices may become pathological as well.

Even in weakly gravitating systems such problems can become so pronounced that the calculations must be halted due to coordinate pathologies. For examples of gauge and slicing conditions and the problems and benefits they bring, see References Bona et al.

(1995) and Balakrishna et al. (1996). These issues will likely require extensive study before successful long term numerical evolutions of 3D orbiting neutron stars and/or black holes can be made.

4 Conclusions

The problem of coalescing neutron stars is a very important one in astrophysics. A careful, fully general relativistic approach to the evolution, coupled with detailed treatment of the microphysics, should lead to quantitative predictions of the gravitational waves emitted from such events. When such detailed studies are compared with observations of the gravitational and electromagnetic waves emitted by these events, many important issues in astrophysics, nuclear physics, and cosmology, will be addressed.

It is clear that a fully relativistic treatment of this problem is needed, but that there are many unsolved problems in developing a 3D code that couples a complete Einstein system solver with a GR hydrodynamics solver. However, the purpose of the NASA Grand Challenge project discussed here is to develop a high performance, parallel code capable of treating such important problems over the next few years. This code will be applied to this problem, and distributed to the relativistic astrophysics community for research in other areas of research.

Acknowledgments: I am grateful to my friends and colleagues for teaching me many things related to this project. In particular, my colleagues on the NASA project are Paul Saylor, Steve Ashby, Ian Foster, James Lattimer, Mike Norman, Douglas Swesty, Wai-Mo Suen, and Clifford Will. This research is supported by the NASA Grand Challenge program, and the Albert-Einstein-Institut.

References

Anninos, P., Massó, J., Seidel, E., Suen, W.-M., and Tobias, M. (1996): Near linear regime of gravitational waves in numerical relativity. *Phys. Rev. D*, **54**, 6544

Anninos, P., Camarda, K., Massó, J., Seidel, E., Suen, W.-M., and Towns, J. (1995): Three-Dimensional Numerical Relativity: The Evolution of Black Holes. *Phys. Rev. D*, **52**, 2059

Balakrishna, J., Daues, G., Seidel, E., Suen, W.-M., Tobias, M., and Wang, E. (1996): Coordinate conditions and their implementation in 3D numerical relativity. *Class. Quant. Grav.*, **13**, L135

Baumgarte, T., Cook, G., Scheel, M., Shapiro, S., and Teukolsky, S. (1997): In: *Proceedings of the Texas Symposium on Relativistic Astrophysics*, in press

Berger, M.J. and Oliger, J. (1984): Adaptive Mesh Refinement for Hyperboli Partial Differential Equations. *J. Comp. Phys.*, **53**, 484

Bernstein, D. and Holst, M. (1997): In: *Proceedings of the Texas Symposium on Relativistic Astrophysics*, in press

Bona, C., Massó, J., Seidel, E., and Stela, J. (1995): New Formalism for Numerical Relativity. *Phys. Rev. Lett.*, **75**, 600

Cook, G.B., Choptuik, M.W., Dubal, M.R., Klasky, S., Matzner, R.A., and Olivera, S.R. (1993): Three-Dimensional Initial Data for the Collision of Two Black Holes. *Phys. Rev. D*, **47**, 1471

Cutler, C., Apostolatos, T.A., Bildsten, L., Finn, L.S., Flanagan, E.E., Kennefick, D., Markovic, D.M., Ori, A., Poisson, E., Sussman, G.J., and Thorne, K.S. (1993): The Last Three Minutes: Issues in Gravitational-Wave Measurements of Coalescing Compact Binaries. *Phys. Rev. Lett.*, **70**, 2984

Davies, M.B., Benz, M.B., Piran, T., and Thielemann, F.K. (1994): Mergin Neutron Stars. I. Initial Results for Coalescence of Noncorotating Systems. *Astrophys. J.*, **431**, 742

Duncan, C., Sczaniecki, L., and Shastri, G. (in prep.): Three-dimensional relativistic hydrodynamics. I. The code, tests, and application to off-center collisions.

Finn, L.S. (1996): Binary inspiral, gravitational radiation, and cosmology. *Phys. Rev. D*, **53**, 2878

Kochanek, C. (1992): Coalescing Binary Neutron Stars. *Astrophys. J.*, **398**, 234

Lattimer, J.M. and Schramm, D.N. (1976): The Tidal Disruption of Neutron Sstars by Black Holes in Close Binaries. *Astrophys. J.*, **210**, 549

Lattimer, J.M. and Swesty, F.D. (1991): A Generalized Equation of State for Hot, Dense Matter. *Nucl. Phys. A*, **535**, 331

Lattimer, J.M., Mackie, F., Ravenhall, D.G., and Schramm, D.N. (1977): The Decompression of Cold Neutron Star Matter. *Astrophys. J.*, **213**, 225

Mathews, G., Bazan, G., and Cowan, J. (1992): Evolution of Heavy-Element Abundances as a Constraint on Sites for Neutron-Capture Nucleosynthesis. *Astrophys. J.*, **391**, 719

Nakamura, T. (1995): in a talk at the Penn State Workshop of Sources of Gravitational Waves, State College, PA, July, 1995

Oohara, K.-I. and Nakamura, T. (1992): Gravitational radiation from coalescing binary neutron stars. In: *Approaches to numerical relativity* (ed. R. d'Inverno). Cambridge University Press, Cambridge

Paczynski, B. (1986): Gamma Ray Bursters at Cosmological Distances. *Astrophys. J.*, **308**, L43

Rasio, F.A. and Shapiro, S.L. (1994): Hydrodynamics of Binary Coalescence. I. Polytropes with Stiff Equations of State. *Astrophys. J.*, **432**, 242

Schutz, B. (1986): In: *Dynamical Spacetimes and Numerical Relativity* (ed. J. Centrella). Cambridge University Press, Cambridge, p. 446

Symbalisty, E.M.D. and Schramm, D.N. (1982): Neutron star collisions and the r-process. *Astrophys. Lett.*, **22**, 143

York, J. (1979): Kinematics and Dynamics of General Relativity. In: *Sources of Gravitational Radiation* (ed. L. Smarr). Cambridge University Press, Cambridge

Zhuge, X., Centrella, J.M., and McMillan, S.L. (1996): Gravitational radiation from the coalescence of binary neutron stars: Effects due to the equation of state, spin, and mass ratio. *Phys. Rev. D* **54**, 7261

A Forty-Year Search for the Hubble Constant

Gustav A. Tammann and Lukas Labhardt

Astronomisches Institut
Universität Basel, Venusstr. 7 , CH-4102 Binningen
labhardt@astro.unibas.ch

Abstract

The calibration of the Hubble constant H_0 to within $\sim 10\,\%$ has been a long and painful process with many setbacks. Thanks to *HST* the process is coming to an end. Almost twenty Cepheid distances to nearby galaxies from *HST* do not themselves yield a useful value of H_0, but they provide the luminosity calibration of seven SNe Ia, put the calibration of the 21cm-line width-luminosity relation on much firmer footing, and discriminate against fallacious distance indicators. The consistent evidence of the large-scale value of the Hubble constant H_0 is reviewed in the light of SNe Ia, the Virgo cluster tied into the cosmic expansion field, and field galaxies. Also the evidence from physical distance determinations is discussed. The conclusion is $H_0 = 55 \pm 8 \ \mathrm{km\,s^{-1}\,Mpc^{-1}}$, corresponding to an expansion age of 15.2 ± 2.2 Gy (if $q_0 = 0.1$) and 12 ± 1.7 Gy (if $q_0 = \ ^1/_2$). Globular clusters, white dwarfs, and radioactive actinides require a time frame of ~ 12 Gy. This and the expansion age agree perfectly if the Universe is open; even for a critical Universe with $q_0 = \ ^1/_2$ the time agreement is as good as one can expect.

1 Introduction

The beginning of the modern calibration of the Hubble constant H_0 is marked by a paper by Sandage (1958). He concluded that $H_0 \approx 75 \ \mathrm{km\,s^{-1}\,Mpc^{-1}}$ and *if* the brightest stars are as bright as $M_B = -9\overset{\mathrm{m}}{.}5$ then $H_0 = 55$. Even brighter stars have been found since, confirming thus that H_0 is "small", i.e. $H_0 < 75$. Subsequent papers by many authors have suggested large excursions with values in the range $18 < H_0 < 110$. In recent years the range has narrowed to $40 < H_0 < 75(80)$. The evidence discussed here leaves a range of $47 < H_0 < 65$ which comes close to the attempted error of only $\pm 10\,\%$.

New Cepheid distances from the Hubble Space Telescope (*HST*) have put the extragalactic distance scale on much stronger footing. Cepheids are the least controversial distance indicators. The zero-point of their period-luminosity (P-L) relation has moved by only $0\overset{\mathrm{m}}{.}07$ over an interval of 30 years (Kraft 1961, Sandage and Tammann (1968), Feast and Walker (1987), Madore and Freedman (1991). The new zero-point from *HIPPARCOS* trigonometric parallaxes (Feast and Catchpole 1997) gives stunning confirmation of the zero-points of Sandage and Tammann (1968) and Madore and Freedman (1991) to within $+0^m\!.02\pm0^m\!.10$ and $+0^m\!.08\pm0^m\!.10$, respectively (Sandage and Tammann 1997).

LMC may serve as an example – although a quite favorable one – for the accuracy of Cepheid distances. The P-L relation of Sandage and Tammann (1968), now corrected by $0^m\!.02$, gives a distance modulus of $(m-M)_{\text{LMC}} = 18.61\pm0.15$, while the purely geometric distance of the ring of the supernova 1987A gives $(m-M)_{\text{LMC}} = 18.56\pm0.05$ (Panagia et al. 1996) and the equally independent distance of the P-L relation of Mira variables, calibrated with $HIPPARCOS$ trigonometric parallaxes, gives $(m-M)_{\text{LMC}} = 18.54\pm0.10$ (van Leeuwen et al. 1997). Finally, Sandage's luminosity calibration (Sandage 1993a) combined with Walker's RR Lyrae star data (Walker 1993) gives $(m-M)_{\text{LMC}} = 18.60\pm0.11$. The slope of the P-L relation taken from LMC is uncritical as long as the available Cepheids cover a sufficient period interval. Metallicity effects on the P-L relation are small (Freedman and Madore 1990, Chiosi et al. 1993, Sandage 1996a); any such effects enter with low weight because no strongly metal-deficient galaxies are considered in the following. Selection bias (Sandage 1988) can be avoided if the Cepheids span a sufficient period interval.

These or similar arguments have led to a general consent that Cepheids, as far as they can reliably be observed, are the most fundamental distance indicators known. Almost twenty Cepheid distances are now available from the ground or mainly from HST. They cannot provide themselves a meaningful value of H_0 because they do not reach – even with HST – beyond 1200 km s^{-1}. Therefore Cepheids are used in two ways to derive H_0: first to determine the distances of nearby galaxies which have produced well observed supernovae of type Ia (SNe Ia) whose absolute maximum magnitude is then used to calibrate the Hubble diagram of SNe Ia out to 30 000 km s^{-1}(Sect. 2), and secondly to derive the distance of the Virgo cluster which is well tied into the cosmic expansion field out to 10 000 km s^{-1}. Because of the important depth effect of the Virgo cluster its Cepheid distance from only four galaxies (two of which are outside the reliable cluster limits) is still unsatisfactory and other distance determinations of the cluster are used as well, several of which are again calibrated through Cepheids (Sect. 3). Section 4 discusses distance determination methods which do not seem to be useful. Tentative results on the distances of the Fornax and Coma clusters are discussed in Sect. 5. Bias-corrected values of H_0 from field galaxies are in Sect. 6. In Sect. 7 purely physical distance determinations are compiled which contribute to H_0. The best value of H_0 and the corresponding expansion age are discussed in Sect. 8. An overview of present ages of the globular clusters, radioactive actinides, and white dwarfs is given in Sect. 9. The discussion is in Sect. 10.

2 H_0 from SNe Ia

The most direct way to the "cosmic" (i.e. large-scale) value of H_0 is now provided by blue SNe Ia. The small scatter of their Hubble diagram proves them to be very useful standard candles at maximum light. An HST program was therefore mounted to calibrate the luminosity of a few nearby SNe Ia through Cepheids. The program team comprises A. Sandage, A. Saha, F.D. Macchetto, N. Panagia, and the present authors. The Hubble diagram of all 56 SNe Ia beyond 1100 km s^{-1}and out to 30 000 km s^{-1}with reasonably well determined maximum magnitudes is shown in Fig. 1 taken from Saha et al. (1997). Only red SNe Ia ($B-V >0^m\!.20$) are omitted to guard against highly absorbed and peculiar objects. All blue SNe Ia with known spectra are "Branch-normal"

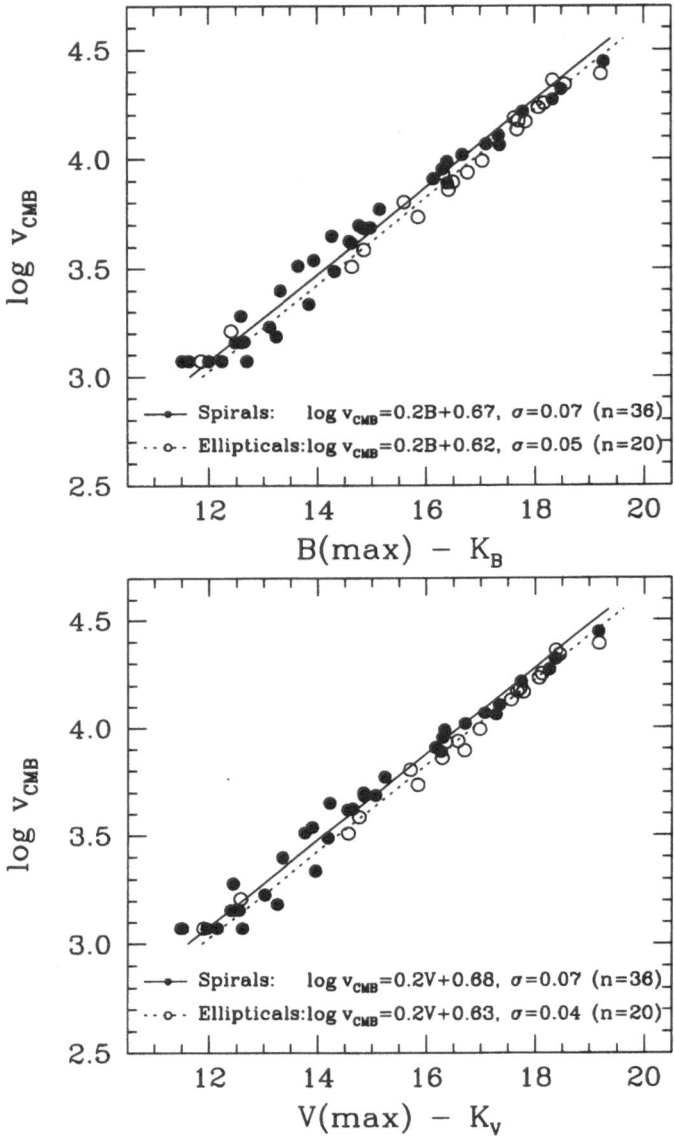

Figure 1 The Hubble diagram of all 56 blue SNe Ia with $1000 < v < 30\,000\,\mathrm{km\,s^{-1}}$ and known $m_B(\mathrm{max})$ and $m_V(\mathrm{max})$. Different symbols are used for SNe Ia in spirals and in E/S0 galaxies. The full line is a linear fit of slope 0.2 to the SNe Ia in spirals, the dotted line is the corresponding fit for SNe Ia in E/S0 galaxies. The velocities $> 3000\,\mathrm{km\,s^{-1}}$ are corrected for the peculiar motion of the "local" volume with respect to the cosmic microwave background (v^{CMB}). The magnitudes are corrected for (very small) redshift effects (K-correction).

(Branch et al. 1993), with the exception of SN 1990T which lies in the Virgo complex and is presumably overluminous. The *intrinsic* scatter of the Hubble diagram must be considerably less than $0^m.35$, because much of the scatter is expected from observational errors and peculiar motions. Indeed the intrinsic scatter must be small because even the most distant SNe Ia lie very close to the theoretical Hubble line of slope 0.2. The argument goes as follows. The most distant SNe Ia occupy a volume about 18 000 times larger than that of the local calibrators. The large volume must contain exceptionally luminous SNe Ia – if they existed – and they have a much enhanced discovery chance for two reasons: their apparent magnitude is brighter than average *and* they stay longer above the detection limit. But still, there are no objects significantly above the Hubble line, not even at large distances. This means: The sample of SNe Ia shown in Fig. 1 constitutes a homogeneous class of very luminous and unabsorbed objects.

In a *linearly* expanding Universe the ridge line of any (unbiased) Hubble diagram has the form

$$\log v = 0.2\, m_\lambda + a_\lambda ,\tag{1}$$

where m_λ is the apparent magnitude in wave band λ and a_λ is a corresponding constant. Equation (1) can easily be transformed into

$$\log H_0 = 0.2\, M_\lambda(\text{max}) + a_\lambda + 5 .\tag{2}$$

In principle H_0 follows from Eq. 2 to within the scatter of Fig. 1 from M_λ (i.e. $M_B(\text{max})$ or $M_V(\text{max})$) of a single SNe Ia. Actually we have calibrated so far the peak luminosity of six SNe Ia. A seventh object has become available through Tanvir's et al. [1995, cf. also Hjorth and Tanvir (1997)] Cepheid distance of the Leo group. (It can be assumed that the member galaxies of this compact group lie practically at the same distance.) The absolute B and V magnitudes of the seven calibrators agree within the observational errors – confirming thus their character as standard candles – and give the mean values (Saha et al. 1997)

$$M_B(\text{max}) = -19.52 \pm 0.07, \qquad M_V(\text{max}) = -19.48 \pm 0.07 .\tag{3}$$

Inserting this calibration into Eq. (2) and taking a_λ from all 56 SNe Ia in Fig. 1 would somewhat underestimate H_0 for the following reason. SNe Ia in elliptical galaxies are $\sim 0^m.25$ *fainter* on average than those in spirals. Five of the calibrating SNe Ia lie, however, in spirals; the remaining two SNe Ia occurred in the Am galaxy NGC 5253, which – because of its Cepheids – has at least some young population. One should therefore compare the calibrators with the Hubble diagram of only SNe Ia in spirals. The 21 best such SNe Ia give Saha et al. (1997)

$$a_B = 0.659 \pm 0.013, \qquad a_V = 0.668 \pm 0.012 .\tag{4}$$

Equations (3) and (4) inserted into Eq. (2) give $H_0(B) = 57 \pm 2$ and $H_0(V) = 59 \pm 3$. These values combined give

$$H_0 = 58 \pm 2 \ \text{km\,s}^{-1}\,\text{Mpc}^{-1} .\tag{5}$$

What are the merits and/or fallacies of this determination?

We first note that the nearer SNe Ia with less than $6300 \, \mathrm{km \, s^{-1}}$ give a somewhat higher value ($H_0 = 61$) and the more distant ones a lower value ($H_0 = 56$). This is because – against all statistical expectations – the distant SNe Ia are somewhat less luminous than the nearer ones. All selection effects cause an opposite trend. In spite of this the decrease of H_0 with distance should not be taken as real because the Hubble diagram of first-ranked cluster galaxies (Sandage and Hardy 1973) shows no sign of such a decrease. Final judgment is left to additional SNe Ia data.

Next, there is the question of the influence of second parameters – mainly the dependence of the SN absolute magnitude on the light curve decline rate Δm_{15}. It has been suggested (Phillips 1993, Hamuy et al. 1995, Hamuy et al. 1996) that this dependence requires a value as high as $H_0 = 65$. However, the suggested steep Δm_{15}-luminosity relation relies also on *red* SNe Ia which are not part of the present experiment. Moreover there is a dependence of Δm_{15} and the Hubble type of the SN parent galaxy. Therefore any effect of Δm_{15} on H_0 is largely accounted for by the above restriction to SNe Ia in spirals. Finally the steep Δm_{15}-luminosity relation, taken at face value, implies that the nearby calibrators are more luminous on average than the distant SNe Ia. This is impossible for statistical reasons, because the *flux-limited* sample of distant SNe Ia must comprise the very most luminous objects while the seven local calibrators, coming from a volume-limited sample, must represent quite closely the average luminosity of blue SNe Ia. The second-parameter problem is discussed in more detail in Saha et al. (1997). Any advocate of $H_0 = 65$ or of an even higher value must accept the consequence to live in a very special part of the Universe in which the SNe Ia are more luminous than elsewhere.

One additional effect could alter the value of H_0, i.e. systematic differences of the internal absorption between the calibrators and the fiducial sample. The blue calibrators have a mean absorption in B of $0^{\mathrm{m}}07$. If the fiducial sample has less absorption on average, the true value of H_0 could be *smaller* by a factor of 1.03 at most. The opposite case, i.e. the fiducial sample having more absorption, is against the expectations of a flux-limited sample which is biased against all dimming effects. In spite of this, Riess et al. (1996) have proposed non-negligible amounts of internal absorption for many SNe Ia in Fig. 1. This suggestion, however, leads to a number of inconsistencies and can be rejected (Saha et al. 1997, Branch et al. 1996).

Equation (5) reflects only the internal error. A discussion of the external errors (Saha et al. 1997) leads to $H_0 = 58^{+7}_{-8}$. The slightly asymmetric error is to reflect the fact that all unaccounted selection effects will lead to an overestimate of H_0.

It should be noted that the empirical luminosity calibration of SNe Ia is in excellent agreement with present *models*. Höflich and Khokhlov (1996) have three SNe Ia in common with the present calibration. Their *model* luminosities are the same to within $0^{\mathrm{m}}12 \pm 0^{\mathrm{m}}21$. Branch's model luminosity (Branch et al. 1997) of SN 1981B agrees fortuitously well with ours, and two SNe Ia of Ruiz-Lapuente (1996) are fainter by only $0^{\mathrm{m}}25 \pm 0^{\mathrm{m}}25$ judging from their late spectra and the inferred $^{56}\mathrm{Ni}$ mass. Any value of H_0 substantially larger than ~ 60 would therefore also seriously question the present theoretical understanding of SNe Ia.

3   The Distance of the Virgo Cluster

3.1   Cepheids

When the first reliable Cepheid distance of a Virgo galaxy (NGC 4321) became available from *HST* (Freedman et al. 1994a) it was precipitately hailed *the* Virgo cluster distance (Mould et al. 1995, Kennicutt et al. 1995), although the value of 17 Mpc was suspiciously low. The next two Virgo galaxies, NGC 4536 (Saha et al. 1996a) and NGC 4496A (Saha et al. 1996b), again had very low distances. Only the fourth galaxy, NGC 4639, revealed the *important depth* of the cluster (note for comparison: the spiral members span $\sim 15°$ in the sky!). Its distance is 25 Mpc ((Sandage et al. 1996) and yet it is a bona fide cluster member; with a *small* velocity of 800 km s^{-1} it cannot be assigned to the background field.

It is no accident that the first three Virgo spirals with Cepheid distances lie on the near side of the cluster. They were selected from the atlas of galaxies (Sandage and Bedke 1988) most suited for *HST*; they were thus biased to begin with. NGC 4639 looks much more difficult and would not have been selected had it not produced the archetypal SN Ia 1990N.

It is now clear that it will take at least a dozen Cepheid distances of a randomly selected sample of Virgo members to obtain a meaningful mean cluster distance.

Meanwhile one may step up the distance of the Leo group out to the Virgo cluster by means of *relative* distance determinations. A discussion of several distance determinations, including Cepheids, of five Leo group members gives a mean modulus of $(m - M) = 30.27 \pm 0.17$ (Hjorth and Tanvir 1997). The best available relative distances between the Leo Group and the Virgo cluster are compiled in Tab. 1. Adding the mean difference of $\Delta(m - M) = 1.25 \pm 0.13$ to the above distance modulus gives a Virgo cluster modulus of $(m - M)_{\text{Virgo}} = 31.52 \pm 0.21$.

Method	$\Delta(m - M)_{\text{Virgo-Leo}}$	Source
Tully-Fisher	1.35 ± 0.20	Federspiel (private communication, 1996)
Globular clusters	1.47 ± 0.42	Harris (1990)
$D_n - \sigma$	0.97 ± 0.29	Faber et al. (1989)
Planetary nebulae	1.15 ± 0.30	Bottinelli et al. (1991)
Velocities	1.30 ± 0.30	Kraan-Korteweg (1986)
mean:	1.25 ± 0.13	

Table 1   The distance modulus difference between the Leo group and the Virgo cluster.

3.2   The Tully-Fisher Relation

There are now 18 spiral galaxies with Cepheid distances (14 of which come from *HST*) which are useful for the calibration of the relation between absolute magnitude and $\log w$ (w = inclination-corrected 21cm-line width). Two close companions of M101 bring the number of useful calibrators to 20. Two galaxies (M101 and M100) are less inclined than the frequently adopted limit of $i = 45°$; yet their inclinations are so well defined by mapping their velocity field that they are still useful as calibrators.

The 20 calibrating galaxies yield the following calibration (Federspiel et al. 1997) of the Tully-Fisher (TF) relation

$$M_B = -6.60 \log w - (3.33 \pm 0.11) \qquad (\sigma = 0.44), \qquad (6)$$

where the slope is taken from the Virgo cluster.

An *objective* and *complete* sample of Virgo spirals is defined by the 48 non-peculiar galaxies of type Sab–Sdm from the Virgo Cluster Catalog (Binggeli et al. 1985) with $i \geq 45°$ and lying within the isopleths of substructures A and B (Binggeli et al. 1993) or, without significantly changing the result, within the X-ray contour of the cluster (Böhringer et al. 1994). This sample together with Eq. (6) gives $(m - M)_{\rm Virgo} = 31.68 \pm 0.12$. For details as to the influence of different input parameters and the difference between (calibrating) field galaxies and Virgo cluster members the reader is referred to Saha et al. (1997).

3.3   Other Distances to the Virgo Cluster

The peak of the luminosity function (LF) of *globular clusters* (GC) has frequently been used as a standard candle. A modern calibration of the GCs in the Galaxy and in M31 combined with a compilation of published GCLFs in five Virgo ellipticals has led to a Virgo modulus of $(m - M)_{\rm Virgo} = 31.75 \pm 0.11$ (Sandage and Tammann 1995). Meanwhile Whitmore et al. (1995) found a very bright peak magnitude in V and I for NGC 4486, which is well determined with *HST* and which corresponds, with our precepts, to a modulus of 31.41 ± 0.28 (Sandage and Tammann 1996). However, the GCs in NGC 4486 have a bimodal color distribution which is suggestive of age differences and possible merger effects (Fritze-von Alvensleben and Gerhard 1994, Elson and Santiago 1996). Turning a blind eye to this problem and averaging over all available GCLFs in Virgo we obtain $(m - M)_{\rm Virgo} = 31.67 \pm 0.15$. We are aware that the method may still face considerable problems.

The $D_n - \sigma$ *method*, normally applied to ellipticals, was extended to the bulges of S0 and spiral galaxies by Dressler (1987). Using the bulges of the Galaxy, M31, and M81 as local calibrators, one obtains $(m - M)_{\rm Virgo} = 31.85 \pm 0.19$ (Tammann 1988).

Novae are potentially powerful distance indicators through their luminosity-decline rate relation. Using the Galactic calibration of Cohen (1985), Capaccioli et al. (1989) have found the apparent distance modulus of M31 to be $(m - M)_{\rm AB} = 24.58 \pm 0.20$ (i.e. somewhat less than indicated by Cepheids). From six novae in three Virgo ellipticals (Pritchet and van den Bergh 1987) concluded that the cluster is more distant by $7^{\rm m}0 \pm 0^{\rm m}4$ than the apparent modulus of M31, implying $(m - M)_{\rm Virgo} = 31.58 \pm 0.45$. The result carries still small weight, but is interesting because it is based on novae exclusively. *HST* observations, although time-consuming, of novae in the Virgo cluster could much improve this *independent* result.

Theoretical *models of SNe Ia* by various authors converge towards $M_B = -19.45 \pm 0.15$ for blue "Branch normal" objects (Branch et al. 1997, Höflich and Khokhlov 1996, Ruiz-Lapuente 1996). It is true that fainter, nearby SNe Ia are known, but being red and spectroscopically peculiar they can easily be singled out. Eight blue SNe Ia which have occurred in the Virgo cluster have $\langle m_B(\max) \rangle = 12^m10 \pm 0^m15$. This value combined with the theoretical calibration gives $(m - M)_{\mathrm{Virgo}} = 31.55 \pm 0.25$. – Had we used instead the empirical calibration in Sect. 2, the Virgo modulus would have become larger by 0^m08. We refrain from using this value because the independent routes towards H_0 in Sect. 2 and 3 are to be kept strictly apart.

3.4 The Adopted Distance of the Virgo Cluster and the Resulting Value of H_0

The distance determinations of the Virgo cluster in Sect. 3.1 to 3.3 are compiled in Tab. 2. All distance moduli are corrected for Galactic absorption by using the absorption given by Burstein and Heiles (1984) for each galaxy. The overall effect of Galactic absorption is small. The resulting mean Virgo distance of 21.6 ± 0.8 Mpc is the same within the errors as the individual determinations. There is no dependence on galaxy type which suggests that early- and late-type Virgo galaxies are at the *same* distance.

The mean recession velocity of the Virgo cluster has given rise to some controversy. The value of Huchra (1988) is significantly too high; Hubble constants based on it are too high by 10-20 % for this single reason! The 364 galaxies with redshifts and listed by Binggeli et al. (1993) to lie within the isopleths of the cluster substructures A and B give $<v_0> = 918 \pm 35$ km s^{-1} (with respect to the centroid of the Local Group). The 361 galaxies within the X-ray contour (Böhringer et al. 1994) give $<v_0> = 983 \pm 39$ km s^{-1}.

Method	$(m - M)_{\mathrm{Virgo}}$	Galaxy type
Cepheids (via Leo)	31.52 ± 0.21	S
Tully-Fisher	31.68 ± 0.12	S
Globular clusters	31.67 ± 0.15	E
$D_n - \sigma$	31.85 ± 0.19	S0 + S
Novae	31.58 ± 0.45	E
Theor. Supernovae	31.55 ± 0.25	E/S0 + S
unweighted mean:	31.64 ± 0.06	
weighted mean:	31.67 ± 0.08	
mean linear distance:	21.6 ± 0.8 Mpc	

Table 2 Distance moduli of the Virgo cluster.

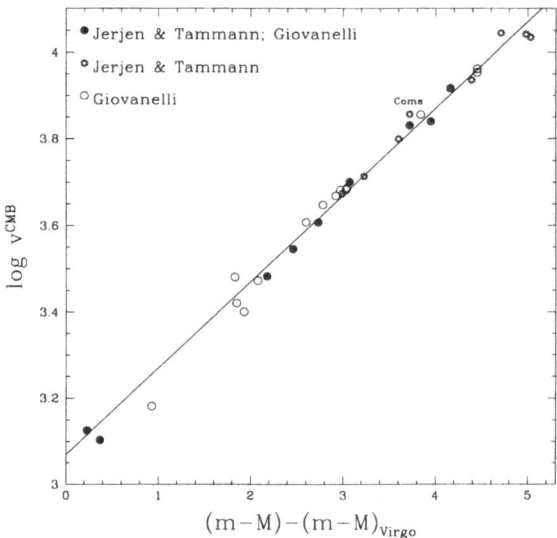

Figure 2 Hubble diagram of 31 clusters with known relative distances. The data were taken from Jerjen and Tammann (1993) (asterisks) and Giovanelli et al. (1997) (open circles). Nine clusters are listed in both sources (filled circles). The abscissa gives the distance modulus relative to the Virgo cluster and the ordinate the log of the recession velocity referred to the CMB.

Taking all 385 galaxies with redshifts in the Virgo survey area (Binggeli et al. 1985), excluding only clear background objects, yields $< v_0 > = 937 \pm 35$ km s^{-1}. From this follows a best cluster velocity of $< v_0 > = 950 \pm 30$ km s^{-1}. There are no significant differences between different Hubble types and between normal and dwarf galaxies. M87, the first-ranked galaxy in substructure A, has a higher velocity (Binggeli et al. 1987), but it should not be given undue weight. It is averaged into the distance determinations of Tab. 2 (where applicable) as well as into the mean cluster velocity.

To obtain the *cosmic* recession velocity of the Virgo cluster the observed value must still be corrected for the deceleration of the Local Group. We adopt $v_{infall} = 220 \pm 50$ km s^{-1} (cf. Tammann 1996) and find for Virgo $v_{cosmic} = 1170 \pm 61$ km s^{-1}. Yet there is another route to derive from the Virgo cluster distance a cosmologically even more meaningful value of H_0 which has the additional advantage that it makes no use of any Virgo cluster velocity.

Cluster distances relative to the Virgo cluster are available for 17 clusters from various methods like the TF method, the $D_n - \sigma$ relation, and first-ranked cluster galaxies [for a compilation cf. Jerjen and Tammann (1993)]. In addition, carefully determined relative TF distances are available for 24 clusters from Giovanelli et al. (1997). The latter list does not include the Virgo cluster, but since eight clusters are in common, the two lists can be combined with a mean error of only 0^{m}05. The double cluster A 2634/66 is not used here because Giovanelli et al. (1997) only gives a distance for A 2634. The resulting Hubble diagram (Fig. 2) contains 31 clusters with distances relative to the Virgo cluster [cf. also Tab. 1 of Tammann and Federspiel (1997)]. Clusters with $v_0 < 3000$ km s^{-1} are corrected for a Virgocentric infall model with a local infall velocity of 220 km s^{-1}. More distant clusters do not partake of the local motion with respect to the CMB. They are therefore corrected for a CMB vector of $630 \, \mathrm{km \, s^{-1}}$. The above dividing limit of $3000 \, \mathrm{km \, s^{-1}}$ is an educated guess [cf. Jerjen and Tammann (1993), Giovanelli et al. (1997)]; the exact choice has no effect on the following conclusions.

The ridge line in Fig. 2 is represented by

$$\log v^{CMB} = 0.2[(m - M) - (m - M)_{Virgo}] + 3.068 \pm 0.024, \qquad \sigma = 0.13 \,. \qquad (7)$$

The slope of 0.2 is enforced. Simple transformation of Eq. 7 gives

$$
\begin{aligned}
\log H_0 &= \log v^{CMB} - \log r_{Mpc} \\
&= -0.2 \, (m - M)_{Virgo} + (8.068 \pm 0.024) \,.
\end{aligned}
\qquad (8)
$$

Inserting the mean distance modulus of the Virgo cluster from Tab. 2 into Eq. (8) immediately yields the value of H_0 out to $11\,000 \, \mathrm{km \, s^{-1}}$

$$H_0 = 55 \pm 4 \, \mathrm{km \, s^{-1} \, Mpc^{-1}} \,. \qquad (9)$$

4   Rejected Distance Indicators

Two methods to determine extragalactic distances – i.e. the maximum luminosity of planetary nebulae (PN) shells and surface brightness fluctuations have yielded in the past unacceptably high values of $H_0 \approx 80$. The reason why these two methods should be rejected is briefly discussed.

The assumption that the luminosity function (LF) of the *shells of planetary nebulae* in the light of the 5007 Å line had a universal cutoff at $M_{5007} = -4^{\mathrm{m}}48$ has led to the proposal that these objects should be used as distance indicators (Jacoby et al. 1990). Yet a comparison of the PN results with Cepheid and other reliable distance determinations is quite disappointing. The relevant data in Tab. 3 suggests that the PN distances can reproduce the nearby Cepheid distances, but that they fall increasingly short as the distance increases to reach a modulus deficiency of $0^{\mathrm{m}}6 - 0^{\mathrm{m}}8$ at the distance of the Virgo and Fornax clusters.

Why are the results of the cutoff magnitude of PN shells so unreliable as distance indicators? A first reason is that the cutoff magnitude depends on the sample size, i.e. on the absolute magnitude of the parent galaxy (Bottinelli et al. 1991). In fact, allowing for this effect it was shown that the available data are perfectly consistent with $H_0 \approx 55$ (Tammann 1993). Numerically simulated LFs of the shell luminosities confirm indeed the dependence on sample size *and* population age (Méndez et al. 1993). Moreover the PN shell luminosities are sensitive to the oxygen content (Richer 1996).

More recently a group of authors (Soffner et al. 1996), including proponents of the universal cutoff magnitude, have developed a new method allowing for sample size *and* other effects to derive distances from PN shells. This method rests on fundamentally different premises and has so far been applied only to the nearby galaxy NGC 300.

Surface brightness fluctuations (SBF) have also been proposed as distance indicators (Tonry and Schneider 1988). The first "test" has remained rather unconvincing, spreading the *elliptical* Virgo cluster members over an interval of 12 to 24 Mpc (Tonry et al. 1990); this interval was interpreted as real although early-type galaxies are known to be concentrated in the *cores* of galaxy clusters. Recently revised SBF distances, allowing for

Mg$_2$ index and color $(V - I)$ effects, agree in four cases with Cepheid distances below $(m - M) = 30.50$. But the three cases beyond this limit where a comparison is possible with reliable distance determinations show very poor agreement, the SBF distances being too short by 0^{m}5 – 0^{m}6 (Tab. 4).

The reasons why the SBFs fail beyond $(m - M) = 30.5$ are not as clear as in the case of the PNe. One can think of the influence of bright stars other than red-giant tip giants, of globular clusters, cosmic rays, internal absorption and metallicity and/or color variations (Sodemann and Thomsen 1996). More fundamental is the question, whether E/S0 galaxies and bulges are really smooth, or whether their internal structure contributes to the SBFs (Lorenz et al. 1993). There is also the serious problem that the observed fluctuation magnitudes must be divided by the surface brightness which is quite dependent in the I-band on the population *age* (Fritze-von Alvensleben, private communication, 1995).

5   The Distances of the Fornax and Coma Clusters

The distance of the Virgo cluster (Sect. 3) is the best known cluster distance at present. Tying the Virgo cluster distance into the Hubble diagram of Fig. 2, which uses only relative distances between the Virgo cluster and other clusters out to 11 000 km s^{-1}, makes therefore optimum use of all relative cluster distances for the determination of the large-scale value of H_0. It may seem hence redundant to discuss here separately the distances of the Fornax and Coma clusters. However, there are a few *direct* distance determinations of these two clusters which can by-step the Virgo cluster entirely. They provide to some extent a check of consistency.

Object	best $(m-M)$	Method	Source	$(m-M)_{PN}$	Source	$\Delta(m-M)$
M81	27.80	Cepheids	[1]	27.72	[2]	+0.08
NGC 5253	28.01	Cepheids	[3]	28.08	[4]	−0.07
M101	29.34	Cepheids	[5]	29.42	[6]	−0.08
Leo group	30.27	Cepheids	[7]	30.02	[8]	+0.25
		and others		29.91	[9]	+0.36
Virgo cluster	31.67	various	§3.4	30.84	[8]	+0.83
Fornax cluster	31.76	SNe Ia	§5	31.09	[8]	+0.67

[1] Freedman et al. 1994b, [2] Jacoby et al. 1989, [3] Saha et al. 1995, [4] Jacoby and Ciardullo 1993, [5] Kelson et al. 1996, [6] Feldmeier et al. 1996, [7] Hjorth and Tanvir 1997, [8] Jacoby and Ciardullo 1993, [9] Feldmeier et al. 1997

Table 3 Comparison of PN distances with "best" distances.

Object	best $(m-M)$	Method	Source	$(m-M)_{SBF}$	Source	$\Delta(m-M)$
NGC 7331	30.95	Cepheids	[1]	30.39	[2]	+0.56
Virgo cluster	31.67	various	§3.4	31.03	[2]	+0.64
Fornax cluster	31.76	SNe Ia	§5	31.23	[2]	+0.53

[1] Hughes (private communication, 1996), [2] Tonry et al. 1997

Table 4 Comparison of SBF distances with "best" distances.

5.1 The Distance of the Fornax Cluster

Three well observed blue SNe Ia have occurred in the Fornax cluster, i.e. SN 1980N and 1981D in NGC 1316 (Sa pec) and SN 1992A in NGC 1380 (S0/a). Their mean magnitudes are $< m_B(\text{max}) >= 12.56 \pm 0.05$ and $< m_V(\text{max}) >= 12.46 \pm 0.07$. These values can be combined with the absolute-magnitude calibration in Eq. 3. However, as noted in Section 2, this calibration rests on SNe Ia in spirals while the Fornax SNe Ia have occurred in early-type galaxies. The latter produce SNe Ia which are on average 0^m23 fainter in B and 0^m26 fainter in V than their counterparts in spirals (Saha et al. 1997). Allowance for this under-luminosity then leads to a mean cluster modulus of $(m - M)_{\text{Fornax}} = 31.76 \pm 0.15$ or 22.5 ± 1.7 Mpc.

The mean recession velocity of the cluster, corrected for Virgocentric infall, is $v_{220} = 1338 \pm 50$ km s^{-1}, which leads to $H_0 = 59 \pm 6$. However, this result has low weight because it makes no allowance for the peculiar motion of the Fornax cluster which is still poorly known and could well be as high as ± 200 km s^{-1}. The difficulty is due to the fact that the *relative* distance determinations between the Fornax cluster and the Virgo cluster (or any other cluster) are quite discrepant [for a discussion see Tammann and Federspiel (1997)]. It is noteworthy that a Cepheid distance of only $(m - M) = 31.30$ has recently been derived for the Fornax Sbc spiral NGC 1365 (Silbermann et al. 1996). If this value is correct at all, it means that NGC 1365 is substantially nearer than the two SN-bearing galaxies of earlier type. This is in fact not unlikely because the late-type spirals are widely distributed about the Fornax cluster and avoid its center. It would be unwise to equate the distance of NGC 1365 with that of the cluster proper.

The literature is abundant with relative distance determinations between the Fornax and the Virgo cluster. Taking 30 determinations between 1973 and 1996 at face value one comes to the conclusion that the late-type galaxies are $0^m35\pm0^m09$ *nearer* than the early-type members. This may be taken as an additional warning against giving too much weight to the distance of NGC 1365.

5.2 The Distance of the Coma Cluster

Taking the Virgo modulus from Tab. 2 and adding the well determined relative modulus Coma-Virgo of $\Delta(m - M) = 3^m72\pm0^m10$ (cf. Sandage and Tammann 1997) gives $(m - M)_{\text{Coma}} = 35.27 \pm 0.13$. The turn-over magnitudes of the GC luminosity function of NGC 4881 (Baum et al. 1995) and IC 4051 (Baum et al. 1997) give $(m - M)_{\text{Coma}} \geq$

35.16 ± 0.23 and 35.15 ± 0.06, respectively. Hjorth and Tanvir (1997) have obtained $(m - M)_{\text{Coma}} = 35.17 \pm 0.24$ from stepping up the mainly Cepheid-based distance of the Leo group by the modulus difference Coma-Leo which they derived from fundamental-plane distance indicators. The combined evidence suggests $(m - M)_{\text{Coma}} = 35.24 \pm 0.15$ or 112 ± 8 Mpc.

The position of the Coma cluster in the Hubble diagram of clusters within $11\,000$ km s^{-1} (Fig. 2) shows the observed cluster velocity of $v_{\text{LG}} = 6912$ km s^{-1}(Zabludoff et al. 1993), after correction for the CMB velocity vector, to be appreciably affected by peculiar motions relative to the CMB frame. Inserting a relative modulus of Coma-Virgo of $\Delta(m-M) = 3^{\text{m}}72 \pm 0^{\text{m}}10$ into Eq. 7 gives $v_{\text{Coma}}^{\text{CMB}} = 6516 \pm 466$ km s^{-1} which is freed from all peculiar motions. But note the non-negligible error of the velocity! This value and the linear Coma distance from above give a large-scale value of $H_0 = 58 \pm 6$ (internal error). The value is compatible with the values in Eq. (5) and (9) to within the errors.

6 H_0 from Field Galaxies

Distance determinations to field galaxies out to 3000 - 5000 km s^{-1} depend on distance indicators like rotation velocities from optical or 21 cm-line widths, the $D_n - \sigma$ relation (diameter versus velocity dispersion), and others. These relations have an intrinsic scatter of $\gtrsim 0^{\text{m}}40$. A scatter thus large invites serious selection effects (Malmquist bias) in all existing *flux-limited* galaxy catalogs. The effect is that the mean luminosity of the cataloged objects increases with distance and that consequently H_0 spuriously increases with distance as well. [Any increase of H_0 by $\gtrsim 5\,\%$ is of course excluded by the linearity of the Hubble diagrams in Figs. 1 and 2 and by that of first-ranked cluster-galaxies (Sandage and Hardy 1973)]. Corrections for Malmquist bias are very difficult because they depend strongly on the size of the *true* scatter (which is difficult to observe), on the fairness of the catalog, on the (unknown) galaxy distribution, and the catalog cutoff. A tutorial on the Malmquist bias is given by Sandage (1995) [cf. also Teerikorpi (1984), Bottinelli et al. (1986, 1988)]. In Tab. 5 only a selection is given of H_0 determinations from field galaxies which are carefully corrected for Malmquist bias.

7   Purely Physical Distance Determinations

Various methods are emerging which allow extragalactic distances to be determined by purely geometrical or physical methods. They are still not competitive mainly due to obvious or possibly hidden systematic error sources. [An exception is the high-precision distance from the ring of SN 1987A (Panagia et al. 1996)]. But there is no doubt that some high-weight determinations of H_0 will become available from these methods in the years to come. It may suffice here to list several methods which hold high promise (Tab. 6). The reader is referred to the original sources for further explanations.

In principle the gravitational lenses overshoot the problem by determining the Hubble constant at high redshifts where one observes a combination of H_0 and q_0. The effect on H_0 is, however, only $\lesssim 10\,\%$. The results of the CMB fluctuation spectrum may eventually not be so much a determination of H_0 than rather a fundamental confirmation of cosmological theories through observations [cf. also Bond et al. (1997)]. The combined evidence in Tab. 6 is consistent with $H_0 \approx 50 - 65$.

8   The Global Value of H_0 and the Expansion Age

Three determinations of the large-scale value of H_0 have been discussed, i.e. the calibration of the Hubble Diagram of SNe Ia through seven SNe Ia with Cepheid distances giving $H_0 = 58 \pm 2$, the distance of the Virgo cluster tied into the rest frame of the CMB yielding $H_0 = 55 \pm 4$, and purely physical methods suggesting $H_0 \approx 50 - 65$. The evidence for $H_0 = 53 \pm 3$ from field galaxies is fully compatible with these determinations, but the range is restricted to distances of $3000 - 5000 \, \mathrm{km \, s^{-1}}$. The present distance information on the Fornax and Coma clusters also agree with these determinations within the considerable errors of their distances and (cosmic) velocities. The conclusion from this is that the best value at present is $H_0 = 55 \pm 8$, where the error is to include also external errors.

The global meaning of this value beyond $5000 \, \mathrm{km \, s^{-1}}$ is assured by the Hubble diagram of first-ranked cluster galaxies (Kristian et al. 1978, Hoessel et al. 1980, Lauer and Postman 1992). The expansion field is linear to within $\sim \pm 5\,\%$ as far as it can be tested; beyond $z = 0.15$ the test becomes ambiguous because of uncontrolled effects of luminosity evolution and q_0. Values larger than $H_0 = 65$ still in the literature can be attributed to a few quite obvious error sources:

1. an unwarrantedly high recession velocity of the Virgo cluster,

2. the unrealistic expectation to fathom the depth of the Virgo cluster with only one well resolved galaxy (NGC 4321); the same objections holds for the assumption of NGC 1365 reflecting the *mean* distance of the Fornax cluster,

3. the myth of a sharp, dispersionless cutoff of the luminosity function of planetary nebulae shells, independent of sample size and other factors,

Method	H_0	Source(s)
Tully-Fisher, distance-limited (local)	48 ± 5	[1] [2]
Tully-Fisher, flux-limited (distant)	< 60	[1] [2]
M101 look-alike diameters	43 ± 11	[3]
M31 look-alike diameters	45 ± 12	[4]
Luminosity classes of spirals	56 ± 5	[5]
M101, M31 look-alike luminosities	55 ± 5	[6]
Tully-Fisher (using magn. + diameters)	55 ± 5	[7]
weighted mean	53 ± 3	

[1] Sandage 1994a, [2] Sandage 1994b, [3] Sandage 1993b, [4] Sandage 1993c, [5] Sandage 1996b, [6] Sandage 1996c, [7] Theureau et al. 1997

Table 5 H_0 determinations from field galaxies corrected for Malmquist bias.

Method	H_0	Source(s)
Radio Remnant of SN 1979C in Virgo (NGC 4321)	54 ± 20	[1]
Expanding-photosphere and ^{56}Ni models of SNe Ia	$55 - 70$	[2] [3] [4]
Expanding-photosphere models of SNe II	73 ± 6	[5]
	< 50	[6]
Sunyaev-Zeldovich effect		
cluster A 2218	45 ± 20	[7] [8] [9]
6 other clusters	60 ± 15	[10] [11]
cluster A 2163	78 ± 30	[12]
2 clusters	42 ± 10	[13]
Gravitational lenses		
QSO 0957 + 561	63 ± 12	[14] [15]
B 0218 + 357	~ 60	[16] [17]
PG 1151 + 080	60 ± 17	[18] [19]
CMB fluctuation spectrum	$30 < H_0 < 50(70)$	[13] [20]

[1] Bartel 1991, [2] Höflich and Khokhlov 1996, [3] Branch et al. 1997, [4] Ruiz-Lapuente 1996, [5] Schmidt et al. 1994, [6] Baron et al. 1995, [7] McHardy et al. 1990, [8] Birkinshaw and Hughes 1994, [9] Lasenby and Hancock 1995, [10] Rephaeli 1995, [11] Herbig et al. 1995, [12] Holzapfel et al. 1996, [13] Lasenby 1997, [14] Kundíc et al. 1997, [15] Falco et al. 1997, [16] Corbett et al. 1995, [17] Nair 1995, [18] Schechter et al. 1996, [19] Keeton and Kochanek 1997, [20] Lineweaver et al. 1997

Table 6 H_0 from physical methods.

4. the reliance on the suspicious surface brightness fluctuation method, which is in open conflict with Cepheid distances, and/or

5. the Malmquist bias which always artificially increases the value of H_0 from flux-limited samples of field galaxies and incomplete catalogs of cluster galaxies.

The final swing of the Hubble constant towards "small" values, i.e. $H_0 \approx 55$, has been largely driven by *HST*. The Cepheid distances from *HST* – calibrating the luminosity of blue SNe Ia as well as the Tully-Fisher relation – have provided the last confirmation. The present highly consistent evidence – if ever proven to be wrong – could only be explained by a conspiracy of nature. Any future suggestions of $H_0 > 65$ have to address this "conspiracy".

With $H_0 = 55 \pm 8$ the Hubble time becomes $1/H_0 = 18 \pm 2.6$ Gigayears (Gy). This is with $\Lambda = 0$ the maximum expansion age. If we adopt $\Omega_0 = 0.2$ as a realistic minimum value of the matter density, the Friedmann time becomes 15.2 ± 2.2 Gy (Sandage 1961). The fashionable minimum Friedmann age comes from $\Omega_0 = 1$ ($q_0 = 1/2$), i.e. 12 ± 1.7. Of course, the age could still be smaller than this if q_0 were larger than $1/2$, but nobody seems to favor this solution.

9   The Age of the Oldest Objects in the Galaxy

The age of our Galaxy is well approximated by the age of the globular clusters, which have formed *before* the formation of the Galactic disk, and by the coolest white dwarfs which record the first deaths of stars with less than $\sim 8\,\mathcal{M}_\odot$ in the Galactic disk. Moreover the onset of the production of radioactive actinides in *short-lived* SNe of type II reflects closely the time of the earliest Galactic star formation.

The fitting of color-magnitude diagrams of globular clusters to theoretical isochrones [e.g. Bergbusch and VandenBerg (1992)] is sensitive to the adopted distance, reddening, the [Fe/H] *and* [O/Fe] ratios, and the mixing length. The distance is usually obtained by fitting the horizontal giant branch (RR Lyr stars) or the main sequence; a new method of fitting the white dwarf cooling sequence has become possible through *HST* (Renzini et al. 1996). Decreasing distances due to brighter RR Lyr stars (Sandage 1993a) and enhanced [O/Fe] ratios have tended to decrease the ages over the last years.

Table 7 lists typical GC ages as of 1996. Incoming trigonometric distances from the *HIPPARCOS* satellite will improve the luminosity calibration of RR Lyr stars and the calibration of the main sequence in function of metallicity. A first analysis (Reid 1997) leads to rather bright main sequences and not only vindicates Sandage's (Sandage 1993a) already bright RR Lyr stars but makes them even brighter by $0^{\mathrm{m}}15$ at [Fe/H]$= -2.2$. This new calibration together with the models of Mazzitelli et al. (1995) yields a *reduced* age of ~ 11 Gy (Sandage and Tammann 1997). It is likely that for the same reason other age determinations in Tab. 7 will come down. From the above evidence it is concluded that a GC age of only 11 Gy is the best value at present.

The number of white dwarfs in the solar neighborhood drops sharply faintwards of $M_V = +16^{\mathrm{m}}0$ (Winget et al. 1987). From this the authors calculated that the faintest, i.e. oldest white dwarfs have a cooling time of 9 Gy which must be increased by the pre-white-dwarf lifetime of about 0.3 Gy. A rediscussion of the white-dwarf luminosity function (Wood 1992), cf. also Kawaler (1997), has given an age range of $7.5 - 11$ Gy. Crystallization processes in white dwarfs, releasing additional energy, tend to increase this age to $9.5 - 12$ Gy or even more (Hernanz et al. 1994). Of course this age determination applies only to the Galactic *disk* as long as there are no available statistics on halo white dwarfs.

If the radioactive actinides in the Solar System were produced in a single type II SN event, it must have occurred ~ 10.6 Gy ago. However, it can be shown that several SNe must have contributed to our chemism; in that case the time since the *first* SN II can only be larger. Allowing for this effect and introducing beta-delayed fission into the r-process network of reactions has increased the age (Cowan et al. 1987). The latest discussion of the input parameters gives an age of 13.8 ± 3 Gy for the onset of heavy-element nucleosynthesis in our Galaxy (Truran 1997). – The Th/Eu age of an ultra-metal-poor halo giant of 17 ± 4 Gy (Cowan et al. 1997) is high but hardly outside the error range. A Galaxy age of $11 - 12$ Gy would seem like a fair assessment of the above triple evidence.

Age [Gy]	Source
14.1 ± 0.3	Sandage 1993d
$11 - 21$ (total range)	Chaboyer 1995
$10 - 14$	Shi 1995
$13(+2, -3)$	Mazzitelli et al. 1995
14.5 ± 1.6	Demarque 1997
< 13	Weiss et al. 1996
$11 - 13$	Caloi et al. 1996
11	Reid 1997

Table 7   Recent age determinations of globular clusters.

To obtain an age of the Universe this number should still be increased by the gestation time of the first stars setting the clocks for globular clusters, white dwarfs, and the Galactic nucleosynthesis. If the first stars formed at $z \approx 10$ and if $0.1 < q_0 < 0.5$ the gestation time was about 0.5 Gy (Sandage 1961). The resulting age of ~ 12 Gy is derived from only Galactic data. Why is this of cosmic relevance? One can make a string of arguments that there was an epoch of preferred galaxy formation and that most galaxies have comparable ages. Even the gas-rich, young-looking LMC has metal-poor globular clusters which are as old as their Galactic counterparts (Brocato et al. 1997). In any case there is no evidence for a galaxy having formed before the Galaxy.

10   Conclusions

The realistic (Friedmann) expansion time of the Universe has been found to be $12 - 15.2$ Gy for Ω_0 between 1 and 0.2. This range may be increased by the uncertainty caused by the external error of the Hubble constant of ± 8. The possible interval becomes then $10.3 - 17.4$ Gy.

The necessary time frame of ~ 12 Gy discussed in Sect. 9 fits astonishingly well into this interval without any necessity of introducing a cosmological constant Λ. If the density parameter is specified to be $\Omega_0 = 1$ ($q_0 = {}^1/_2$) the Friedmann time interval is narrowed down to $10.3 - 13.7$ Gy. The upper limit here corresponds to $H_0 = 47$, which is not impossible but at present quite unpopular. Yet a contradiction-free Friedmann time of 12 Gy is obtained with the most likely value of $H_0 = 55$. Only if $H_0 > 60$ and $q_0 = {}^1/_2$ the Friedmann time decreases to < 11 Gy, and the time agreement becomes marginal.

In any case it would be absurd to speak of a "time scale crisis". The nearly perfect agreement of the expansion age of the Universe and the age of its constituents to within the observational errors is in fact so impressive that simple Friedmann models are as viable as ever.

Acknowledgments: The authors thank Dres. A. Sandage, A. Saha, F.D. Macchetto and N. Panagia for a most joyful collaboration in the *HST* project of the calibration of H_0. They acknowledge the support of the Swiss National Science Foundation.

References

Baron, E., Hauschildt, P.H., Branch, D., Austin, S., Garnavich, P., Ann, H.B., Wagner, R.M., Filippenko, A.V., Matheson, T., and Liebert, J. (1995): Non-LTE spectral analysis and model constraints on SN 1993J. *Astrophys. J.*, **441**, 170

Bartel, N. (1991): An estimate of the distance to M100 in the Virgo cluster via VLBI of SN 1979C: Updates and prospects. In: *Supernovae* (ed. S.E. Woosley). Springer, New York, 760

Baum, W.A., Hammergren, M., Thomsen, B., Groth, E.J., Faber, S.N., Grillmair, C.J., and Ajhar, E.A. (1997): Distance to Coma cluster and a value for H_0. *Astrophys. J.*, in press

Baum, W.A., Hammergren, M., Groth, E.J., Ajhar, E.A., Lauer, T.R., O'Neil, E.J., Jr., Lynds, C.R., Faber, S.M., Grillmair, C.J., Holtzman, J.A., and Light, R.M. (1995): Globular clusters in Coma galaxy NGC 4881. *Astron. J.*, **110**, 2537

Bergbusch, P.A. and VandenBerg, D.A. (1992): Oxygen-enhanced models for globular cluster stars. II. Isochrones and luminosity functions. *Astrophys. J. Suppl.*, **81**, 163

Binggeli, B., Popescu, C.C., and Tammann, G.A. (1993): The kinematics of the Virgo cluster revisited. *Astron. Astrophys. Supp. Ser.*, **98**, 275

Binggeli, B., Sandage, A., and Tammann, G.A. (1985): Studies of the Virgo cluster. II. A catalog of 2096 galaxies in the Virgo cluster area. *Astron. J.*, **90**, 1681

Binggeli, B., Tammann, G.A., and Sandage, A. (1987): Studies of the Virgo cluster. VI. Morphological and kinematical structure of the Virgo cluster. *Astron. J.*, **94**, 251

Birkinshaw, M. and Hughes, J.P. (1994): A measurement of the Hubble constant from the X-ray properties and the Sunyaev-Zel'dovich effect of Abell 2218. *Astrophys. J.*, **420**, 33

Böhringer, H., Briel, U.G., Schwarz, R.A., Voges, W., Hartner, G., and Trümper, J. (1994): The structure of the Virgo cluster of galaxies from Rosat X-ray images. *Nature*, **368**, 828

Bond, J.R., Efstathiou, G., and Tegmark, M. (1997): Forecasting cosmic parameter errors from microwave background anisotropy experiments. *Mon. Not. Roy. Astron. Soc.*, submitted

Bottinelli, L., Gouguenheim, L., Patural, G., and Teerikorpi, P. (1986): The Malmquist bias and the value of H_0 from the Tully-Fisher relation. *Astron. Astrophys.*, **156**, 157

Bottinelli, L., Gouguenheim, L., Paturel, G., and Teerikorpi, P. (1988): The Malmquist bias in the extragalactic distance scale: Controversies and misconceptions. *Astrophys. J.*, **328**, 4

Bottinelli, L., Gouguenheim, L., Paturel, G., and Teerikorpi, P. (1991): A systematic effect in the use of planetary nebulae as standard candles. *Astron. Astrophys.*, **252**, 550

Branch, D., Fisher, A., and Nugent, P. (1993): On the relative frequencies of spectroscopically normal and peculiar type Ia supernovae. *Astron. J.*, **106**, 2383

Branch, D., Nugent, P., and Fisher, A. (1997): Type Ia supernovae as extragalactic distance indicators. In: *Thermonuclear Supernovae* (eds. P. Ruiz-Lapuente, R. Canal, and J. Isern). Kluver, Dordrecht, 715

Branch, D., Fisher, A., Nugent, P., and Baron, E. (1996): On H_0 from type Ia supernovae. In: *The Extragalactic Distance Scale*, Poster papers from the STScI Symposium (eds. M. Livio, M. Donahue, and N. Panagia). STScI, Baltimore, 4

Brocato, E., Castellani, V., Ferraro, F.R., Piersimoni, A.M., and Testa, V. (1997): The age of the old Magellanic Cloud clusters. II. NGC 1786, NGC 1841, and NGC 2210 as evidence for an old coeval population of LMC and Galactic globular clusters. *Mon. Not. Roy. Astron. Soc.*, in press

Burstein, D. and Heiles, C. (1984): Reddening estimates for galaxies in the Second Reference Catalog and the Uppsala General Catalog. *Astrophys. J. Suppl.*, **54**, 33

Caloi, V., D'Antona, F., and Mazzitelli, I. (1996): Horizontal branch models and the distances to globular clusters. In: *The Extragalactic Distance Scale*, Poster Papers from the STScI Symposium (eds. M. Livio, M. Donahue, and N. Panagia). STScI, Baltimore, 9

Capaccioli, M., Della Valle, M., D'Onofrio, M., and Rosino, L.A. (1989): Properties of the nova population of M31. *Astron. J.*, **97**, 1622

Chaboyer, B. (1995): Absolute ages of globular clusters and the age of the Universe. *Astrophys. J.*, **444**, L9

Chiosi, C., Wood, P., and Capitanio, N. (1993): Theoretical models of Cepheid variables and their $BVI(c)$ colors and magnitudes. *Astrophys. J. Suppl.*, **86**, 541

Cohen, J.G. (1985): Nova shells. II. Calibration of the distance scale using novae. *Astrophys. J.*, **292**, 90

Corbett, E.A., Browne, I.W.A., Wilkinson, P.N., and Patniak, A.R. (1996): Radio measurement of the time delay in 0218+357. In: *Astrophysical Applications of Gravitational Lensing* (eds. C.S. Kochanek and J.N. Hewitt). Kluwer, Dordrecht, 37

Cowan, J.J., Thielemann, F.-K., and Truran, J.W. (1987): Nuclear chronometers from the r-process and the age of the Galaxy. *Astrophys. J.*, **323**, 543

Cowan, J.J., McWilliam, A., Sneden, C., and Burris, D.L. (1997): The thorium chronometer in CS 22892-052: Estimates of the age of the Galaxy. *Astrophys. J.*, in press

Demarque, P. (1997): The ages of globular clusters. In: *The Extragalactic Distance Scale* (eds. M. Livio, M. Donahue, and N. Panagia). Baltimore: STScI, in press

Dressler, A. (1987): The $D_n - \sigma$ relation for bulges of disk galaxies - A new, independent measure of the Hubble constant. *Astrophys. J.*, **317**, 1

Elson, R.A.W. and Santiago, B.X. (1996): The M87 globular cluster system revisited. *Mon. Not. Roy. Astron. Soc.*, **280**, 971

Faber, S.M., Wegner, G., Burstein, D., Davies, R.L., Dressler, A., Lynden-Bell, D., and Terlevich, R.J. (1989): Spectroscopy and photometry of elliptical galaxies. VI. Sample selection and data summary. *Astrophys. J. Suppl.*, **69**, 763

Falco, E.E., Shapiro, I.I., Moustakas, L.A., and Davis, M. (1997): An estimate of H_0 from Keck spectroscopy of the gravitational lens system 0957 + 561. *Astrophys. J.*, **484**, 70

Feast, M.W. and Catchpole, R.M. (1997): The Cepheid PL zero-point from *HIPPARCOS* trigonometrical parallaxes. *Mon. Not. Roy. Astron. Soc.*, **286**, L1

Feast, M. and Walker, A.R. (1987): Cepheids as distance indicators. *Ann. Rev. Astron. Astrophys.*, **25**, 345

Federspiel, M., Tammann, G.A., and Sandage, A. (1997): The Virgo cluster distance from 21 cm-line widths. *Astron. Astrophys.*, submitted

Feldmeier, J.J., Ciardullo, R., and Jacoby, G.H. (1996): The planetary nebula distance to M101. *Astrophys. J.*, **461**, L25

Feldmeier, J.J., Ciardullo, R., and Jacoby, G.H. (1997): Planetary nebulae as standard candles. XI. Application to spiral galaxies. *Astrophys. J.*, in press

Freedman, W.L. and Madore, B.F. (1990): An empirical test for the metallicity sensitivity of the Cepheid period-luminosity relation. *Astrophys. J.*, **365**, 186

Freedman, W.L., Madore, B.F., Mould, J.R., Ferrarese, L., Hill, R., Kennicutt, R.C., Jr., Saha, A., Stetson, P.B., Graham, J.A., Ford, H., Hoessel, J.G., Huchra, J., Hughes, S.M., and Illingworth, G.D. (1994a): Distance to the Virgo cluster galaxy M 100 from *HubbleSpaceTelescope* observations of Cepheids. *Nature*, **371**, 757

Freedman, W.L., Hughes, S.M., Madore, B.F., Mould, J.R., Lee, M.G. Stetson, P., Kennicutt, R.C., Turner, A., Ferrarese, L., Ford, H., Graham, J.A., Hill, R., Hoessel, J.G., Huchra, J., and Illingworth, G.D. (1994b): The *HubbleSpaceTelescope* extragalactic distance scale key project. I. The discovery of Cepheids and a new distance to M81. *Astrophys. J.*, **427**, 628

Fritze-von Alvensleben, U. and Gerhard, O.E. (1994): The star formation history of NGC 7252. *Astron. Astrophys.*, **285**, 775

Giovanelli, R., Haynes, M.P., Herter, T., Vogt, N.P., Da Costa, L.N., Freudling, W., and Salzer, J.J. (1997): The I band Tully-Fisher relation for cluster galaxies: A template relation, its scatter and bias corrections. *Astron. J.*, **113**, 53

Hamuy, M., Phillips, M.M., Maza, J., Suntzeff, N.B., Schommer, R.A., and Aviles, R. (1995): A Hubble diagram of distant type Ia supernovae. *Astron. J.*, **109**, 1

Hamuy, M., Phillips, M.M., Suntzeff, N.B., Schommer, R.A., Maza, J., and Aviles, R. (1996): The Hubble diagram of the Calan/Tololo type Ia supernovae and the value of H_0. *Astron. J.*, **112**, 2398

Harris, W.E. (1990): CCD photometry of globular clusters in NGC 3377. *Publ. Astron. Soc. Pacific*, **102**, 966

Herbig, T., Lawrence, C.R., Readhead, A.C.S., and Gulkis, S. (1995): A measurement of the Sunyaev-Zel'dovich effect in the Coma cluster of galaxies. *Astrophys. J.*, **449**, L5

Hernanz, M., Garcia-Berro, E., Isern, J., Mochkovitch, R., Segretain, L., and Chabrier, G. (1994): The influence of crystallization on the luminosity function of white dwarfs. *Astrophys. J.*, **434**, 652

Hjorth, J. and Tanvir, N.R. (1997): Calibration of the fundamental plane zero-point in the Leo I group and an estimate of the Hubble constant. *Astrophys. J.*, **482**, 68

Hoessel, J.G., Gunn, J.E., and Thuan, T.X. (1980): The photometric properties of brightest cluster galaxies. I. Absolute magnitudes in 116 nearby Abell clusters. *Astrophys. J.*, **241**, 486

Höflich, P. and Khokhlov, A. (1996): Explosion models for type Ia supernovae: A comparison with observed light curves, distances, H_0, and q_0. *Astrophys. J.*, **457**, 500

Holzapfel, W.L., Arnaud, M., Ade, P.A.R., Church, S.E., Fischer, M.L., Mauskopf, P.D., Rephaeli, Y., Wilbanks, T.M., and Lange, A.E. (1996): Measurement of the Hubble constant from X-ray and 2.1 millimeter observations of Abell 2163. *Astrophys. J.*, **480**, 449

Huchra, J.P. (1988): On infall into the Virgo cluster. In: *The Extragalactic Distance Scale* (eds. S. van den Bergh and C.J. Pritchet). Publ. Astron. Soc. Pacific, San Francisco, 257

Jacoby, G.H. and Ciardullo, R. (1993): Luminosity functions of planetary nebulae. In: *Planetary Nebulae* (eds. R. Weinberger and A. Acker). Kluwer, Dordrecht, 503

Jacoby, G.H., Ciardullo, R., and Ford, H.C. (1990): Planetary nebulae as standard candles. V. The distance to the Virgo cluster. *Astrophys. J.*, **356**, 332

Jacoby, G.H., Ciardullo, R., Ford, H.C., and Booth, J. (1989): Planetary nebulae as standard candles. III. The distance to M81. *Astrophys. J.*, **344**, 704

Jerjen, H. and Tammann, G.A. (1993): The Local group motion towards Virgo and the microwave background. *Astron. Astrophys.*, **276**, 1

Kawaler, S.D. (1997): White dwarf stars. In: *Stellar Remnants* (eds. G. Meynet and D. Schaerer). Springer, Berlin, 1

Keeton, C.R. and Kochanek, C.S. (1997): Determining the Hubble constant from the gravitational lens PG 1115 + 080. *Astrophys. J.*, in press

Kelson, D.D., Illingworth, G.D., Freedman, W.F., Graham, J.A., Hill, R., Madore, B.F., Saha, A., Stetson, P.B., Kennicutt, R.C., Jr., Mould, J.R., Hughes, S.M., Ferrarese, L., Phelps, R., Turner, A., Cook, K.H., Ford, H., Hoessel, J.G., and Huchra, J. (1996): The extragalactic distance scale key project. III. The discovery of Cepheids and a new distance to M101 using the *HubbleSpaceTelescope*. *Astrophys. J.*, **463**, 26

Kennicutt, R.C., Freedman, W.L., and Mould, J.R. (1995): Measuring the Hubble constant with the *HubbleSpaceTelescope*. *Astron. J.*, **110**, 1476

Kraan-Korteweg, R.C. (1986): A catalog of 2810 nearby galaxies - The effect of the Virgocentric flow model on their observed velocities. *Astron. Astrophys. Supp. Ser.*, **66**, 255

Kraft, R.P. (1961): Color excesses for supergiants and classical Cepheids. V. The period-color and period-luminosity relations: a revision. *Astrophys. J.*, **134**, 616

Kristian, J., Sandage, A., and Westphal, J.A. (1978): The extension of the Hubble diagram. II. New redshifts and photometry of very distant galaxy clusters - First indication of a deviation of the Hubble diagram from a straight line. *Astrophys. J.*, **221**, 383

Kundić, T., Turner, E.L., Colley, W.N., Gott, J.R., III, Rhoads, J.E., Wang, Y., Bergeron, L.E., Gloria, K.A., Long, D.C., Malhorta, S., and Wambsganns, J. (1997): A robust determination of the time delay in 0957+561 A, B and a measurement of the global value of Hubble's constant. *Astrophys. J.*, **482**, 75

Lasenby, A.N. (1997): The Sunyaev Zel'dovich effect. In: *The Extragalactic Distance Scale* (eds. M. Livio, M. Donahue, and N. Panagia). STScI, Baltimore, in press

Lasenby, A. and Hancock, S. (1995): Detection of primary and secondary anisotropies in the CMB. In: *Current Topics in Astrofundamental Physics: The Early Universe* (eds. N. Sanchez and A. Zichichi). Kluwer, Dordrecht, 327

Lauer, T.R. and Postman, M. (1992): The Hubble flow from brightest cluster galaxies. *Astrophys. J.*, **400**, L47

Lineweaver, C.H., Barbosa, D., Blanchard, A., and Bartlett, J.G. (1997): Constraints on Hubble's constant, Ω_{baryon} and Λ from cosmic microwave background observations. *Astron. Astrophys.*, in press

Lorenz, H., Böhm, P., Capaccioli, M., Richter, G.M., and Longo, G. (1993): A new technique to gauge luminosity fluctuations in galaxies. *Astron. Astrophys.*, **277**, L15

Madore, B.F. and Freedman, W. L. (1991): The Cepheid distance scale. *Publ. Astron. Soc. Pacific*, **103**, 933

Mazzitelli, I., D'Antona, F., and Caloi, V. (1995): Globular cluster ages with updated input physics. *Astron. Astrophys.*, **302**, 382

McHardy, I.M., Stewart, G.C., Edge, A.C., Cooke, B., Yamashita, K., and Hatsukade, I. (1990): Ginga observations of Abell 2218 – Implications for H_0. *Mon. Not. Roy. Astron. Soc.*, **242**, 215

Méndez, R.H., Kudritzki, R.P., Ciardullo, R., and Jacoby, G.H. (1993): The bright end of the planetary nebula luminosity function. *Astron. Astrophys.*, **275**, 534

Mould, J.R., Huchra, J.P., Bresolin, F., Ferrarese, L., Ford, H.C., Freedman, W.L., Graham, J., Harding, P., Hill, R., Hoessel, J.G., Hughes, S.M., Illingworth, G.D., Kelson, D., Kennicutt, R.C., Jr., Madore, B.F., Phelps, R., Stetson, P.B., and Turner, A. (1995): Limits on the Hubble constant from the *HST* distance of M100. *Astrophys. J.*, **449**, 413

Nair, S. (1995): A lens model for B0218+357. In: *Astrophysical radiomeasurements of the time delay in 0218+357* (eds. C.S. Kochanek and J.N. Hewitt). Kluwer, Dordrecht, 197

Panagia, N., Gilmozzi, R., Sonneborn, G., Cassatella, A., Fransson, C., Kirshner, R.P., Lundquist, P., and Wamsteker, W. (1996): Improved distance determination to SN 1987A. In: *The Extragalactic Distance Scale*, Poster papers from the STScI Symposium (eds. M. Livio, M. Donahue, and N. Panagia). STScI, Baltimore, 54

Phillips, M.M. (1993): The absolute magnitudes of type Ia supernovae. *Astrophys. J. Lett.*, **413**, L105

Pritchet, C. J. and van den Bergh, S. (1987): Observations of novae in the Virgo cluster. *Astrophys. J.*, **318**, 507

Reid, N.I. (1997): Younger and brighter – New distances to globular clusters based on *HIPPARCOS* parallax measurements of local subdwarfs. *Astrophys. J.*, in press

Renzini, A., Bragaglia, A., Ferraro, F.R., Gilmozzi, R., Ortolani, S., Holberg, J.B., Liebert, J., Wesemael, F., and Bohlin, R.C. (1996): The white dwarf distance to the globular cluster NGC 6752 (and its age) with the *HubbleSpaceTelescope*. *Astrophys. J.*, **465**, L23

Rephaeli, Y. (1995): Comptonization of the cosmic microwave background: The Sunyaev-Zeldovich Effect. *Ann. Rev. Astron. Astrophys.*, **33**, 541

Richer, M.G. (1996): Planetary nebulae: their use as a tool to probe the evolution of galaxies. *J. Roy. Astron. Soc. Can.*, **90**, 246

Riess, A.G., Press, W.H., and Kirshner, R.P. (1996): A precise distance indicator: type Ia supernova multicolor light-curve shapes. *Astrophys. J.*, **473**, 88

Ruiz-Lapuente, P. (1996): The Hubble constant from ^{56}Co-powered nebular candles. *Astrophys. J.*, **465**, 83

Saha, A., Sandage, A., Labhardt, L., Tammann, G.A., Macchetto, F.D., and Panagia, N. (1996a): Cepheid calibration of the peak brightness of SNe Ia. V. SN 1981B in NGC 4536. *Astrophys. J.*, **466**, 55

Saha, A., Sandage, A., Labhardt, L., Tammann, G.A., Macchetto, F.D., and Panagia, N. (1996b): Cepheid calibration of the peak brightness of type Ia supernovae. VI. SN 1960F in NGC 4496A. *Astrophys. J. Suppl.*, **107**, 693

Saha, A., Sandage, A., Labhardt, L., Tammann, G.A., Macchetto, F.D., and Panagia, N. (1997): Cepheid calibration of the peak brightness of SNe Ia. VIII. SN 1990N in NGC 4639. *Astrophys. J.*, in press

Saha, A., Sandage, A., Labhardt, L., Schwengeler, H., Tammann, G.A., Panagia, N., and Macchetto, F.D. (1995): Discovery of Cepheids in NGC 5253: Absolute peak brightness of SN Ia 1895B and SN Ia 1972E and the value of H_0. *Astrophys. J.*, **438**, 8

Sandage A. (1958): Current problems in the extragalactic distance scale. *Astrophys. J.*, **127**, 513

Sandage, A. (1961): The ability of the 200-inch telescope to discriminate between selected world models. *Astrophys. J.*, **133**, 355

Sandage, A. (1988): Cepheids as distance indicators when used near their detection limit. *Publ. Astron. Soc. Pacific*, **100**, 935

Sandage, A. (1993a): Temperature, mass, and luminosity of RR Lyrae stars as functions of metallicity at the blue fundamental edge. II. *Astron. J.*, **106**, 703

Sandage, A. (1993b): $H_0 = 43 \pm 11$ km s^{-1} Mpc^{-1} based on angular diameters of high-luminosity field spiral galaxies. *Astrophys. J.*, **402**, 3

Sandage, A. (1993c): H_0 found by comparing linear diameters of M31 with similar field galaxies. *Astrophys. J.*, **404**, 419

Sandage, A. (1993d): Globular cluster ages determined from the Oosterhoff period-metallicity effect using oxygen-enhanced isochrones. III. *Astron. J.*, **106**, 719

Sandage, A. (1994a): Bias properties of extragalactic distance indicators. 1: The Hubble constant does not increase outward. *Astrophys. J.*, **430**, 1

Sandage, A. (1994b): Bias properties of extragalactic distance indicators. 2: Bias corrections to Tully-Fisher distances for field galaxies. *Astrophys. J.*, **430**, 13

Sandage, A. (1995): Practical cosmology: Inventing the past. In: *The Deep Universe* (eds. B. Binggeli and R. Buser). Springer, Berlin, 1

Sandage, A. (1996a): Annual Report 1994/95 of the Carnegie Observatories. *Bull. Amer. Astron. Soc.*, **28**, 51

Sandage, A. (1996b): Bias properties of extragalactic distance indicators. V. H_0 from luminosity functions of different spiral types and luminosity classes corrected for bias. *Astron. J.*, **111**, 1

Sandage, A. (1996c): Bias properties of extragalactic distance indicators. VI. Luminosity functions of M31 and M101 look-alikes listed in the RSA2: H_0 therefrom. *Astron. J.*, **111**, 18

Sandage, A. and Bedke, J. (1988): *Atlas of Galaxies*. NASA, Washington, D.C.

Sandage, A. and Hardy, E. (1973): The redshift-distance relation. VII. Absolute magnitude of the first three ranked cluster galaxies as functions of cluster richness and Bautz-Morgan cluster type: The effect on q_0. *Astrophys. J.*, **183**, 743

Sandage, A. and Tammann, G.A. (1968): A composite period-luminosity relation for Cepheids at mean and maximum light. *Astrophys. J.*, **151**, 531

Sandage, A. and Tammann, G.A. (1995): Steps toward the Hubble constant. X. The distance of the Virgo cluster core using globular clusters. *Astrophys. J.*, **446**, 1

Sandage, A. and Tammann, G.A. (1996): An alternate calculation of the distance to M87 using the Whitmore et al. luminosity function for its globular clusters: H_0 therefrom. *Astrophys. J.*, **464**, L51

Sandage, A. and Tammann, G.A. (1997): Stunning confirmation of the ground-based Cepheid P-L zero-point using *HIPPARCOS* trigonometric parallaxes. *Mon. Not. Roy. Astron. Soc.*, submitted

Sandage, A., Saha, A., Tammann, G.A., Labhardt, L., Panagia, N., and Macchetto, F.D. (1996): Cepheid calibration of the peak brightness of type Ia supernovae: calibration of SN 1990N in NGC 4639 averaged with six earlier type Ia supernova calibrations to give H_0 directly. *Astrophys. J.*, **460**, L15

Schechter, P.L., Bailyn, C.D., Barr, R., Barvainis, R., Becker, C.M., Bernstein, G.M., Blakeslee, J.P., Bus, S.J., Dressler, A., Falco, E.E., Fesen, R.A., Fischer, P., Gebhardt, K., Harmer, D., Hewitt, J.N., Hjorth, J., Hurt, T., Jaunsen, A.O., Mateo, M., Mehlert, D., Richstone, D.O., Sparke, L.S., Thorstensen, J.R., Tonry, J.L., Wegner, G., Willmarth, D.W., and Worthey, G. (1996): The quadruple gravitational lens PG 1115 + 080: Time delays and models. *Astrophys. J. Lett.*, **475**, L85

Schmidt, B.P., Kirshner, R.P., Eastman, R.G., Phillips, M.M., Suntzeff, N.B., Hamuy, M., Maza, J., and Aviles, R. (1994): The distances to five type II supernovae using the expanding photosphere method, and the value of H_0. *Astrophys. J.*, **432**, 42

Shi, X. (1995): The uncertainties in the age of globular clusters from their helium abundance and mass loss. *Astrophys. J.*, **446**, 637

Silbermann, N.A., Harding, P., Madore, B.F., Kennicutt, R.C., Jr., Saha, A., Stetson, P.B., Freedman, W.L., Mould, J.R., Graham, J.A., Hill, R.J., Turner, A., Bresolin, F., Ferrarese, L., Ford, H., Hoessel, J.G., Han, M., Huchra, J., Hughes, S.M.G., Illingworth, G.D., Phelps, R., and Sakai, S. (1996): The *HubbleSpaceTelescope* key project on the extragalactic distance scale. VI. The Cepheids in NGC 925. *Astrophys. J.*, **470**, 1

Sodemann, M. and Thomsen, B. (1996): Variation in the surface brightness fluctuations of M32. *Astron. J.*, **111**, 208

Soffner, T., Mendez, R.H., Jacoby, G.H., Ciardullo, R., Roth, M.M., and Kudritzki, R.P. (1996): Planetary nebulae and HII regions in NGC 300. *Astron. Astrophys.*, **306**, 9

Tammann, G.A. (1988): The distance of the Virgo cluster – A review. In: *The Extragalactic Distance Scale* (eds. S. van den Bergh and C.J. Pritchet). Publ. Astron. Soc. Pacific, San Francisco, 282

Tammann, G.A (1993): Why are planetary nebulae poor distance indicators? In: *Planetary Nebulae*, IAU Symp. 155 (eds. R. Weinberger and A. Acker). Kluwer, Dordrecht, 515

Tammann, G.A. (1996): Why is there still controversy on the Hubble constant? *Rev. Mod. Astron.*, **9**, 139

Tammann, G.A. and Federspiel, M. (1997): Focussing in on H_0. In: *The Extragalactic Distance Scale* (eds. M. Livio, M. Donahue, and N. Panagia). STScI, Baltimore, in press

Tanvir, N.R., Shanks, T., Ferguson, H.C., and Robinson, D.R.T. (1995): Determination of the Hubble constant from observations of Cepheid variables in the galaxy M96. *Nature*, **377**, 27

Teerikorpi, P. (1984): Malmquist bias in a relation of the form $M = a \cdot p + b$. *Astron. Astrophys.*, **141**, 407

Theureau, G., Hanski, M., Ekholm, T., Bottinelli, L., Gouguenheim, L., Paturel, G., and Teerikorpi, P. (1997): Kinematics of the local universe V. The value of H_0 from the Tully-Fisher B and $\log D_{25}$ relations for field galaxies. *Astron. Astrophys.*, **322**, 730

Tonry, J.L. and Schneider, D.P. (1988): A new technique for measuring extragalactic distances. *Astron. J.*, **96**, 807

Tonry, J.L., Ajhar, E.A., and Luppino, G.A. (1990): Observations of surface-brightness fluctuations in Virgo. *Astron. J.*, **100**, 1416

Tonry, J.L., Blakeslee, J.P., Ajhar, E.A., and Dressler, A. (1997): The SBF survey of galaxy distances. I. Sample selection, photometric calibration, and the Hubble constant. *Astrophys. J.* **475**, 399

Truran, J.W. (1997): Nucleochronology. In: *The Extragalactic Distance Scale* (eds. M. Livio, M. Donahue, and N. Panagia). STScI, Baltimore, in press

van Leeuwen, F., Feast, M.W., Whitelock, P.A., and Yudin, B. (1997): First results from *HIPPARCOS* trigonometrical parallaxes of Mira-type variables. *Mon. Not. Roy. Astron. Soc.*, **287**, 955

Walker, A.R. (1993): The Large Magellanic Cloud clusters NGC 1835 – Photometry of the RR Lyrae stars. *Astron. J.*, **105**, 527

Weiss, A., Salaris, M., and Degl'Innocenti, S. (1996): Lower ages for the oldest clusters. In: *The Extragalactic Distance Scale*, Poster Papers from the STScI Symposium (eds. M. Livio, M. Donahue, and N. Panagia). STScI, Baltimore, 71

Whitmore, B.C., Sparks, W.B., Lucas, R.A., Macchetto, F.D., and Biretta, J.A. (1995): *Hubble Space Telescope* observations of globular clusters in M87 and an estimate of H_0. *Astrophys. J. Lett.*, **454**, L73

Winget, D.E., Hansen, C.J., Liebert, J., van Horn, H.M., Fontaine, G., Nather, R.E., Kepler, S.O., and Lamb, D.Q. (1987): An independent method for determining the age of the Universe. *Astrophys. J.*, **315**, L77

Wood, M.A. (1992): Constraints on the age and evolution of the Galaxy from the white dwarf luminosity function. *Astrophys. J.*, **386**, 539

Zabludoff, A.I., Geller, M.J., Huchra, J.P., and Vogeley, M.S. (1993): The kinematics of dense clusters of galaxies. I. The data. *Astron. J.*, **106**, 1273

Experimental Gravity

Michael Soffel

Lohrmann Observatorium
Technische Universität Dresden, Mommsenstr. 13, D-01062 Dresden
`soffel@rcs.urz.tu-dresden.de`

Abstract

The present status of experimental gravity is reviewed. We discuss experimental tests of Newton's law of gravity (the $1/r^2$-law, $\dot{G}/G$ and the numerical value of G), of various forms of the equivalence principle (the weak equivalence principle, Einstein's equivalence principle and the strong equivalence principle related with the Nordtvedt effect) as well as various tests of metric theories of gravity. Among the latter we describe the search for gravito-magnetism (Lense-Thirring effects), measurements of the geodetic precession and light-deflection (signal retardation) effects. The important topics of binary pulsars and gravitational wave physics are beyond the scope of that article.

1 Introduction

Since many years Einstein's theory of gravity (general relativity) has evolved far from being merely an esoteric, mathematically oriented discipline, although a huge number of researchers are still mainly attracted by its mathematical beauty. Indeed, the pure mathematical framework still exerts a magic force upon a huge number of people that are willing to devote their lives for a deeper understanding of that gorgeous theoretical building. Fortunately, even in Germany the number of those people has steadily increased who enjoy the field of tension between pure theory and experimental confirmation in general relativity, i.e., who take general relativity seriously as a branch of physics that has to be verified by experiments. It is remarkable that in the last decades a large number of new techniques have been developed that were used to perform tests of Einstein's theory of gravity. Among these we find laser ranging to retroreflectors on the lunar surface (Lunar Laser Ranging, LLR), Very Long Baseline Interferometry (VLBI) and radar measurements to planets or spacecrafts. Today the number of experiments that test some aspects of Einstein's theory of gravity is enormous indeed. For that reason it is useful to put some spiritual order into this wealth of experiments. Here it is advantageous that in our solar system gravitational fields are weak and orbital velocities of massive bodies (Sun, Moon, planets) are small. In dimensionless units, the gravitational potential, U/c^2, and the squares of astronomical orbital velocities, $(v/c)^2$, in our solar system are as small as about 10^{-6}. This implies that in our cosmic neighborhood relativistic effects are small and for many purposes Newton's theory of gravity is quite sufficient. If one starts with the Newtonian framework and wants to move towards Einstein's general relativity the socalled 'equivalence principle' plays a crucial role. Roughly speaking this principle says that all test-bodies of sufficiently small size fall freely at the same rate in some external gravitational field if they are not influenced by some other non-gravitational forces. Theoretically, the equivalence principle leads to the possibility of interpreting the action of gravity as phenomenon of space-time curvature. If one considers the results from Special Relativity then one arrives at the conclusion that gravity can be described by some 'metric theory' (Will 1993), where a test body moves along a geodesic in some curved space-time. Beyond Newton's theory, in the framework of a first post-Newtonian theory a large class of posssible metric theories of gravity can be described by means of several selected PPN-parameters. Such a parameterized post-Newtonian (PPN) scheme may serve as test-theory for most gravitational experiments in the solar system. Beyond its post-Newtonian core every metric theory of gravity makes its own predictions for the strong-field regime or large orbital velocities. Under certain conditions some of them are accessible to measurements even today.

In the following I will review some important tests of Einstein's theory of gravity, involving solar system objects only. A discussion of experimental gravity by means of binary pulsars or gravitational waves is beyond the scope of this article.

2 Newton's Law of Gravity

Let us start with Newton's law for the free-fall of a test-mass in the form

$$m_I \ddot{\mathbf{r}} = -G \frac{M m_G}{r^2} \frac{\mathbf{r}}{r},$$

where M denotes the field generating mass and $m_I (m_G)$ the inertial (gravitational) mass of the testbody. In the late eighties some groups had claimed to have found some evidence for a violation of the inverse square law, like gravimetric measurements in mines (e.g., Stacey et al. 1987) or a reanalysis of the classical Eötvös experiments by Fischbach et al. (1986, 1988), due to a new fundamental force (5th force) in nature, leading to a deviation from the $1/r^2$-law according to

$$G(r) = G_\infty \left[1 + \alpha(1 + r/\lambda)e^{-r/\lambda} \right]$$

with $\alpha \neq 0$. Some older measurements seemed to support this suspicion: accelerometer measurements in the vicinity of a cliff by Thieberger (1987), torsion balance measurements in the vicinity of a cliff by Boynton et al. (1987), gravimetric measurements at the WTVD-tower in Clayton, NC by Eckhard and et al. (1988) and gravimetric measurements in a 2033 m deep borehole in the ice of Greenland by Ander et al. (1989). However, considering systematic errors from neighboring mass distributions (small hillsites and ridges) most results were later found to be in agreement with Newton's law of gravity (Parker and Zumberge 1989, Bartlett and Tew 1989). More recent tower experiments also at the WABG tower ($h \sim 610\,\mathrm{m}$) in Inverness, MS, showed no evidence for a violation of the inverse square law; instead, in the geophysical window for λ between 10 m and 10 km they now provide the best upper limits for the strength α of a possible 5th force (Fig. 1). A group from the University of Zurich (Cornaz et al. 1994, Hubler et al. 1995) tested the inverse square law for the range of a few centimeters to some hundred meters using a single balance that compared the weights of two 1 kg steel masses located above and below

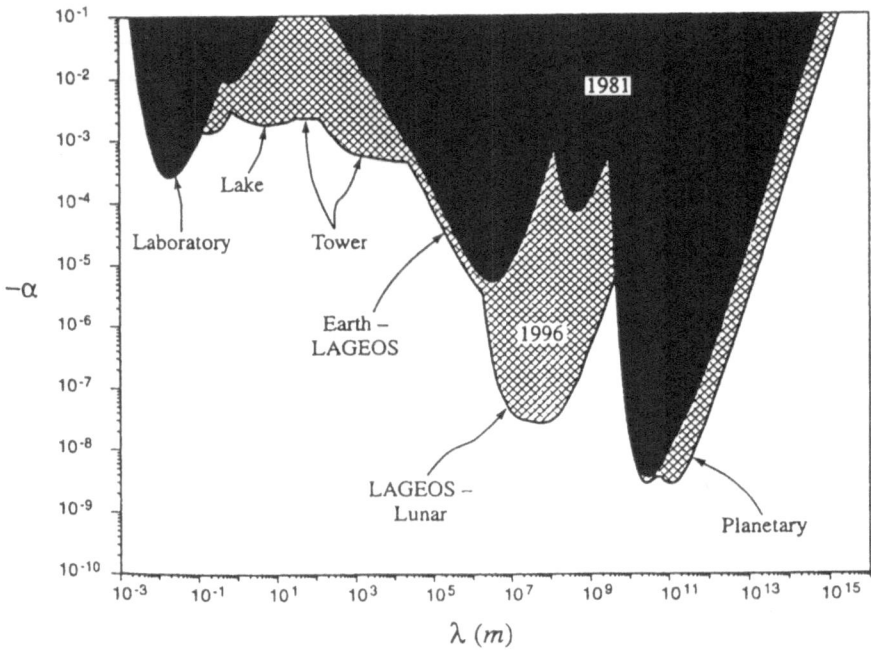

Figure 1 The constraints on α as a function of λ in 1981 (dark region) and again in 1996 (hatched region) (from Romaides et al. 1997). Not all of the latest results are considered.

the variable water level of a pumped-storage lake (the Gigerwald lake experiment); they obtained an upper limit for α of the order 0.003. For even smaller distances considerable progress was achieved with the Eöt-Wash experiments carried out at the University of Washington in Seattle (Heckel et al. 1989, Adelberger et al. 1990, 1991). These experiments were mainly sensitive to a composition-dependent 5th force (depending, e.g., upon baryon numbers). These experiments use a mass-quadrupole torsion balance mounted on a turntable that rotates with a multiple of the oscillation period of the suspension fibre. For distances larger than a few km results from laser ranging to satellites like LAGEOS can be used to derive upper limits for α, lunar laser ranging data provide the best upper limit for λ of the order of the Earth-Moon distance ($\alpha < 10^{-11}$; note that this result was not included in Fig. 1). Finally, for even larger distances planetary data can be employed (Fig. 1).

Table 1 shows that the relative temporal change of the gravitational constant, $\dot{G}/G$, cannot be larger than a few parts in 10^{-12} per year. One sees that Lunar Laser Ranging yields the smallest upper limit for $\dot{G}/G$. These LLR-measurements determine the Earth-Moon distance by means of travel-times of laser-pulses of about 5 cm pulse-length. For common pulse powers of about 10^9 W and wavelengths of 532 nm such a pulse consists of about 10^{19} photons. This photon disk of 5 cm thickness alluminates an area of about $20\,\mathrm{km}^2$ on the lunar surface. If it hits one of several laser retroreflectors that has been placed on the Moon by Apollo astronautes or unmanned Soviet probes about one photon per pulse finds its way back into the receiving system. By means of several filtering techniques one tries to discriminate the correct photons from background noise. The

$\dot{G}/G$ in 10^{-11}/yr	Reference
radar measurements to planets and spacecrafts	
< 40	Shapiro et al. 1971
< 1.5	Reasenberg and Shapiro 1976, 1978
< 14	Anderson et al. 1978
Lunar laser ranging	
< 3.0	Williams et al. 1978
< 1.0	Müller et al. 1991
< 0.8	Williams et al. 1996
< 0.5	Schneider et al. 1997
PSR 1913+16	
< 2.3	Damour et al. 1988

Table 1 Measurements of $\dot{G}/G$ using different methods.

accuracy of LLR meanwhile is at the 1 cm level and one should keep in mind that the average distance to the Moon is about 385000 km. In the Earth-Moon system the value for $\dot{G}/G$ is highly correlated with the value $\dot{n}_M$ for the tidal friction. However, because of the influence from the Sun there is enough sensitivity for an excellent measurement of $\dot{G}/G$.

Recently, there have been interesting developments as far as the numerical value of G is concerned. The official Committee on Data for Science and Technology (CODATA) value, $G = (6.67259 \pm 0.00085) \times 10^{-11}$ m^3 kg^{-1} s^{-2}, was mainly derived from the measurements of Luther and Towler (1982) and was assigned an uncertainty of 128 ppm. It is still by far the worst determination of a fundamental constant of nature. Recently this value has been questioned by several groups: the German PTB (Michaelis et al. 1995) obtained a value 0.6% ($\sim$ 40 standard deviations) higher, a New Zealand group (Fitzgerald and Armstrong 1995) reported a value 0.1% lower, while a Wuppertal group (Walesch et al. 1995) obtained a value 0.06% lower than the CODATA value. Now, apart from the PTB measurements all groups employed the time-of-swing method with a torsion balance assuming the spring constant of the torsion fibre to be independent of frequency. According to Kuroda (1995) this assumption might in fact be wrong and recent tests of the Kuroda hypothesis supported his conjecture (Bagley and Luther 1997). So it seems likely that with a different experimental setup (e.g. Gundlach et al. 1996) or control of the spring constant improved values for G will appear in the near future.

3 The Equivalence Principle

3.1 The Weak Equivalence Principle

The weak equivalence principle says that all uncharged test bodies whose gravitational influence can be neglected fall freely at the same rate in some external gravitational field and this acceleration is independent of mass, geometry, composition, temperature, etc. of the test body. If we write Newton's law of free-fall as above then the weak equivalence principle claims the equality of inertial mass, $m_\mathcal{I}$, and passive gravitational mass, m_G, of the test body.

Rumours relate free-fall experiments from the leaning tower of Pisa with the name of Galileo, but, being aware of friction problems, likely he instead performed experiments with inclined planes. However, Galileo certainly was not the first who had performed such kind of experiments. Already in the 5th century the Byzantinian scholar Ioannis Philoponus reported:

> For if you let fall from the same height two weights of which one is many times as heavy as the other, you will see that the .... difference in time is a very small one (Clagett 1955).

The first careful measurements for the weak equivalence principle, however, obviously go back to Newton. He writes:

> ... I tried the thing in gold, silver, lead, glass, sand, common salt, wood, water, and wheat. I provided two equal wooden boxes. I filled the one with wood, and suspended an equal weight of gold (as exactly as I could) in the centre of oscillation of the other. The boxes, hung by equal threads of 11 feet, made a couple of pendulums perfectly equal in weight and figure, and equally exposed to the resistance of air: and, placing the one by the other, I observed them to play together forwards and backwards for a long while, with equal vibrations. And therefore the quantity of matter in the gold was to the quantity of matter in the wood as the action of the motive forces upon all the gold to the action of the same upon all the wood; that is, as the weight of the one to the weight of the other.
>
> And by these experiments, in bodies of the same weight, one could have discovered a difference of matter less than the thousandth part of the whole (Newton, I., Principia)

Figure 2 provides an overview of the historically important tests of the weak equivalence principle. Here, the Eötvös ratio

$$\eta_\mathrm{E}(1,2) \equiv \left(\frac{m_\mathrm{G}}{m_\mathcal{I}}\right)_1 - \left(\frac{m_\mathrm{G}}{m_\mathcal{I}}\right)_2,$$

displayed on the vertical axis, is the difference of the ratios of gravitational to inertial mass for body 1 and 2. Remarkable progress was achieved in the late 19th and early 20th century by baron Lorand von Eötvös and colleagues by means of a torsion balance. If the weak equivalence principle were violated for the two masses located at the arms of a torsion balance then a torque would act on the balance since, because of inertial forces, the suspension fibre does not point precisely towards the center of gravitational attraction. If the balance is then rotated this torque becomes time dependent and the resulting oscillation can be detected. Eötvös et al. analyzed a variety of materials like water, talcum, copper, asbestos or such strange materials as snakewood together with platinum. The accuracy of these measurements was about 10^{-8}. Later it was improved

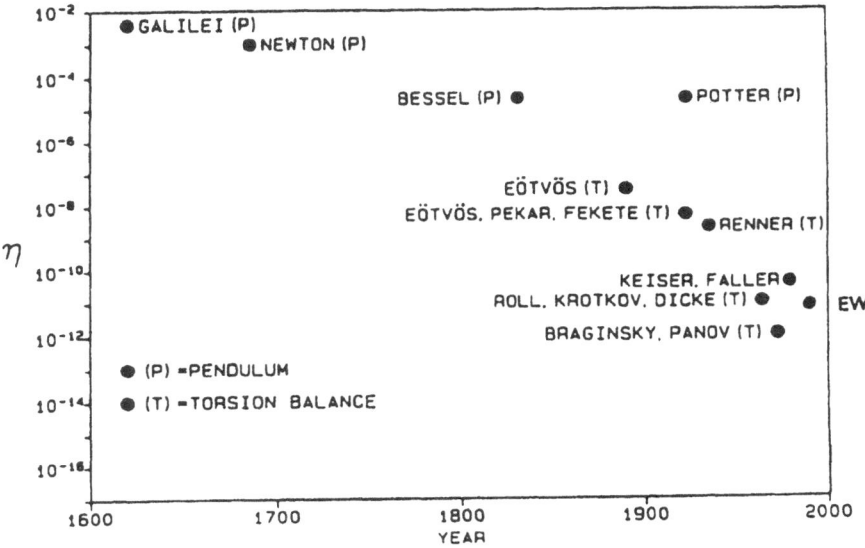

Figure 2 Upper limits for the Eötvös-parameter η_E.

by Pekar and Fekete (Eötvös et al. 1922) to about 3×10^{-9}. This accuracy is already remarkable and, strictly speaking, not easy to understand. Roll et al. (1964) write:

> One mystery connected with Eötvös' original experiment is how he could achieve a precision of 3×10^{-9} if the effect of his own body on his apparatus (through gravitational field gradients) was as much as 80 times (as may be easily calculated). The answer probably is that he left the balance undisturbed until it came to rest, and then he observed for only a short period of time before the instrument would respond to his own gravitational field. Also, he may have observed in such a standardized fashion that this effect may have always been nearly the same, tending to cancel out in the data reduction. One may easily calculate that with Eötvös' apparatus the observation time (total time sitting at apparatus) would need to be as short as 30 sec if errors from the field gradient effect were not to exceed 3×10^{-9}.

Maybe Eötvös had some slim assistent of low weight responsible for the readings. The technique using a torsion balance was substantially improved in the early 60ies by Roll et al. (1964), who achieved an accuracy of about 10^{-11}. The principle of the Princeton experiment differed from that of baron Eötvös and his colleagues in that they looked for possible torques resulting from the gravitational attraction of the Sun and the centrifugal acceleration caused by the orbital motion of the Earth about the Sun. In Eötvös-type experiments the torsion balance had to be rotated around the fibre axis, causing uncontrollable errors; in the Princeton experiment this rotation was automatically induced by the Earth's rotation in space. In the design of the Princeton experiment a lot of perturbations were checked: those by gravity field gradients, varying magnetic fields caused by magnetic impurities of the materials, variable electrostatic fields, effects from gas pressure, from Brownian motion, temperature variations, vibrations of the underground, etc. Figure 3 shows the essential parts of the apparatus used by Roll et al. (1964). The central part of the torsion balance is an equilateral triangle formed by quartz bars, 6 cm on a

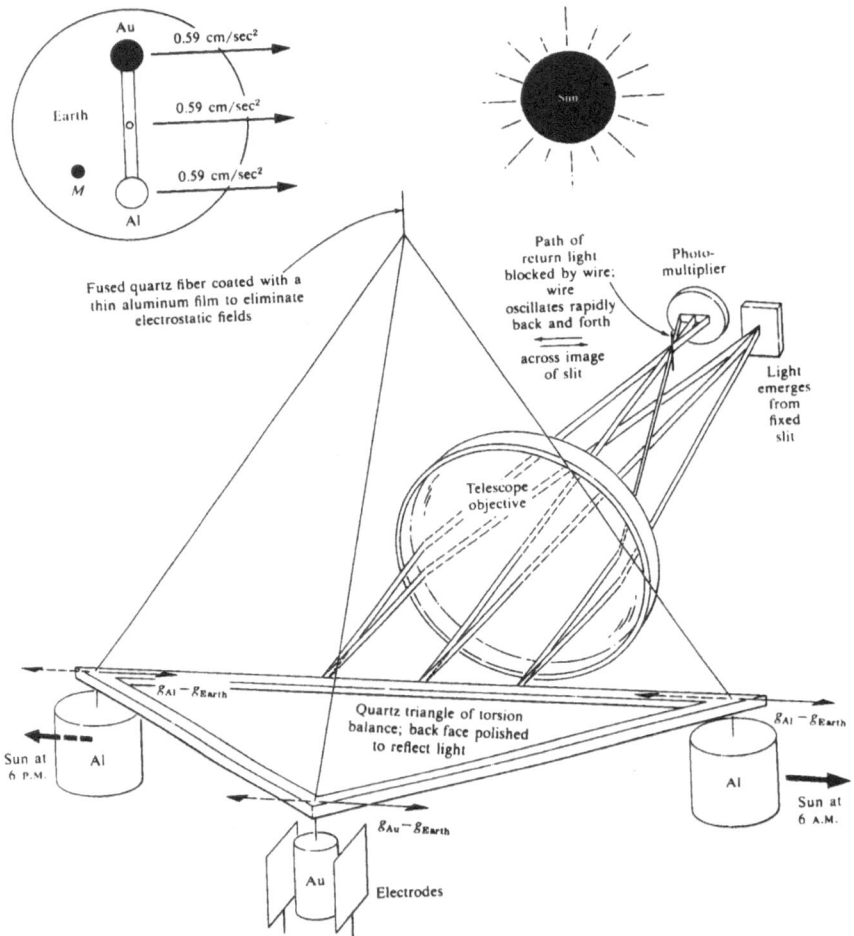

Figure 3 Essential parts of the apparatus used by Roll et al. (1964) to test the weak equivalence principle (from Misner et al. 1973).

side. Equal masses of gold and aluminum hang at the corners of this supporting construction. If the aluminum and gold masses would fall at different rates in the gravitational field of the Sun they would exert a torque on the balance that, according to the Earth's rotation, would vary with diurnal period. Thus perturbations with a different period can be eliminated. The triangular shape of the balance was chosen to minimize perturbations due to gravity field gradients. For further reduction of these gradients the experiment was performed underground by remote control. By means of electrodes around the gold mass and a feedback loop the position of the balance was always kept in its zero position. The D.C. voltage required to restore the balance to its zero position was recorded, Fourier-analyzed and the magnitude of the Fourier-component of 24-hour period finally gave an upper bound for the Eötvös ratio for the gold-aluminum pair (Misner et al. 1973).

Braginsky and Panov (1972) from Moscow University performed a similar experiment with aluminum and platinum. They claimed an accuracy of about 10^{-12}; however, a reanalysis of their data by F. Everitt and colleagues raised doubts about this phantastic accuracy.

There are reasons to believe that a significant improvement of these experiments can only be achieved in a spacecraft orbiting the Earth. One reason are the various noise sources that limit the accuracy of such Earth-bound experiments. The accuracy of Eötvös type experiments is limited by thermal- and gravity-field gradients; that of Dicke type experiments by frequency conversion of seismic noise. These perturbations can be reduced significantly in a spacecraft with drag-free control. Even a free-fall experiment with highest accuracy seems to be feasible in orbit.

In the last years a Satellite Test of the Equivalence Principle (STEP 1996) was extensively discussed which should provide a test of the weak equivalence principle at the 10^{-17} level, an improvement of about six orders of magnitude with respect to Earth-bound experiments.

The project STEP was proposed to ESA in November 1989 by an international team of scientists chaired by C.W.F. Everitt and P.W. Worden, Jr., from Stanford University. In principle, STEP is a free-fall experiment on board a drag-free satellite. Central part is a differential accelerometer consisting of two hollow cylindrical test masses placed inside of each other that, because of magnetic forces, can move only along their common z-axis. The relative position of these two masses can be measured with an accuracy of about 10^{-13} cm by means of squid-magnetometers. The two masses are centered by means of positioning electrodes so they experience the same gravitational field. Various perturbations can be eliminated because they act equally onto the two masses. In one design a total of three accelerometers with five different materials will be employed.

STEP failed to become a medium size mission of ESA in the last years; presently the future of STEP is unclear.

3.2   Einstein's Equivalence Principle

Simply speaking the weak equivalence principle says that in some sufficiently small freely falling laboratory (Einstein's elevator) uncharged test-bodies behave as if they were in an inertial frame. Einstein had generalized that to what is now called Einstein's equivalence principle (EEP): in such a freely falling elevator all non-gravitational laws of physics (Maxwell's laws of electrodynamics, the laws of weak and strong interaction, etc.) take their usual form from Special Relativity where gravitational fields play no role.

One important test of Einstein's equivalence principle is provided by measuring the gravitational redshift of photons. If a photon with initial frequency ν_E traverses a potential difference of ΔU, then an observer at rest in the gravitational field measures a frequency shift that is given by

$$z = \frac{\nu_R - \nu_E}{\nu_R} = \frac{\Delta U}{c^2},$$

where ν_R denotes the frequency after traversing the potential difference. This follows indeed from Einstein's equivalence principle as is, e.g., shown in Will (1993). Allowing for a violation of EEP, we may write

$$z = (1 + \alpha)\frac{\Delta U}{c^2}.$$

During the years 1960 to 1965, R.V. Pound and colleagues (Pound and Rebka 1960, Pound and Snider 1965) from Harvard University performed the first precise gravitational redshift experiments with gamma-rays of extremely small linewidth employing the Mössbauer effect in Fe^{57}. The ratio of linewidth to frequency was of the order of 10^{-18}. Precise frequency-shifts of the gamma-rays over a vertical distance of 25 m (the height of the tower of Jefferson laboratory, Harvard University) were determined by moving the absorber material with a small relative velocity with respect to the emitter and measuring ν_R by means of the Doppler-effect. These early measurements had an accuracy of about 1%. Other experiments measured the frequency shift of spectral lines in the gravitational field of the Sun. In subsequent ones the rate of atomic clocks on board of aircrafts or rockets were investigated. One effect that is directly related with the gravitational redshift of electromagnetic signals concerns the rates of clocks: clocks are slowed down in a gravitational field. This is not difficult to understand since electromagnetic signals are the tool to compare the rates of different clocks if they are well separated. In October 1971, J. Hafele and R. Keating (1972), equipped with four cesium clocks, were flying around the Earth, once in westward direction and once in eastward direction. Theoretically one expected a time difference between the clocks onboard and a reference clock in Washington D.C. after one flight around the Earth because of the gravitational redshift and the quadratic Doppler effect. Theoretically the onboard clocks after the eastward trip should be behind by about 40 ns and after the westward trip about 275 ns ahead of time. The experiment gave for the eastward flight a time difference of $-(59 \pm 10)$ ns and for the westward flight (273 ± 7) ns, leading to an upper limit for $|\alpha|$ of 0.1.

During the years 1975 and 1976, C.O. Alley (1979) conducted a series of interesting clock experiments by means of rubidium clocks onboard of high flying aircrafts. Here, laser pulses were send from the ground to the aircraft and back for time comparisons and precise orbital determination. Until now the most precise test of the EEP is the rocket experiment of Vessot and Levine (1979) (see Vessot et al. 1980, Vessot 1984), NASA's Gravity Probe A (GP-A) experiment from June 18, 1976. Here a Scout rocket carried a hydrogen-maser clock with a stability of 10^{-14} to a height of about 10 000 km. During the duration of the flight of about two hours the frequency of this hydrogen-clock was compared with the frequencies of two Earth-bound atomic clocks (also hydrogen-masers). To this end this orbit of the rocket was carefully monitored from five different ground stations. With great care first order Doppler effects were experimentally eliminated. An analysis of GP-A data gave

$$\alpha < 2 \times 10^{-4}.$$

Several null gravitational redshift experiments have been performed by several groups (Turneaure et al. 1983, Demidov et al. 1992, Godone et al. 1995) by comparing the readings of several 'nonidentical clocks' (clocks that are constructed differently). Writing

$$\nu_A/\nu_B = (\nu_{A_0}/\nu_{B_0}) \left[1 + (\beta_A - \beta_B) \frac{U}{c^2} \right],$$

these experiments gave $|\beta_{Cs} - \beta_H| < 10^{-4}$ and $|\beta_{Cs} - \beta_{Mg}| < 7 \times 10^{-4}$, where Cs, H and Mg refers to Cs-clocks, H-masers and Mg-clocks.

EEP has many consequences that can be checked experimentally. E.g., Local Lorentz Invariance (LLI) is implied by EEP (Will 1993) and the Hughes-Drever experiment (Hughes et al. 1960, Drever 1961) presents a well-known test of it. In the Glasgow version, the experiment examined the $J = 3/2$ ground state of the ^{7}Li nucleus in an external magnetic field. The state is split into four levels with equal spacing. Any perturbation related with some preferred direction of space, obviously a violation of LLI, should destroy this equality. By means of nuclear magnetic resonance techniques stringent contraints for possible deviations from equality were derived.

In some alternative theories of gravity like Moffat's nonsymmetric gravitation theory (Moffat 1991), in the presence of gravity space becomes anisotropic and birefringent. The gravity-induced birefringence of space implies that the propagation through a gravitational field can alter the polarization of light. Haugan and Kauffmann (1995) have used data from polarization measurements of extragalactic sources to constrain birefringence induced by the gravity field of our Galaxy.

3.3   The Strong Equivalence Principle

The strong equivalence principle (SEP) extends the EEP by considering also the motion of self-gravitating bodies in some external gravitational field. A violation of the SEP would imply that test-bodies of different gravitational self-energy Ω would fall at different rates in some external gravitational field. Let us write the acceleration of such a test body as

$$\mathbf{a} = \left(\frac{m_G}{m_\mathcal{I}}\right)\mathbf{g}_{\text{ext}} \equiv \left(1 + \eta_N \frac{\Omega}{mc^2}\right)\mathbf{g}_{\text{ext}}.$$

Here, the so-called Nordtvedt-parameter η_N describes the degree of possible violation of the SEP. For bodies of a few meters in size the ratio of gravitational self-energy to rest-energy is of the order of 10^{-26}, hence in Eötvös-type laboratory experiments the gravitational self-energy plays no role. This, however, is not the case for astronomical bodies; for the Earth this ratio is about 5×10^{-10}, for the Moon about 2×10^{-11}. Let us consider specifically the Earth-Moon system in orbit around the Sun. If

$$\delta_N \equiv \left(\frac{m_G}{m_\mathcal{I}}\right)_M - \left(\frac{m_G}{m_\mathcal{I}}\right)_E \neq 0,$$

then Earth and Moon would fall at different rates towards the Sun with an anomalous relative acceleration $\delta\mathbf{a}$ between Earth and Moon of

$$\delta\mathbf{a} = GM_S \left[\frac{(\mathbf{R} - \mathbf{r})}{|\mathbf{R} - \mathbf{r}|^3}\left(\frac{m_G}{m_\mathcal{I}}\right)_M - \frac{\mathbf{R}}{R^3}\left(\frac{m_G}{m_\mathcal{I}}\right)_E\right] \simeq \frac{GM_S}{R^3}\mathbf{R}\,\delta_N,$$

leading to anomalous oscillations of the Earth-Moon distance of

$$\delta r \simeq 13\,\eta_N \cos(\omega_M - \omega_S)\,t \;[\text{m}]$$

(Nordtvedt 1988, Williams et al. 1996). Here, $\mathbf{R}$ denotes the vector from the Sun to the Earth, $\mathbf{r}$ that from the Earth to the Moon; $\omega_M(\omega_S)$ is the angular velocity of the Moon (Sun) around the Earth. In case the SEP would be violated the lunar orbit around the Sun would be polarized in the direction of the Sun (Nordtvedt 1968a,b, Nordtvedt and Will 1972, Will and Nordtvedt 1972, Will 1973, Nordtvedt 1973). Lunar laser ranging provides an ideal tool for a test of the SEP by looking for a possible Nordtvedt effect. Different groups (Williams et al. 1996, Schneider et al. 1997) have analyzed LLR data to determine the value for the Nordtvedt parameter η_N with the result:

$$\eta_N < 10^{-3},$$

strongly supporting the validity of Einstein's theory of gravity, where $\eta_N = 0$. One should note, that basically in every other reasonable alternative theory of gravity this strong version of the equivalence principle is violated in some form or another. Formulating the strong equivalence principle rigorously is not an easy matter. In some sense it is equivalent with the validity of Einstein's theory of gravity.

4   Tests of Metric Theories of Gravity

Einstein's equivalence principle has a profound consequence: the gravitational interaction can be understood to result from space-time curvature. It is quite obvious that if EEP were violated such an interpretation becomes impossible. Einstein's theory of gravity is the simplest of all metric theories of gravity that is compatible with Special Relativity. It is formulated with one single g-field, the metric field $g_{\mu\nu}$, describing the infinitesimal distance ds between neighboring points in space-time:

$$ds^2 = g_{\mu\nu}dx^\mu dx^\nu = g_{00}(c\,dt)^2 + 2g_{0i}\,cdt\,dx^i + g_{ij}dx^i dx^j,$$

where $x^\mu = (ct, x^i), i = 1,2,3$ are the space-time coordinates. To lowest order

$$g_{00} = -1 + \frac{2U}{c^2} + O(c^{-4}),$$

i.e., the Newtonian potential U appears in the time-time component of the metric. The Newtonian field equations (the Poisson equations)

$$\Delta U = -4\pi G\rho$$

are contained as a limit in Einstein's field equations

$$\Phi(g_{\mu\nu}, \partial g_{\mu\nu}) = -4\pi G\mathcal{F}(\text{matter variables}).$$

These relate the components of the metric tensor und its first and second derivatives with those matter variables that act as sources of the gravitational field. In the framework of a post-Newtonian approximation the metric, e.g., for one isolated body, can be written in the form

$$g_{00} = -1 + \frac{2U}{c^2} - 2\beta\left(\frac{U}{c^2}\right)^2 ; \quad g_{0i} = -(\gamma + 1)G\frac{(\mathbf{J} \times \mathbf{x})^i}{c^3 r^3}; \quad g_{ij} = \left(1 + \gamma\frac{2U}{c^2}\right)\delta_{ij},$$

where we have introduced the PPN-parameters β and γ. $\mathbf{J}$ is the angular momentum of the isolated body. In Einstein's theory of gravity $\beta = \gamma = 1$. The Newtonian potential U is related with 'electric-type' gravity, $\mathbf{J}$, or more generally any mattter current, with gravito-magnetism.

It is well known that Einstein's theory of gravity leads to an anomalous perihelion precession of planets. In our PPN-formalism it is described by

$$(\Delta\varpi)_{\text{rev.}} = 2\pi(2\gamma + 2 - \beta)\frac{GM_S}{c^2 a(1 - e^2)},$$

where a and e are the semi-major axis and eccentricity of the planetary orbit. The solar quadrupole moment also contributes to such a perihelion precession and there was a time when people thought this contribution could be significant. Helioseismological measurements of the solar normal modes, however, lead to a solar quadrupole moment of about 2×10^{-7} and a negligible contribution to perihelia precessions of planets. For quite a while the perihelion precession of Mercury can be measured by means of radar signals. Presently one has achieved an accuracy of about a few per mille,

$$(2\gamma + 2 - \beta)/3 = 1.000 \pm 0.002,$$

in accordance with Einstein's theory.

There is indirect evidence of the existence of gravity-magnetism (Lense-Thirring effects), e.g., in the motion of artificial Earth's satellites. If one writes the metric in the barycentric system (with origin at the barycenter of the solar system) it contains a term of gravito-magnetic origin,

$$g_{0i} = \Delta \frac{GM_E v_E^i}{c^2 r},$$

where the quantity Δ takes the value 4 in Einstein's theory. After transforming the metric into a geocentric system this term is precisely cancelled by another one. Formally one finds anomalous distance oscillations of a satellite orbit proportional to $\Delta - 2\gamma - 2$ that vanishes identically in Einstein's theory because of the presence of gravito-magnetism. Assuming the orbit of the lasergeodynamical satellite LAGEOS to be known to better than about 10 cm one can derive an upper limit of .004 for $\Delta - 2\gamma - 2$, a strong evidence of the existence of gravito-magnetism in nature. An even stronger evidence of this existence results from the periastron advance of the pulsar in the binary system PSR 1913+16 (Nordtvedt 1993, private communication). In this binary system the masses of both components are comparable and hence the gravito-magnetic interaction contributes significantly to the observed periastron advance. The direct measurement of gravito-magnetism is one goal of the well-known Stanford gyroscope experiment (NASA's gravity probe B, GP-B; Everitt 1974, Lipa and Everitt 1978, Bardas et al. 1992, Xiao et al. 1992), that was proposed already in 1959 by Pugh and in 1960 by Schiff (Fig. 4). According to Einstein's theory the spin-axis of a torque-free gyroscope in some (drag-free) Earth's satellite should precess with respect to the fixed-stars according to

$$\frac{d\mathbf{S}}{d\tau} = \frac{3}{2} \frac{\mathbf{v} \times \nabla U}{c^2} + 2 \frac{\nabla \times \mathbf{g}}{c^2}$$

with $g_i \equiv g_{0i}$. Here, the first term on the right hand side is of gravito-electrical origin and describes geodetic precession. The second gravito-magnetical term describes the socalled 'Lense-Thirring precession'. For the GP-B experiment the geodetic precession amounts to $6.9''$ per year and the Lense-Thirring precession to $0.05''$ per year. Goal of GP-B is a determination of the geodetic (Lense-Thirring) precession with an accuracy of 10^{-5} (0.3%). In this experiment the gyros consist of almost perfectly round quartz spheres of 2 cm radius, rotating with 2000 Hz. They are coated with a thin layer of superconducting niobium that because of the rotation induces a magnetical (London) moment perfectly aligned with the rotation axis. The motion of the spin axis can then be monitored by means of a squid magnetometer. Recently the test facilities on Earth were completed and first tests were successful. First tests in orbit are expected to be carried out soon.

For the orbit of the satellite LAGEOS the gravito-magnetic Lense-Thirring effect leads to an additional drift of the node by three arcsec per century. Now, the even zonal harmonics

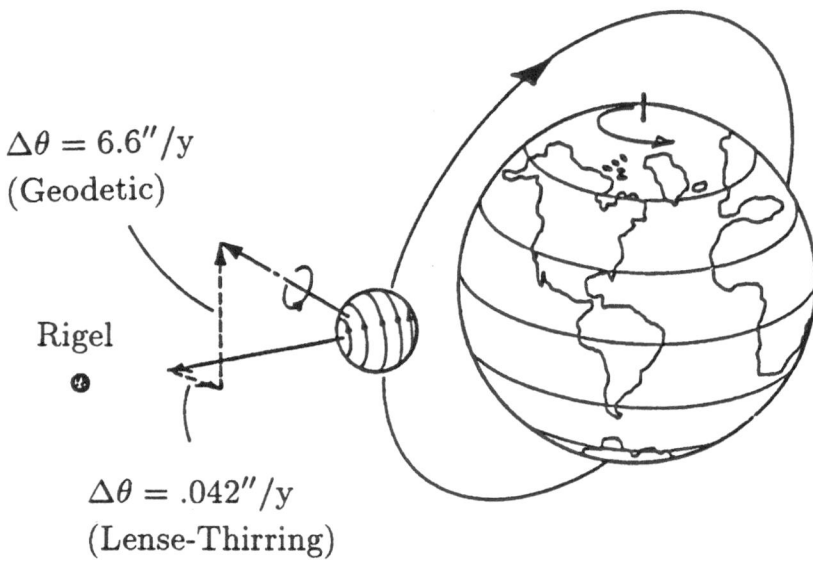

$\Delta\theta = 6.6''/\mathrm{y}$
(Geodetic)

Rigel

$\Delta\theta = .042''/\mathrm{y}$
(Lense-Thirring)

Figure 4 The Stanford gyroscope experiment. The relativistic precession rates of gyroscope whose spin vector is parallel to the line of sight to a guide star and located in the orbital plane of a 650 km polar orbit. The geodetic precession is in the plane of the orbit and has a predicted value in Einstein's theory of 6.6 arc sec/y. The Lense-Thirring precession is in the plane of the celestial equator and in the same sense as the Earth's rotation; its predicted value is 0.042 arc sec/y.

of the Earth lead to a nodal drift that by far exceeds the Lense-Thirring one; the Lense-Thirring effect is roughly comparable with that from the $l = 12$ multipole moment of the Earth. Since the accuracy by which these multipole moments can be determined is relatively poor the gravito-magnetic drift cannot be determined by analyzing the data from one single satellite. However, to first order the nodal drift due to the multipole moments is of the form

$$[\Delta\Omega(Y_l^m)]_{\text{sec.}}^{(1)} = \cos I \cdot \Psi(a, e, \text{even powers of } \sin I),$$

where I denotes the inclination of the satellite orbit. If one considers a second satellite orbit with the same values for a and e but with inclination $I' = 180° - I$, then in the sum of the two nodes the contributions from the Newtonian multipole moments cancel to a high degree. Since the Lense-Thirring precession always acts in the direction of the Earth's rotation it has the same value for both orbits. In this manner it should be possible to measure directly the Lense-Thirring effect from the data of such two complementary satellite orbits. This idea goes back to Ignazio Ciufolini (1989, 1991); the experiment meanwhile is called LAGEOS-3 experiment (LAGEOS-2 is in orbit since 1992; the inclination of the LAGEOS-2 orbit is not complementary to that of LAGEOS-1). Two studies and an evaluation by NASA came to the conclusion, that the Lense-Thirring precession with such an experiment can be measured with a precision of about 10%, given three years of data. With an improved Earth's model, that became available recently, the accuracy could even be of the order of 3%. However, the launch date for LAGEOS-3 still is a mystery. Likely, the competition with GP-B plays a role here, causing unforseeable delays for the launch. Recently Ciufolini (1997, unpublished notes) has thought about the possibility to measure the Lense-Thirring effect from LAGEOS-1 and LAGEOS-2 data by looking into the drifts of the nodes and the arguments of perigee. To understand the error budgets and correlations more clearly, however, more work has to be done.

The geodetic precession in Einstein's theory of gravity leads to an additional precession of the lunar orbit in space by about $2''$ per century. To measure this effect one may write

$$\dot{\omega}_M = 2''/100\,y \times (1 + h)$$

and may determine the value for h from observational data. If one experimentally finds evidence for some nonzero value of h one might argue against Einstein's theory. Shapiro et al. (1988) were the first who in this way determined the geodetic precession by means of LLR data. They used data from the years between 1970 and 1986 and estimated a total of 335 parameters, about 250 of them were time dependent Earth's rotation parameters. It is amusing to note that simply by analyzing the Earth-Moon distance they were able to determine the mass of Jupiter with a precision of about 10%. The analysis of Shapiro et al. (1988) gave

$$h = 0.019 \pm .020,$$

where systematic errors were taken into account. Two similar determinations of the geodetic precession of the lunar orbit by the JPL-group (Dickey et al. 1994) and a German group (Müller et al. 1991) also found no deviation from the prediction of Einstein's theory. Müller et al. (1991) determined the value of h from about 6300 LLR-data from the time between 1969 and 1990; they found $h = 0.002 \pm 0.010[0.002]$, where the formal 1σ error is indicated in square brackets.

One well-known prediction of General Relativity is the light-deflection in the gravitational field of the Sun. At the limb of the Sun it amounts to $1.75''$ and decreases like $1/d$ with increasing impact parameter d. In the PPN-formalism this deflection angle is proportional to $(\gamma + 1)/2$, so the space-curvature parameter γ can be determined from light-deflection measurements. The light deflection in the gravitational field of the Sun was first detected by the British expeditions to Sobral (Brazil) and Principe (Gulf of Guinea) during the total eclipse on May 29, 1919. An analysis of the data gave for the extrapolated deflection angle at the limb of the Sun $1.''98 \pm 0.''16$ (Sobral) and $1.''61 \pm 0.''40$ (Principe) in accordance with Einstein's theory (Dyson et al. 1920). Until now photographic plates of star fields were taken during at least nine total solar eclipses and compared with corresponding night plates. None of these measurements did contradict predictions from Einstein's theory but the accuracy was never better than about 10%. Often poor seeing conditions contributed to this fact. Now, light signals are not only deflected from some gravitational center but in some gravity field their propagation is also slowed down. This is the gravitational time delay, which was first predicted by I. Shapiro (Shapiro effect). Time delay measurements using radar data to VIKING (Reasenberg et al. 1979) gave

$$\gamma = 1. \pm 0.002.$$

The gravitational time delay plays an important role in Very Long Baseline interferometry (VLBI). In VLBI one receives radio signals of distant extragalactic radio sources (quasars) with sufficiently large radio antennas. These antennas typically are separated by thousands of kilometers without any direct connection. At each station the radiosignals are transformed from the GHz into the MHz region and together with time-tags from some local oscillator (H-maser) recorded on video tape. The time-tags serve as phase reference, a crucial tool for the working of VLBI. The video tapes of several (at least two) radio stations are then brought together and by means of some correlator the cross-correlation of the data produces an interference pattern from which the observables (like the group delay) can be determined. Presently, with VLBI angular separations can be determined with an accuracy of better than 1 mas. That implies that the light deflection (more precisely the gravitational time delay) in the gravitational field of the Sun can be measured even for radio sources lying almost 180° from the Sun. An analysis of VLBI data by Robertson et al. (1991) gave

$$\gamma = 1. \pm .002[.00096].$$

In this study a total of 342819 VLBI-data from the geodetic programs POLARIS (polar motion analysis), IRIS (International Radio Interferometric Surveying) and the Crustal Dynamics Program (CDP) were analysed.

Recent VLBI observations of the extragalactic radio sources 3C273B and 3C279 by Lebach et al. (1995) gave $\gamma = 0.9996 \pm 0.0017$. Using all VLBI data available Eubanks et al. (1997) found $\gamma - 1 = \pm 3 \times 10^{-4}$, where again a realistic error is given.

References

Adelberger, E.G., Heckel, B.R., Stubbs, C.W., and Rogers, W.F. (1991): Searches for New Macroscopic Forces. *Ann. Rev. Nucl. Particle Sci.*, **41**, 269

Adelberger, E.G., Heckel, B.R., Smith, G., Su, Y., and Swanson, H.E. (1990): Eötvös experiments, lunar ranging and the strong equivalence principle. *Nature*, **347**, 261

Alley, C.O. (1979): Relativity and clocks. In: *Proc. of the 33rd Annual Symposium on Frequency Control.* Electronic Industries Association, Washington, D.C.

Ander, M.E., Zumberge, M.A., Lautzenhiser, T., Parker, R.L., Aiken, C.L.V., Gorman, M.R., Nieto, M.M., Cooper, A.P.R., Ferguson, J.F., Fisher, E., McMechan, G.A., Sasagawa, G., Stevenson, J.M., Backus, G., Chave, A.D., Greer, J., Hammer, P., Hansen, B.L., Hildebrand, J.A., Kelty, J.R., Sidles, C., and Wirtz, J. (1989): Test of Newton's Inverse-Square Law in the Greenland Ice Cap. *Phys. Rev. Lett.*, **62**, 985

Anderson, J.D., Keesey, M.S., Lau, E.L., Standish, E.M., and Newhall, X.X. (1978): Tests of General Relativity using astrometric and radiometric observations of the planets. *Acta Astronautica*, **5**, 43

Bagley, C.H. and Luther, G.G. (1997): Preliminary Results of a Determination of the Newtonian Constant of Gravitation: A Test of the Kuroda Hypothesis. *Phys. Rev. Lett.*, **78**, 3047

Bardas, D., Taber, M.A., Buchman, S., DeBra, D.B., Everitt, C.W.F., Gill, D., Green, G., Gutt, G., Kasdin, N.J., Keiser, G.M., Lipa, J.A., Lockhart, J.M., Muhlfelder, B., Parkinson, B.W., Turneaure, J.P., Van Patten, R.A., Xiao, Y., Zhou, P., Parmley, R., Reynolds, G., Calhoon, S., Clappier, R., Frank, D., Grady, J., Grammer, J., Read, D., Salmon, J., and Vassar, R. (1992): Gravity Probe B: II. Hardware Development; Progress Towards the Flight Instrument. In: *Proc. of the Sixth Marcel Grossmann Meeting* (eds. H. Sato and T. Nakamura). World Scientific, Singapore, p. 382

Bartlett, D.F. and Tew, W.L. (1989): Possible Effect of the Local Terrain on the North Carolina Tower Gravity Experiment. *Phys. Rev. Lett.*, **63**, 1531

Boynton, P.E., Crosby, D., Ekstrom, P., and Szumilo, A. (1987): Search for an Intermediate-Range Composition-Dependent Force. *Phys. Rev. Lett.*, **59**, 1385

Braginsky, V.B. and Panov, V.I. (1972): Verification of the Equivalence of Inertial and Gravitational Mass. *Sov. Phys. JETP*, **34**, 463

Ciufolini, I. (1989): A Comprehensive Introduction to the LAGEOS Gravitometric Experiment: From the Importance of the Gravitomagnetic Field in Physics to Preliminary Error Analysis and Error Budget. *Int. J. Mod. Phys. A.*, **4**, 3083

Ciufolini, I. (1991): New Class of Theories of Gravity Not Described by the Parametrized Post-Newtonian (PPN) Formalism. *Int. J. Mod. Phys. A.*, **6**, 5511

Clagett, M. (1955): *Greek Science in Antiquity.* Abelard Schumann, New York.

Cornaz, A., Hubler, B., and Kündig, W. (1994): Determination of the Gravitational Constant at an Effective Interaction Distance of 112 m. *Phys. Rev. Lett.*, **72**, 1152

Damour, T., Gibbons, G.W., and Taylor, J.H. (1988): Limits on the Variability of G Using Binary-Pulsar Data. *Phys. Rev. Lett*, **61**, 1151

Demidov, N.A., Ezhov, E.M., Sakharov, B.A., Uljanov, B.A., Bauch, A., and Fischer, B. (1992): Investigations of the frequency instability of CH1-75 hydrogen masers. *Proceedings of the 6th European Frequency and Time Forum.* Noordwijk, NL, ESA

Dickey, J.O, Bender, P.L., Faller, J.E., Newhall, X.X., Ricklefs, R.L., Ries, J.G., Shelus, P.J., Veillet, C., Whipple, A.L., Wiant, J.R., Williams, J.G., and Yoder, C.F. (1994): Lunar Laser Ranging: A Continuing Legacy of the Apollo Program. *Science*, **265**, 482

Drever, R. (1961): A Search for Anisotropy of Inertial Mass using a Free Precession Technique. *Phil. Mag.*, **6**, 683

Dyson, F., Eddington, A.S., and Davidson, C. (1920): A determination of the deflection of light by the Sun's gravitational field from observations made at the total eclipse of May 29, 1919. *Phil. Trans. Roy. Soc.*, **220A**, 291

Eckhardt, D.H., Jekeli, C., Lazarewicz, A.R., Romaides, A.J., and Sands, R.W. (1988): Tower Gravity Experiment: Evidence for Non-Newtonian Gravity. *Phys. Rev. Lett.*, **60**, 2567

Eötvös, R.V., Pekar, D., and Fekete, E. (1922): Beiträge zum Gesetze der Proportionalität von Trägheit und Gravität. *Ann. Phys.*, **68**, 11

Eubanks, T.M., Matsakis, D.N., Martin, J.O., Archinal, B.A., McCarthy, D.D., Klioner, S.A., Shapiro, S., and Shapiro, I.I. (1997): Advances in Solar System Tests of Gravity. Abstract submitted for the APR97 Meeting of The American Physical Society, Session K11: Gravitation Experiment and Theory. 1 p.

Everitt, C.W.F. (1974): In: *Experimental Gravitation: Proc. of Course 56 of the International School of Physics "Enrico Fermi"* (ed. B. Bertotti). Academic Press, New York.

Fischbach, E., Sudarsky, D., Szafer, A., Talmadge, C., and Aronson, S.H. (1986): Reanalysis of the Eötvös Experiment. *Phys. Rev. Lett.*, **56**, 3

Fischbach, E., Sudarsky, D., Szafer, A., Talmadge, C., and Aronson, S.H. (1988): Long-Range Forces and the Eötvös Experiment. *Ann. Phys. (N.Y.)*, **182**, 1

Fitzgerald, M. and Armstrong, T. (1995): Newton's Gravitational Constant with Uncertainty Less Than 100 pm. *IEEE Trans. Instrum. Meas.*, **44**, 494

Godone, A., Novero, C., and Tavella, P. (1995): Null gravitational redshift experiment with nonidentical atomic clocks. *Phys. Rev. D*, **51**, 319

Gundlach, J.H., Adelberger, E.G., Heckel, B.R., and Swanson, H.E. (1996): New technique for measuring Newton's constant G. *Phys. Rev. D*, **54**, R1256

Hafele, J.C. and Keating, R.E. (1972): Around-the-World Atomic Clocks: Predicted Relativistic Time Gains. *Science*, **177**, 166

Hafele, J.C. and Keating, R.E. (1972): Around-the-World Atomic Clocks: Observed Relativistic Time Gains. *Science*, **177**, 168

Haugan, M.P. and Kauffmann, T.F. (1995): New test of the Einstein equivalence principle and the isotropy of space. *Phys. Rev. D*, **52**, 3168

Heckel, B.R., Adelberger, E.G., Stubbs, C.W., Su, Y., Swanson, H.E., Smith, G., and Rogers, W.F. (1989): Experimental Bounds on Interactions Mediated by Ultralow-Mass Bosons. *Phys. Rev. Lett.*, **63**, 2705

Hubler, B., Cornaz, A., and Kündig, W. (1995): Determination of the gravitational constant with a lake experiment: New constraints for non-Newtonian gravity. *Phys. Rev. D*, **51**, 4005

Hughes, V.W., Robinson, H.G., and Beltran-Lopez, V. (1960): Upper limit for the anisotropy of inertial mass from nuclear resonance experiments. *Phys. Rev. Lett.*, **4**, 342

Kuroda, K. (1995): Does the Time-of-Swing Method Give a Correct Value of the Newtonian Gravitational Constant? *Phys. Rev. Lett.*, **75**, 2796

Lebach, D.E., Corey, B.E., Shapiro, I.I., Ratner, M.I., Webber, J.C., Rogers, A.E., Davis, J.L., and Herring, T.A. (1995): Measurement of the Solar Gravitational Deflection of Radio Waves Using Very-Long-Baseline Interferometry. *Phys. Rev. Lett.*, **75**, 1439

Lipa, J.A. and Everitt, C.W.F. (1978): The role of cryogenics in the gyroscope experiment. *Acta Astronautica*, **5**, 119

Luther, G.G. and Towler, W.R. (1982): Redetermination of the Newtonian Gravitational Constant G. *Phys. Rev. Lett.*, **48**, 121

Michaelis, W., Haars, H., and Augustin, R. (1995): A new precise determination of Newton's gravitational constant. *Metrologia*, **32**, 267

Misner, C., Thorne, K.S., and Wheeler, J.A. (1973): *Gravitation*. Freeman, San Francisco

Moffat, J.W. (1991): In: *Gravitation 1990*, Proceedings of the Banff Summer Institute, Banff, Canada (eds. R. Mann and P. Wesson). World Scientific, Singapore

Müller, J., Schneider, M., Soffel, M., and Ruder, H. (1991): Testing Einstein's theory of gravity by analyzing lunar ranging data. *Astrophys. J. Lett.*, **382**, L101

Nordtvedt, K. (1968a): Equivalence Principle for Massive Bodies. I. Phenomenology. *Phys. Rev.*, **169**, 1014

Nordtvedt, K. (1968b): Equivalence Principle for Massive Bodies. II. Theory. *Phys. Rev.*, **169**, 1017

Nordtvedt, K. (1973): Post-Newtonian Gravitational Effects in Lunar Laser Ranging. *Phys. Rev. D*, **7**, 234

Nordtvedt, K. (1988): Gravitomagnetic Interaction and Laser Ranging to Earth Satellites. *Phys. Rev. Lett.*, **61**, 2647

Nordtvedt, K. and Will, C.M. (1972): Conservation laws and preferred frames in relativistic gravity. II. Experimental evidence to rule out preferred-frame theories of gravity. *Astrophys. J.*, **177**, 775

Parker, R.L. and Zumberge, M.A. (1989): An analysis of geophysical experiments to test Newton's law of gravity. *Nature*, **342**, 29

Pound, R.V. and Rebka, G.A., Jr. (1960): Apparent Weight of Photons. *Phys. Rev. Lett.*, **4**, 337

Pound, R.V. and Snider, J.L. (1965): Effect of Gravity on Gamma Radiation. *Phys. Rev.*, **140**, B788

Pugh, G.E. (1959): *WSEG Research Memo*, **11**. U.S. Dept. of Defence

Reasenberg, R.D. and Shapiro, I.I. (1976): In: *Atomic Masses and Fundamental Constants*, Vol. 5 (eds. J.H. Sanders and A.H. Wapstra). Plenum, New York

Reasenberg, R.D. and Shapiro, I.I. (1978): In: *On the Measurement of Cosmological Variations of the Gravitational Constant* (ed. L. Halpern). University Press of Florida, Gainesville

Reasenberg, R.D., Shapiro, I.I., MacNeil, P.E., Goldstein, R.B., Breidenthal, J.C., Brenkle, J.P., Cain, D.L., Kaufmann, T.M., Komarek, T.A, and Zygielbaum, A.I. (1979): Viking relativity experiment: verification of signal retardation by solar gravity. *Astrophys. J.*, **234**, L219

Robertson, D.S., Carter, W.E., and Dillinger, W.H. (1991): New measurement of solar gravitational deflection of radio signals using VLBI. *Nature*, **349**, 768

Roll, P.G., Krotkov, R., and Dicke, R.H. (1964): The Equivalence of Inertial and Passive Gravitational mass. *Ann. Phys. (N.Y.)*, **26**, 442

Romaides, A.J., Sands, R.W., Fischbach, E., and Talmadge, C.L. (1997): Final results from the WABG tower gravity experiment. *Phys. Rev. D*, **55**, 4532

Schiff, L.I. (1960): Possible new experimental test of general relativity theory. *Phys. Rev. Lett.*, **4**, 215

Schneider, M., Müller, J., Schreiber, U., and Egger, D. (1997): Hochpräzisionsvermessung der Mondbewegung. *Astronomie und Raumfahrt*, **34**, 4

Shapiro, I.I., Smith, W.B., Ash, M.B., Ingalls, R.P., and Pettengill, G.H. (1971): Gravitational Constant: Experimental Bound on Its Time Variation. *Phys. Rev. Lett.*, **26**, 27

Shapiro, I.I., Reasenberg, R.D., Chandler, J.F., and Babcock, R.W. (1988): Measurement of the de Sitter Precession of the Moon: A Relativistic Three-Body-Effect. *Phys. Rev. Lett.*, **61**, 2643

Stacey, F.D., Tuck, G.J., Moore, G.I., Holding, S.C., Goodwin, B.D., and Zhou, R. (1987): Geophysics and the law of gravity. *Rev. Mod. Phys.*, **59**, 157

STEP, Satellite Test of the Equivalence Principle (1996): Testing the Equivalence Principle in Space. Proc. of an International Symposium held in Pisa, Italy, 6-8 April 1993 (ed. R. Reinhard). *ESA Publications Division*, WPP-115

Thieberger, P. (1987): Search for a Substance-Dependent Force with a New Differential Accelerometer. *Phys. Rev. Lett.*, **58**, 1066

Turneaure, J.P., Will, C.M., Farrell, B.F., Mattison, E.M., and Vessot, R.F.C. (1983): Test of the principle of equivalence by a null gravitational red-shift experiment. *Phys. Rev. D*, **27**, 1705

Vessot, R.F.C. (1984): Tests of gravitation and relativity. *Contemp. Phys.*, **25**, 355

Vessot, R.F. and Levine, M.W. (1979): A Test of the Equivalence Principle Using a Space-Borne Clock. *Gen. Rel. and Grav.*, **10**, 181

Vessot, R.F.C., Levine, M.W., Mattison, E.M., Blomberg, E.L., Hoffman, T.E., Nystrom, G.U., Farrel, B.F., Decher, R., Eby, P.B., Baugher, C.R., Watts, J.W., Teuber, D.L., and Wills, F.D. (1980): Test of Relativistic Gravitation with a Space-Borne Hydrogen Maser. *Phys. Rev. Lett.*, **45**, 2081

Walesch, H., Meyer, H., Piehl, H., and Schurr, J. (1995): The Gravitational Force at Mass Separations from 0.6 m to 2.1 m and the Precise Measurement of *G*. *IEEE Trans. Instrum. Meas.*, **44**, 491

Will, C.M. (1973): Relativistic gravity in the solar system. III. Experimental disproof of a class of linear theories of gravitation. *Astrophys. J.*, **185**, 31

Will, C.M. (1993): *Theory and experiment in gravitational physics* (2nd revised edition). Cambridge University Press, Cambridge

Will, C.M. and Nordtvedt, K. (1972): Conservation laws and preferred frames in relativistic gravity. I. Preferred-frame theories and an extended PPN formalism. *Astrophys. J.*, **177**, 757

Williams, J.G., Newhall, X.X., and Dickey, J.O. (1996): Relativity parameters determined from lunar laser ranging. *Phys. Rev. D*, **53**, 6730

Williams, J.G., Sinclair, W.S., and Yoder, C.F. (1978): Tidal acceleration of the Moon. *Geophys. Res. Lett.*, **5**, 943

Xiao, Y.M., Bardas, D., Buchman, S., Cohen, C., Everitt, C.W.F., Gill, D., Keiser, G.M., Van Patten, R.A., Taber, M., Turneaure, J.P., Van Hooydonk, T., Walter, T., and Zhou, P. (1992): Gravity Probe B: III. The Precision Gyroscope. In: *Proc. of the Sixth M. Grossmann Meeting on General Relativity* (eds. H. Sato and T. Nakamura). World Scientific, Singapore, p. 394